CURRENT STATUS OF NANO MATERIALS FOR MULTI FUNCTIONAL APPLICATIONS

VOLUME 2: NANO STRUCTURES TOWARD BIOMEDICAL APPLICATION

CURRENT STATUS OF NANO MATERIALS FOR MULTI FUNCTIONAL APPLICATIONS

VOLUME 2: NANO STRUCTURES TOWARD BIOMEDICAL APPLICATION

Edited by:

Dr. Aneeya Kumar Samantara and Dr. Satyajit Ratha

ARCLER PRESS

www.arclerpress.com

Current Status of Nano Materials for Multi functional Applications

Volume 2: Nano structures toward Biomedical Application

Dr. Aneeya Kumar Samantara and Dr. Satyajit Ratha

Arcler Press
2010 Winston Park Drive,
2nd Floor
Oakville, ON L6H 5R7
Canada
www.arclerpress.com
Tel: 001-289-291-7705
001-905-616-2116
Fax: 001-289-291-7601
Email: orders@arclereducation.com

ISBN: 978-1-77361-540-0 (Hardcover)

ABOUT THE EDITORS

Dr. Aneeya Kumar Samantara is a post-doctorate fellow at National Institute of Science Education and Research, Odisha, India. He did his Ph.D. in Chemistry from CSIR-Institute of Minerals and Materials Technology, Odisha, India. Before joining Ph.D., he completed Master in Science in Advanced Organic Chemistry from Ravenshaw University, and Master of philosophy (M.Phil.) from Utkal University. Aneeya's research interest includes the synthesis of metal chalcogenides and graphene composites for energy storage and conversion applications and designing of electrochemical sensors for detection of different bioanalytes. To his credit, he has authored in 18 peer reviewed international journals, one Book (Springer Nature) and four book chapters.

Dr. Satyajit Ratha has pursued his Ph.D. at the Indian Institute of Technology Bhubaneswar, India. Prior to joining IIT Bhubaneswar, he received his Bachelor of Science, First Class Honours from Utkal University, in 2008 and Master of Science from Ravenshaw University in 2010. Satyajit's research interests include two dimensional semiconductors, nanostructure synthesis, applications, energy storage devices and supercapacitors. He has authored and co-authored about 21 peer reviewed international journals and one Book (Springer Nature).

Dedication

Dr. Aneeya K. Samantara would like to dedicate this Book to,
His parents, Mr. Braja Bandhu Dash, Mrs. Swarna Chandrika Dash and his beloved wife Elina

Dr. Satyajit Ratha would like to dedicate this Book to,
His Parents, Mrs. Prabhati Ratha and Mr. Sanjaya Kumar Ratha

TABLE OF CONTENTS

LIST OF FIGURES

chitosan-based nanocapsules, and its ability for drug release upon enzymatic (protease) degradation (From Zhou et al., 2, 1021, 2013).

Figure 4.1. Schematic presentation of different approaches of gold nanoparticles mediated cancer ablation.

Figure 4.2. (A) Illustration summarizing cubic Au nanoframe formation. Coloration indicates the conversion of a Ag nanocube into a Au/Ag nanocage and then a predominately Au nanoframe. Beginning with Ag nanocubes, Au/Ag nanoboxes are prepared by galvanic replacement (step 1). Then a wet etchant removes the remaining Ag to form a porous nanocage (step 2), which with more etchant evolves into a cubic nanoframe (step 3). (B-E) TEM and SEM (insets) of (B) 50 nm Ag nanocubes, (C) Au/Ag nanoboxes prepared by galvanic replacement, and (D) nanocages and (E) nanoframes prepared with $Fe(NO_3)_3$ as a Ag etchant (From Skrabalak et al., Accounts of chemical research, 41, 1587, 2008).

Figure 4.3. Apoptosis and necrosis are the two mechanisms of cell death induced by photothermal therapy. A cell undergoing apoptosis maintains its membrane integrity and produces "eat me" signals to mark the cell for phagocytosis without incurring inflammation. Apoptosis may lead to secondary necrosis in which the cell experiences loss of membrane integrity and release of damage-associated molecular patterns (DAMPs), but without activating phagocytosis. Conversely, primary necrosis destroys plasma membrane integrity to cause the release of DAMPs, leading to an inflammatory response (From Melamed et al., ACS Nano, 9, 6, 2015).

Figure 4.4. Left: Experimental UV-vis absorption spectrum of a gold nanorod sample with an average aspect ratio of 3.3. The band at 525 nm is referred to as the transverse plasmon resonance, while the one centered at 740 nm is called the longitudinal plasmon absorption. Right: TEM image of the same solution (From Link et al., Journal of Physical Chemistry B, 103, 3073, 1999).

Figure 4.5. (A) TEM and UV-vis spectra of RGD-[64Cu]Au NR808. (B) Temperature mapping during the 808 nm laser irradiation (2 W/cm^2) of an RGD-[64Cu]Au NR808 suspension in distilled water by real-time thermal imaging using a thermal camera (FLIR). (C) Representative whole-body coronal PET images of U87MG tumor-bearing mice at 4, 16, 24, and 45 h after intravenous injection of gold nanorods (From Sun et al., ACS Nano, 8, 8438, 2014).

Figure 4.6. (A) SEM and TEM images of plasmonic gold vesicles (GVs) self-assembled from GNPs, the Size distribution of GVs by DLS and UV-vis spectra of BCP-tethered GNPs and GVs. (B) MDA-MB-435 cell viability at different concentrations of GVs, Ce6, and GV-Ce6 for 12 h at 37 C with or without. Fluorescence images of Calcein AM and Ethidium homodimer-1 containing

LIST OF TABLES

LIST OF CONTRIBUTORS

Dr. Aneeya K. Samantara
Post-doctorate fellow, School of Chemical Sciences
National Institute of Science Education and Research, Bhubaneswar, Odisha, 752050

Dr. Satyajit Ratha
PhD, School of Basic Sciences
Indian Institute of Technology, Bhubaneswar, Odisha, 752050

Mr. Lakkoji Satish
Department of Chemistry
National Institute of Technology, Rourkela – 769008, Odisha, India

Mrs. Ayonbala Baral
PhD Hydro & Electrometallurgy Department
CSIR-Institute of Minerals and Materials Technology, Bhubaneswar – 751013, Odisha, India
AcSIR CSIR-Institute of Minerals and Materials Technology, Bhubaneswar – 751013, Odisha, India

Dr. Malay Kumar Ghosh
Senior Principal Scientist, Head of the Department
Hydro & Electrometallurgy Department
CSIR-Institute of Minerals and Materials Technology, Bhubaneswar – 751013, Odisha, India

Mrs. Diptipriya Sethi
Ph.D, Environment and sustainability department
CSIR-Institute of Minerals & Materials Technology, Bhubaneswar-751013, India

AcSIR CSIR-Institute of Minerals and Materials Technology, Bhubaneswar – 751013, Odisha, India

Mrs. Geetanjali Mishra
Ph.D, Hydro & Electrometallurgy Department
CSIR-Institute of Minerals & Materials Technology, Bhubaneswar-751013, India
AcSIR CSIR-Institute of Minerals and Materials Technology, Bhubaneswar – 751013, Odisha, India

Dr. Barsha Dash
Scientist, Hydro & Electrometallurgy Department
CSIR-Institute of Minerals and Materials Technology, Bhubaneswar – 751013, Odisha, India

Dr. Ramasamy Sakthivel
Senior Principal Scientist, Environment and sustainability department
CSIR-Institute of Minerals & Materials Technology, Bhubaneswar-751013, India

Mr. Ashok K. Nayak
PhD, DNA Nanotechnology & Application Laboratory
CSIR-Institute of Minerals & Materials Technology, Bhubaneswar, 751013, India
Department of Biotechnology, Ravenshaw University, Cuttack, 753003, India

Ms. Alfa S. Mohanty
DNA Nanotechnology & Application Laboratory
CSIR-Institute of Minerals & Materials Technology, Bhubaneswar, 751013, India
Academy of Scientific and Innovative Research, New Delhi, India

Dr. Arun Kumar Pradhan
Assistant Professor, Centre for Biotechnology, Siksha O Anusandhan University, Bhubaneswar, India
M/S Odisha NET Academy & Research Centre, Bhubaneswar, Odisha, India-751021

Ms. Tanushree Mahapatra
M/S Odisha NET Academy & Research Centre, Bhubaneswar, Odisha, India-751021
Utkal University, Bhubaneswar-751004, India

Dr. Minati Behera
PhD, Department of Environmental and Sustainability
CSIR-Institute of Minerals and Materials Technology, Bhubaneswar-751013, India

Mrs. Sanghamitra Mohapatra
PhD, Department of Environmental and Sustainability
CSIR-Institute of Minerals and Materials Technology, Bhubaneswar-751013, India

Mrs. Upasana Mohapatra
PhD, Department of Environmental and Sustainability
CSIR-Institute of Minerals and Materials Technology, Bhubaneswar-751013, India

PREFACE

The nano-world has taken the human society by surprise due to its altogether different aspects and capabilities beyond natural senses. The transition from the conventional macro-world is not easy as there are neither sufficient concordant physical representations nor any phenomenology that could simplify the rules followed in the nano-world. Extensive research in recent times have revealed the multifaceted prospects of nano-materials, therefore posing stiff challenges for the development of techniques which could bring desired modification strategies in order to engineer these nano-materials for specific application which is highly unlikely in the case of systems at macro scale. Nevertheless, the unprecedented success story of nanotechnology has come at a time when the human society desire to control almost everything at atomic/molecular level which could provide great control over a wide range of applications starting from basic scientific analysis to high-end industrial research. Operating at sub-micrometer level, nanotechnology has revolutionized biological, chemical and pharmaceutical industries. Besides, there are various other promising aspects of nanotechnology, e.g., catalyst development for wastewater treatment and energy conversion, synthesis of high-performance nano-materials/composites for energy storage, photovoltaics, optoelectronics, and sensors, etc. Thus, nanotechnology (or nanotechnologies) provides an extended range of methodologies which could not only help building nano-materials with desired structures and morphologies but also reveal their true multidimensional aspects. With rapid evolution in synthesis and characterization techniques (thanks to nanotechnology), there has been a constant surge of interest from the research community to check the feasibility and effectiveness of several nanoparticles (both metallic and non-metallic) in various applications. Their high surface to volume ratio results in exceptional catalytic and surface kinetics in contrast to the bulk system. Also, due to the extremely small dimension, few of these nano-materials (e.g., carbon quantum dots) show unique features (due to quantum confinement and size dependent physicochemical properties) that have been found to be of quantum nature. Furthermore, engineered nano-materials as biomarkers and targeted drug delivery for effective diagnosis and therapeutics have been the highlight of the cancer research. There is, however, ambiguity due to conflicting/contradictory results and an overdose of information from numerous sources, which require immediate categorization that could provide an opportunity to correlate the relevant research yields to streamline the whole process, starting from synthesis to the potential application of nano-materials.

This book highlights the true and diverse aspects of a broad spectrum of nano-materials with clear emphasis on the various synthesis methodologies that have gone under painstakingly slow optimization process, their application in various emerging biomedical fields of research (e.g., bio-imaging, targeted drug delivery, antimicrobial activity, tissue engineering, wastewater treatment, etc.) and the potential they hold for the future.

SUMMARY

This book/volume emphasizes the enormous potential/possibilities that the nano-materials possess by unfolding their history, recent/present trends with a brief discussion about future endeavor. The multifaceted application of these nano-materials covering a wide range of critical domains, e.g., health, food, and pharmaceuticals have been included in this volume through dedicated chapters in a highly organized manner to capture the attention of readers from diverse sections. Be it antimicrobial development, wastewater treatment or complex conjugate systems comprising biomolecule (e.g., protein) or effective therapeutic for cancerous tumors, nano-materials have showed great potential in all these fields. Six chapters have been included in this volume/book through contribution from authors affiliated to distinguished research laboratories and academic institutions, covering a myriad of nano-materials of distinctive features and their application in various fields.

The first chapter entitled "Functionalized Carbon Quantum Dots for Biomedical Application" compiled by Samantara and Ratha, have discussed the emergence of carbon-based quantum dots and their dominating properties over the traditional fluorescent dyes and transition metal-based quantum dots. The different synthetic methodologies, properties, and functionalization is discussed. Moreover, a brief discussion on their physicochemical properties, higher bio/hemocompatibility, and tunable photoluminescence behavior is presented which make these quantum dots as a potential candidate for advanced therapeutic/biomedical applications.

The second chapter, "Reconnaissance of protein-nanoparticle conjugates: A mini-review on development and application," compiled by Lakkoji Satish et al., stresses upon the implementation of sophisticated nanotechnology for the amalgamation of various nanoparticles with biomolecules like protein to form bioconjugates which could immensely benefit the biomedical research and also food, chemical industries as well. The authors have provided a vivid explanation regarding the effect of nanoparticles on the structure and stability of the protein molecules which could reveal better understanding of various vital protein immobilization techniques. In the next chapter entitled "Metal Nanoparticles as Potential Antimicrobial Agents," compiled by Diptipriya Sethi et al., have discussed the current challenge faced by the microbicides market. Though most of the industry grade antimicrobials are quite effective, it, however, remains a daunting task to tackle the severity of few infections caused by multi-drug resistant bacterial strains and those associated with biofilms. The authors have further discussed about the biocompatibility of several promising nanoparticles and have compared the same with the microbicides that are available in the pharmaceutical market.

Chapter Four entitled "Multi-Utility of Gold Nanoparticles in Cancer," compiled by Ashok K. Nayak et al., discuss about the superiority of gold nanoparticles towards cancer therapeutics because of their features such as ligand targeting, bioimaging, as carriers of antisense drugs and integration with other surface modifiers. The authors have also discussed the green synthetic routes to obtain gold nanoparticles which could propel their practical implementation ineffective cancer therapeutics.

The fifth chapter entitled "Application of Silver nanoparticles on various types of cancerous cells" thrusts emphasis on the implementation of silver nanoparticles, known for their excellent disinfecting and antimicrobial properties, in the diagnosis and potential treatment of various cancerous growths in the human body. The chapter compiled by Arun Kumar Pradhan et al., further proposed the preparation and implementation of nanoparticles of other elements and extends their usage to diagnose and treat several other chronic diseases besides cancer.

In Chapter Six, "Nanomaterials as future prospects for wastewater treatment," Minati Behera et al., have discussed about the effectiveness of various nanomaterials in treating wastewater to make it portable so as to check the depleting fresh water reserve and prevent drinking water scarcity in future.

1
CHAPTER

FUNCTIONALIZED CARBON QUANTUM DOTS FOR BIOMEDICAL APPLICATION

Aneeya K. Samantara[1] and Satyajit Ratha[2]

[1] National Institute of Science Education and Research, Bhubaneswar, Odisha, 752050

[2] Indian Institute of Technology, Bhubaneswar, Odisha, 752050

CONTENTS

ABSTRACT

The carbon nanoparticles of lateral dimension ~10 nm regarded as carbon quantum dots have drawn significant attention from materials researchers since few decades. These quantum dots can be categorized into two types such as carbon quantum dots (CQDs) and graphene quantum dots (GQDs) depending upon the source from which they are derived and their lateral size. Their ease of preparation, intriguing physicochemical properties, tunable photoluminescence behavior, and excellent bio/hemo compatibility strongly suggest their potential candidature for implementation in many fields of biomedicine and diagnosis. Furthermore, the availability of oxygen functional groups on their surface allows forming conjugate with a variety of biomolecules facilitating the drug delivery and biosensing applications. In this chapter, a brief introduction on the carbon-based quantum dots followed by their synthesis, functionalization, and application are presented.

Keywords: Carbon quantum dots, Graphene quantum dots, Functionalization, Top-down approach, Bottom-up approach

1.1. INTRODUCTION

Interestingly, for a long period of time, the fluorescent organic dye molecules were used as the biological markers and also for the detection of metal ions in real samples (Aldeek et al., 2011; Bourlinos et al., 2008). Thereafter, a new material named as quantum dot (QDs) entered the world of fluorescent materials. These are generally the transition metal-based nanomaterials (CdSe, PbTe, ZnS, etc.) and their composites (Jaiswal et al., 2003; Jie et al., 2008; Alivisatos et al., 2005; Gao and Xu, 2009). These materials are preferred over the conventional organic dye molecules due to some of their fascinating behaviors like high optical brightness, sensitive towards imaging, photostability, resistance towards metabolic degradation, high photoluminescence lifetimes, high power conversion efficiency, etc. For these fascinating properties, these QDs were used in designing of photovoltaics, solar cells, development of fluorometric sensors and broadly used as biomarkers (Tang et al., 2014; Qian et al., 2010). Although the QDs are used in so many emerging fields, the presence of hazardous metal (like cadmium), high cost and strict synthesis conditions limit their wide applications (Gao et al., 2009). Therefore, the further finding of a new material having such type of fluorescent behaviors along with low toxicity and high biocompatibility is desired. After the passage of a long duration of time, carbon-based quantum dots come in the research field (Xu et al., 2004). Since then, these carbon-based quantum dots have created a separate field of interest over the traditional fluorescent organic dye molecules and the transition metal-based quantum dots mainly in biomedical applications.

Generally, there are two types of carbon-based quantum dots, i.e., carbon quantum dot (CQDs) and graphene quantum dot (GQDs) that are named according to the source from which they have been derived. These are zero-dimensional, quasi-spherical, discrete carbon nanoparticles with a high concentration of oxygen (the oxygen comes from surface oxidation during the synthesis). Along with these properties, the CQDs contain sp^2 graphitic π bonds showing the nanocrystalline behaviors (Shen et al., 2012; Baker et al., 2010; Li et al., 2011; Wang et al., 2010; Bourlinos et al., 2008). Whereas, the GQDs are generally derived from the graphene-based starting materials or by following some univocal wet chemical methods exhibiting a graphene lattice inside them along with the sizes less than 100 nm. Since it is mostly derived from the Graphene-based materials, it also comprises a few layers of graphene sheets (<10) (Lee et al., 2012; Philpott et al., 2012; Li and Yan, 2010; Lin and Zhang, 2012; Kim et al., 2012). The CQDs/GQDs have many dominating properties over the traditional QDs along with negligible cytotoxicity, non-blinking behavior, better biocompatibility, diverse biochemical characteristics, etc. Due to these remarkable properties, these CQDs/GQDs are used for the development of fluorescent-based biosensors, optoelectronic devices, drug carriers for various anticancer drugs, imaging materials for the in vitro and in vivo cell imaging, etc. (Cao et al., 2007; Wei et al., 2012; Wang et al., 2011; Guo et al., 2012; Li et al., 2010; Cao et al., 2011; Yu and Kwak, 2012).

Again, in order to enhance the photoluminescence behavior, fluorescent quantum yield, improve their compatibility and solubility in solvents other than water; the surfaces of CQDs/GQDs are functionalized, that can be carried out by different strategies enabling to modify with foreign molecules. During synthesis, many oxygenated functional groups develop on the surface of these QDs. The carboxylic, hydroxyl, carbonyl groups, mostly present at the edges of the QDs whereas the core of the particles comprises of mainly sp^2 bonded carbon skeletons, so that the edges are functionalized by mainly treating amino-containing compounds. On the other hand, the surfaces of these QDs also get functionalized by doping with hetero atoms rather than carbon. This type of heteroatom doping generally carried out during the synthesis, and the heteroatom selection based on the composition of the starting material taken. Mostly, the surface doping of carbon quantum dot have carried out by incorporating nitrogen (N), sulfur (S), boron (B), etc. atoms into the carbon skeleton (Dong et al., 2013; Qu et al., 2013; Jahan et al., 2013).

In this chapter, we sum up various synthetic protocols for the development of carbon-based quantum dots, through which we aimed to give a brief idea to the readers and encourage them to go insight these new emerging materials. This also outlines the fundamental concepts regarding the fluorescent enhancement of Carbon quantum dots, which will germinate a research interest to explore different biomedical applications.

1.2. SYNTHESIS OF CARBON QUANTUM DOTS

The synthesis schemes and starting precursors have a promising role to determine the optical (luminescence, quantum yield, etc.), chemical (elemental composition, functional groups on their surface, etc.) and physical (crystallinity, size, solvent compatibility, etc.) properties of the QDs.

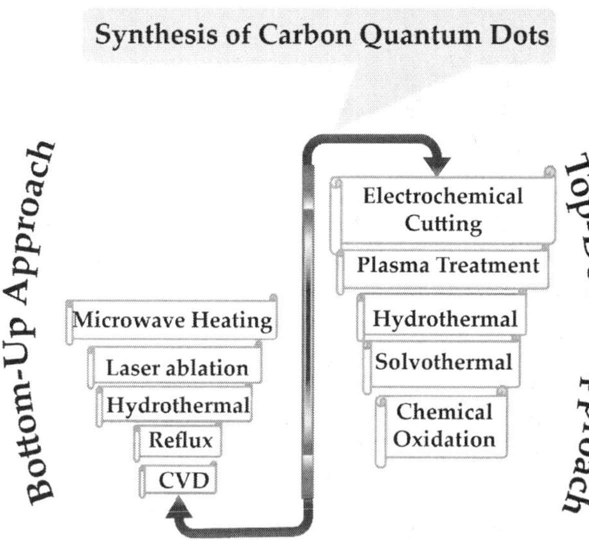

Figure 1.1: Different synthetic approaches for the synthesis of CQDs.

Although a variety of synthetic strategies have been developed, they can be categorized into two major approaches such as, Top-down approach and Bottom-up approach. In the case of the Top-down approach, large sized carbon precursors (graphene sheets, carbon fiber and carbon nanotubes, graphite, etc.) are broken down into small particles having fluorescence properties. Whereas the starting materials like small carbon-containing molecules (urea, citric acids, carbohydrates, proteins, amino acids, etc.) undergo polymerization, carbonization, and functionalization to form such particles in a bottom-up approach. However the synthetic methods like arc discharge, electrochemical cutting, hydrothermal cutting, etc. are taken under the top-down and the hydrothermal, microwave irradiation, solvothermal, etc. are categorized under the bottom-up approach (Figure 1.1).

1.2.1. Top-Down Approach

Arc-discharge is the first method used for the preparation of CQDs. Xu et al. (2004) tried to separate the single-walled carbon nanotube prepared from the arc discharge method. At first, the arc-discharge shoot was oxidized with 3.3 N HNO_3 followed by washing with NaOH solution. During this oxidation process, a large number of carboxyl groups got developed on the surface which was confirmed by the Fourier transform infrared (FTIR) spectral analysis.

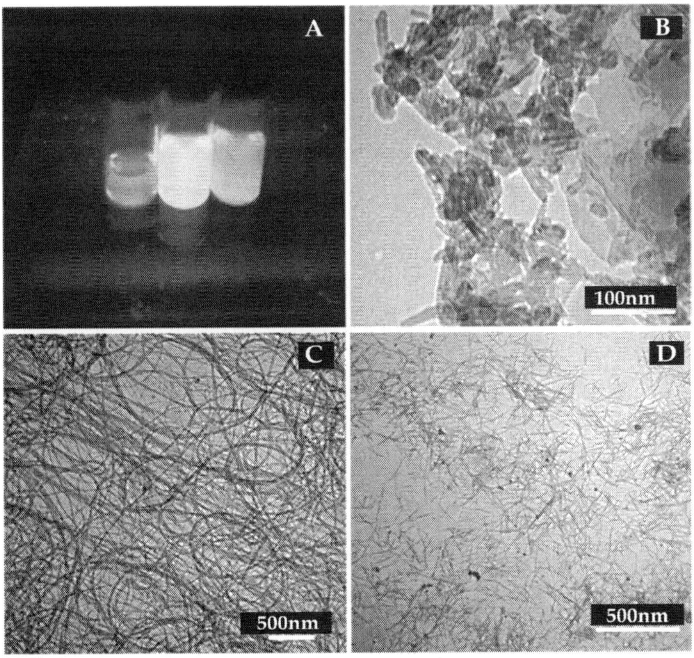

Figure 1.2: TEM images of (A) SWNTs, (B) short tubular material and (C)

fluorescent carbon particles, (D) optical images of fluorescent carbon materials under 365 nm UV-lamp, (From Xu et al., J. Am. Chem. Soc., 2004, 126, 12736–12737).

Then by gel electrophoresis using agarose gel and glass beads, the solution mixture have been separated. During the separation process, three types of materials were found on the illumination of 365 nm ultraviolet light; (a) long nanotubes, which are unable to penetrate into the matrix, (b) a short, irregular tubular material and (c) a fast-moving band of highly fluorescent material. Out of them, the fast-moving band of highly fluorescent material (i.e., regarded as the CQDs) has been collected and that give three fluorescent color bands (greenish blue, yellow and red) under 365 nm UV-lamp exposures. Again the mixture has been separated into three different parts according to their size, by means of Centricon filtration devices of the nominal molecular weight limit. The quantum yield of this yellow fluorescent material was 0.016 under excitation of 366 nm (Figure 1.2) (Xu et al., 2004).

In many of the cases, by taking graphene sheets as the starting material GQDs were derived following different methods like hydrothermal/ solvothermal processes, electrochemical cutting, nanolithography, etc. are followed for the scalable preparation (Li et al., 2012; Pan et al., 2010; Zhu et al., 2011). All these processes of synthesis involve the cutting of graphene sheets to generate the fluorescent GQDs (Figure 1.3). It is simply like a cutting of the Graphene sheets by a scissor into small pieces.

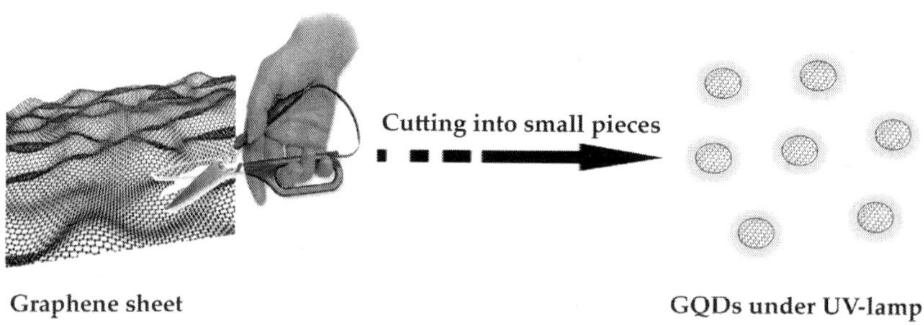

Figure 1.3: A schematic presenting the cutting of graphene sheets to produce the GQDs.

Besides, the solvothermal method is a very simple one to synthesize the GQDs from graphene sheets. Zhu et al., have prepared highly fluorescent GQDs by taking Graphene oxide (GO, prepared by modified Hummer's method) as the starting material (Zhu et al., 2011). A suspension of GO in dimethylformamide (27 mg/ml) has been prepared, poured in a Teflon lined stainless steel autoclave and placed in a hot air oven at 200°C for 5 hour. After the reaction, two types of materials are found in the autoclave, (a) a black precipitate and (b) brown colored transparent suspension. Out of these two, the black precipitate was throughout and the GQDs were collected from the brown suspension. The solvent of the suspension was removed by evaporation through a rotary evaporator and the GQDs were purified by column chromatography using silica gel and water as the eluent. It was observed that the GQDs give very strong green fluorescent under UV lamp with 11.8% PLQY at 360 nm excitation. They show very well size distribution with an average diameter of 5.3 nm with of 2–3 layers of graphene sheets. The GQDs shows better solubility in a number of polar solvents including water along with very good biocompatibility and less cytotoxicity, for which it was successfully used as the biomarker for the in vitro imaging of MG-63 cells.

Pan et al. (2010) prepared the GQDs through the hydrothermal synthetic method by taking the rippled graphene sheets (GSs) obtained from the thermal reduction of graphene oxide. Before the hydrothermal treatment, these graphene sheets were oxidized by using a mixture of conc. H_2SO_4 and conc. HNO_3 in 1:3 ratios. So that the GSs cut into smaller sheets and simultaneously huge numbers of oxygen functional groups get developed on the surface, making the GSs more hydrophilic. The GQDs are found to be uniform in size with average diameter 9.6 nm and more than 85% of them consist of 1–3 layers. A generalized mechanism has been assumed for the GQDs formation by the solvothermal/hydrothermal method and their luminescence behavior. Actually, during the oxidation (when treated with the 1:3 mixture of conc. H_2SO_4 and HNO_3), the epoxy groups tend to form a line on the carbon lattice and then their cooperative alignment induces a rupture of the underlying C–C bonds.

Once the epoxy groups appear on the surface, it immediately transfers to the epoxy pairs and then to stable carbonyl pairs which were confirmed from the FTIR analysis. The presences of such linear defects make the GSs more fragile and easily broken up during the hydrothermal/solvothermal deoxydation process. During which, the oxygen atoms in the epoxy line get removed producing the GQDs. They also derived the structural models of these GQDs both in acidic and alkali media which is based on the reversible protonation at the carbine like zig-zag sites and of oxygen-containing functional groups. During the protonation, reversibility was maintained between the zigzag site of GQDs and H^+, so the emissive triplet carbene gets broken up and becomes inactive towards photoluminescence. But in the case of alkaline medium, a strong response towards the photoluminescence was observed due to restoration of free zigzag sites.

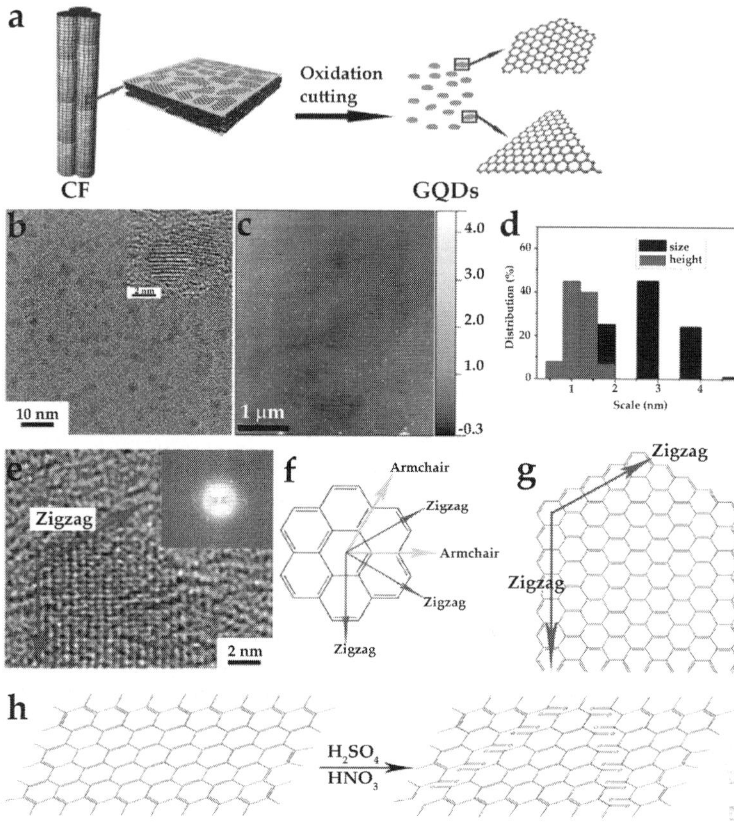

Figure 1.4: (a) Representation scheme of oxidation cutting of CF into GQDs. (b) TEM images of GQDs (synthesized reaction temperature at 120°C), inset of (b) is the HRTEM of GQDs. (c) AFM image of GQDs. (d) Size and height distribution of GQDs. (e) HRTEM image of the edge of GQD, inset is the 2D FFT of the edge in (e). (f) Schematic illustration showing the orientation of the hexagonal graphene network and the relative zigzag and armchair directions. (g) Schematic representation of the edge termination of the HRTEM image in (e). (h) Proposed mechanism for the chemical oxidation of CF into GQDs, (From Peng et al., Nano Lett. 2012, 12, 844–849).

The chemical oxidation method is a powerful and convenient method for the scalable preparation of GQDs. Generally, acids like, conc. H_2SO_4, conc.

HNO_3 and their mixtures with an optimized ratio are used as the oxidizing agent. These acids simultaneously cut down the large-sized graphene layers and develop oxygen functionalities on the QDs surfaces. By using the chemical oxidation method and etching, a quantitative amount of GQDs has been derived from the pitch-based carbon fiber (CF) (Peng et al., 2012). Here, the carbon fiber has been treated with a mixture of H_2SO_4/HNO_3 in 3:1 ratio for 24 hours at three different temperatures (80, 100, 120°C) producing variable sizes of GQDs (Figure 1.4). The formation of variable sized GQDs (1 and 4 nm in size with 1−3 layers of graphene) was found to be reaction temperature dependent and emits blue, green and yellow luminescence under 365 nm light irradiation. Here also the oxidative unzipping of the sp^2 carbon frame taking place through the breakdown of the lined epoxy/carbonyl functional groups as taking place in the case of the hydrothermal cutting method.

In another work, Dong et al. (2012) have prepared single and multi-layered GQDs for cell imaging and optoelectronics application. The CX-72 (commercially available carbon black as the carbon source) has been treated with 6 M HNO_3 which made the carbon precursor to undergo oxidative cutting and produce GQDs. By this process, both single and multi-layered GQDs were produced and are separated out by means of simply centrifugation with HNO_3 solution. These single layered GQDs present high penetration efficiency into the cell, so that they are applied as the effective agents for targeted drug delivery and intracellular imaging. Qiao et al. (2010) develop the CQDs by chemically oxidizing the activated carbon (commercially available). The CQDs were synthesized by treating the activated carbons (as the starting material, i.e., wood activated carbon, charcoal activated carbon and coal activated carbon) with HNO_3. Here the oxidative etching process plays a key role to produce CQDs. These particles are about 2–6 nm in size and gives bright blue fluorescence under 365 nm UV-light illumination. These CQDs shows excellent properties towards biological imaging due to its high luminescence, easy penetration into the live cells, photostability and non-toxicity like promising properties.

Kwon et al. (2014) have prepared the size controlled GQDs and apply them as well depot's for OLED giving a white light with an external quantum efficiency of 0.1% (Figure 1.5). Here a two-step process has been followed; (a) tattering of graphite, followed by (b) amidative cutting. In the first step, the tattered graphite has been produced by means of HNO_3 treatment on which so many oxygenated functional groups get developed. After that, different sizes of GQDs were obtained by amidation process followed by removal of excess oxygen functionalities through in situ hydrazine treatment. The size of the synthesized GQDs (2, 4, 7, 10 nm) is manipulated by simply varying the concentration of olylamine and observed that the size is inversely proportional to the concentration of olylamine used during amidation. These GQDs produce colorful photoluminescence ranging from blue to brown depending upon their size. The defect sites formed during these processes plays a key role for low energy emission and reduction in exciton lifetime.

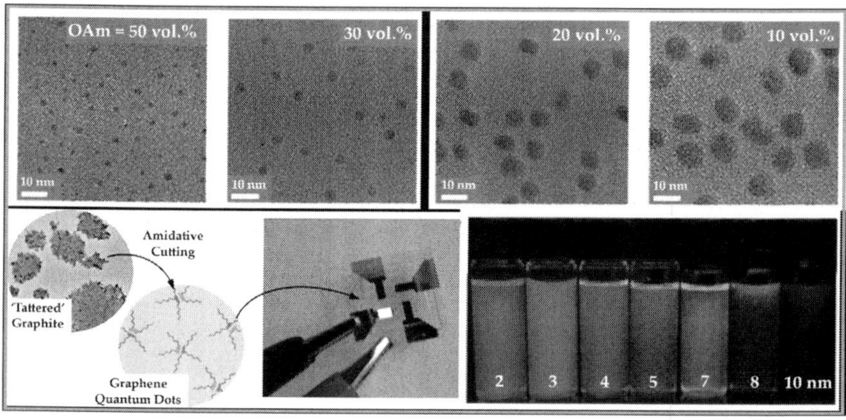

Figure 1.5: (a) TEM images of the GQDs with varied sizes of 2, 4, 7, and 10 nm (left-to-right). The text insets represent the concentration of olylamine, (b) schematic presentation for the formation of GQDs from tattered graphene, (c) fluorescence color of the different sized GQDs under 365 nm UV-lamp, (From Kwon et al., Nano Lett., 2014, 14, 3, 1306).

Although the above synthesis methodologies have been broadly adopted for the preparation of carbon-based quantum dots, but are associated with limitations like the requirement of peculiar equipment and expensive chemicals. In order to overcome the above shortcomings, the electrochemical cutting method has been chosen as efficient one. Generally, Graphite rod/Multi-walled carbon nanotube (despite of graphite oxide as used in other methods whose synthesis requires toxic chemicals) are used as the working electrode and mostly the non-aqueous solvents are taken as the electrolyte. Shinde et al. (2012) prepared the GQDs by taking multi-walled carbon nanotubes (MWCNTs) as the starting material. In the arrangement of electrochemical setup, glassy carbon electrode (MWCNTs drop casted on it), Pt foil and Pt wire are taken as the working, reference and counter electrode respectively and 3 mM $LiClO_4$ in propylene carbonate as the supporting electrolyte. Two-step electrochemical scanning was carried out by applying 1V anodic potential that causes the oxidation of the MWCNTs and –1V cathodic potential cause the reduction of the oxidized materials. The oxidation and reduction process makes the lateral unzipping of MWCNTs and produces GQDs. It has been observed that, in presence of aqueous electrolyte, graphene nanoribbons are produced instead of GQDs. The size of GQDs is inversely proportional to the time of oxidation (step-1) i.e., small sized particles obtained on increasing the oxidation time. They are very well dispersed showing narrow size distribution with 8.2±0.3, 5±0.3, 3±0.3 nm for 7, 11, and 15 hour of oxidation at 90°C and 23±2 nm at 30°C. The particles produced at 15-hour oxidation are about 1–3 atomic layers thick as observed from the AFM analysis.

Instead of organic solvents, aqueous medium can also be used as the supporting electrolyte for production of GQDs. Zhang et al. (2012) have prepared GQDs by pursuing electrochemical oxidation of graphite rod followed by hydration reduction. The electrolysis was carried out with graphite rod as the working electrode and a platinum foil counter electrode in a NaOH supporting electrolyte within a current intensity range of 80–200 mA/cm^2. The black color solution obtained from electrolysis (for 3 hours) was reduced by 80% (1 ml) hydrazine hydrate followed stirring for 8.5 hours. After removal of the black precipitates, pure GQDs were collected from the supernatant by means of centrifugation (10,000 rpm) and dialysis. The GQDs are highly water-soluble small sheets of diameter 5–10 nm and constitute single graphene sheets (<5 nm). They show excitation wavelength independent emission with PLQY of 14%, giving yellow fluorescence under 365 nm UV-lamp irradiation. The plasma treatment is one of the efficient methods used for the large-scale, single step production of highly pure CQDs. In these cases, the plasma is used for both production of QDs and their surface functionalization. Wang et al. (2012) Prepared highly fluorescent CQDs by adopting the oxygen plasma treatment with the chicken egg. The egg yolk and egg white separated nicely by an egg separator and placed between the quartz slides of the plasma generator for 3 minutes. The black color products thus formed were dissolved in water that emits bright blue fluorescence with PLQY of 2–8% and were employed for the development of fluorescent ink for printing purpose.

1.2.2. Bottom-Up Approach

As discussed in the above sections, by using different synthetic methods, the carbon-based quantum dots can be derived from the molecular precursors.

Out of others, the microwave-assisted route is a powerful one for the scalable preparation of CQDs in a cost-effective way. It provides consistent heating throughout the reaction solution, making the formation of uniform sized particles. Tang et al. (2012) Prepared highly photoluminescent GQDs by taking the glucose as the carbon precursor in deionized water via the microwave assisted hydrothermal method (Figure 1.6).

The as-formed GQDs were observed to be of uniform in size (3.4±0.5 nm) with few-layer thickness (height= ~3.2 nm) giving unique ultraviolet luminescence. The mechanism for formation of GQDs from glucose by this microwave method can be generalized for all the carbohydrates of empirical formulae 1:2:1 (C, H, and O). According to them, on the beginning of the reaction, the glucose molecule undergoes dehydration forming the nucleus (composed of C=C skeleton), then with the passage of time the side growth takes place for which the spherical GQDs are obtained. It was also verified that, on long-term heating, larger sized particles are formed. So by manipulating heating time one can manipulate the size of GQDs. These GQDs act as a very efficient light converter for the conversion of blue light to white light when coated in a blue light emitting diode.

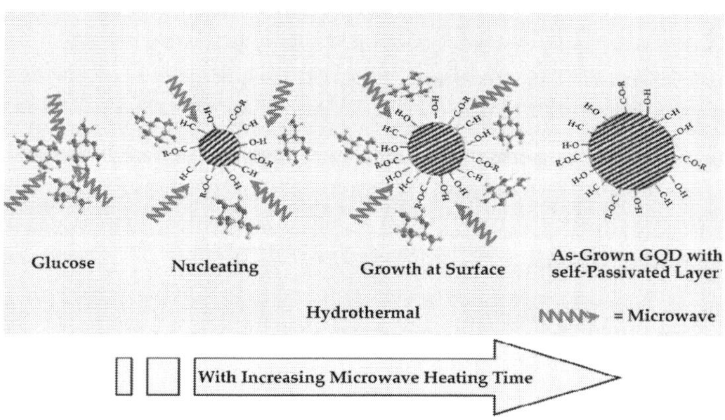

Figure 1.6: Synthetic mechanism for formation of GQDs from glucose, Color converter of the GQDs. Luminescence spectra of the blue LED with and without GQDs coating. Left inset: Photographs of the GQD coated LED without (top) and with applied voltage (bottom). Right inset: The CIE chromaticity coordinates for the illuminating blue LED with and without GQD layer, (From Tang et al., ACS Nano, 2012, 6, 5102).

In another report, Zhai et al. (2012) Prepared highly fluorescent CQDs by taking citric acid as carbon precursor and various N-containing molecules as the source for N-doping by means of microwave irradiation. The particles were found to be 2.2–3 nm in size with a PLQY of 30.2%. Along with the hydrophilic CQDs, organophilic CQDs were also developed having good dispersibility in organic solvents [Stassinopoulos et al., 2008; Bourlinos et al., 2008; Wang et al., 2010; Wang et al., 2011). Mitra et al. (2012) prepared the hydrophobic CQDs with both blue and green luminescence depending upon the excitation wavelength by using microwave techniques. Here, apoloxamer (PF-68) have taken as the carbon source in phosphoric acid and the mixture was exposed to microwave irradiation (450W) only for 4 minutes. During this microwave treatment, fragmentation of PF-68 takes place, forming the carbon core, then the passivation of the core carried out by condensation of hydroxyl groups with the fragmented/non-fragmented PF-68 (Wang et al., 2010). The as-prepared CQDs were found to be organophilic in nature, having very good dispersibility in a number of organic solvents. The small size (5–20 nm) and bright photoluminescence property (PLQY=7%) made them efficient material for development of fluorescent hydrophobic surfaces. Not only the aqueous medium, but also the organic solvent can be employed for the synthesis of CQDs by this microwave irradiation method. It is well reported that the photoluminescence property of CQDs depends upon their size which can be modulated by manipulating the carbonization time (i.e., by increasing the time, size of CQDs increases) (Fang et al., 2012). But in case of Microwave synthetic route, it is limited due to the use of water as the solvent. So in order to manipulate the size of CQDs some alternative solvent other than water should be chosen. Liu et al. (2014) develop such type of synthetic procedure by adopting sucrose as the carbon precursor in diethylene glycol solvent. The CQDs were having very narrow size distribution (average size 5 nm) and highly hydrophilic in nature. The green photoluminescence and negligible cytotoxicity enable them to be an efficient biomarker for in vitro cell imaging.

Many researchers have been adopted the hydrothermal method for the large-scale production of CQDs from a verity of sources. It is also regarded as a green approach due to use of water instead of any organic solvents. In particular, Hsu et al. (2012) prepared high-quality CQDs from hydrophilic compounds like ethylenediaminetetraacetic acid (EDTA), 2-amino-2-hydroxymethyl-propane-1, 3-diol (TRIS), glycerine, and cadaverine by following hydrothermal route.

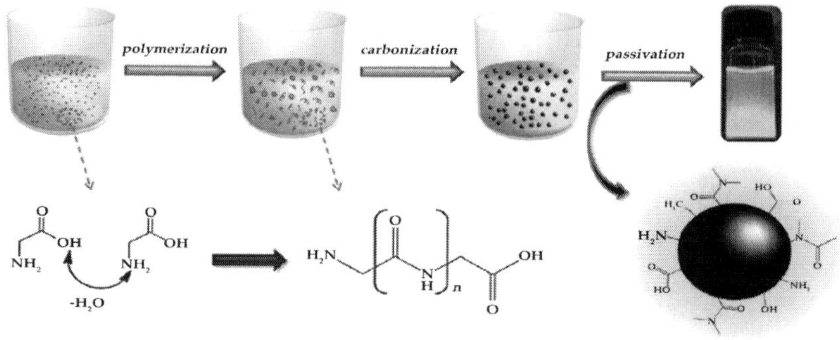

Figure 1.7: Growth model of CQDs from Glycine by following dehydration, polymerization, carbonization, and passivation, (From Hsu et al., Chem. Commun. 2012, 48, 3984).

Here, the starting materials were dissolved in water and put in a Teflon lined autoclave at 300°C for 2 hours, after that the hydrophilic CQDs were obtained. The particles show the very narrow size distribution with diameter ranging from 2.6 to 7.9 nm and high photoluminescence behavior having quantum yield 26.6%, 26.0%, 30.6%, 5.4% respectively. Among the different starting materials, CQDs derived from Glycine show very high PLQY (30.6%) due to the presence of both amine and carboxyl functional groups. Such type of CQDs (from Glycine) formation by this hydrothermal route proceeds through four steps (Figure 1.7). Initially, the molecules get associated by means of hydrogen bonding, and then on heating dehydration take place. On continuation of heating, polymerization take place followed by carbonization process. After these three steps; the carbon core is formed, then immediately diffusion of solute particles towards the surface of core takes place making the core grow in size.

This growth model was also adopted by Yang et al. (2012) for the formation of CQDs by taking Chitosan as the carbon precursor. In this work, luminescent CQDs were developed having very less cytotoxicity and high biocompatibility, so that they were broadly applied for the imaging of A549 cells. Similarly, Liang et al. (2013) synthesized the highly luminescent CQDs (PLQY 31.6%) from gelatin (as the carbon precursor) without using any surface passivating chemicals. Yang et al. (2011) prepared the intrinsically fluorescent carbon dots from glucose and interestingly manipulate the fluorescent color by varying the concentration of monopotassium phosphate (KH_2PO_4). The intrinsic fluorescence behavior results from the simultaneous formation and surface passivation during the hydrothermal treatment. On the other hand, the green and blue fluorescent carbon dots (C-Green and C-Blue) were collected by varying the concentration of KH_2PO_4 with constant glucose concentration. Here the presence of KH_2PO_4 prevents the growth of carbon nuclei directing the formation of small-sized particles (Liang et al., 2013; Demir-Cakan et al., 2009). Also, the selection of carbon precursor is a crucial task towards development of multicolor fluorescent CQDs.

As we know the surfactants are generally used for stabilization and shape modulation of the metal nanoparticles, but they can be used as the carbon precursor for CQDs synthesis due to the existence of high carbon content (Antonietti et al., 2001; Pileni et al., 2007). Kozak et al. (2013) for the first time take the cetylpyridinium chloride as the carbon precursor and hydrothermally prepared the multicolor fluorescent CQDs (blue-green-yellow photoluminescence). They assumed that this multicolor luminescence behavior depends upon the solvent used, excitation wavelength, reaction condition, etc. Here the surfactant acts as the carbon donor, contribute the emission by employing its π–electron system, and stabilizes the formed carbon nanoparticles. Due to their multiple photoluminescent property, higher amphiphilic nature resistant to the photobleaching; these CQDs can be employed as the suitable alternative to the traditional transitional metal quantum dots. Very recently, Samantara et al. (2016) have prepared the CQDs from different Good's buffer by following the hydrothermal reaction. The as-obtained CQDs shows blue color luminescence with PLQY of 30.5−47.39% and are employed for the in vitro imaging of HEK-293 and H357 cells and targeted delivery of Doxorubicin.

Like other synthetic protocols, reflux reaction is one of the important methodologies for the production of CQDs. As from the above discussions, the microwave-assisted synthesis of CQDs is an efficient one in terms of quick synthesis, but the quality of the CQDs is inferior to that obtained from the reflux reaction. Generally, the reflux reaction is carried out either by pyrolysis of carbonaceous molecules (normally carbohydrates) in the presence of surface passivating agents or dehydration followed by surface passivation (Peng et al., 2009; Qiao et al., 2010). Shen et al. (2013) have prepared the pH-sensitive photoluminescent carbon nanoparticles by refluxing the polyethyleneimine in 6M HNO_3. In this type of reaction, the reflux time directs the size of carbon nanoparticles formed i.e., on increasing the time small-sized particles are obtained (particles of diameter 12.8 and 9.3 nm were obtained from 12- and 24-hour reaction). The particles give constant photoluminescent behavior in high ionic concentration, but very sensitive towards the pH of the solution. Bhunia et al. (2014) have prepared highly fluorescent carbon nanoparticles by taking the Thiamine (Vitamin-B1) as the carbon precursor and modulate their photoluminescence property by simply manipulating the reaction temperature and phosphate salt. At lower reflux temperature and by taking Na_3PO_4 highly fluorescent CQDs (PLQY of 76%) has been prepared, but on increasing the temperature (130°C) comparative less luminescent (PLQY=25%) CQDs obtained. According to them, the high PL behavior originates from the heteroatom (N, P, S, and O) doping in the CQDs. The presence of such heteroatoms in smallest dimension (1–10 nm) of carbon controls the ratio of sp^2 and sp^3 carbon atoms optimizing the length of conjugated double bond (Bhunia et al., 2013). This causes the fluorescence enhancement in the doped CQDs.

The solution processable methods were so attractive towards the large-scale preparation of GQDs. Still, some limitations are there like, difficult to produce the size controlled GQDs, low yield and strict synthetic process, for which the development of a new synthetic method is required (Sun et al., 2006; Zheng et al., 2009; Yang et al., 2012; Liu et al., 2009). Ding et al. (2014) prepared GQDs by chemical vapor deposition method, which is a relatively economical and versatile route for large-scale production (Ding et al., 2014). A great advantage of this technique is, not only size controlled, but also the thickness controlled GQDs can be easily derived from variations in reaction temperature, pressure, and gas flow rate. The GQDs were produced from the decomposition of methane onto the hexagonal boron nitride plate without the use of any metal catalysts. The GQDs are of 80 nm in diameter and 1–10 atomic layers thick. These materials can be broadly used in the development of various optoelectronic devices like; touch screen, LED, flat panel displays, etc.

1.3. PROPERTIES OF CQDS

The carbon-based quantum dots are tiny carbon particles with quasi-spherical in shape having lateral dimension of ≤10 nm. These small dimensions make them to pass the cell membrane and to enter into the cytoplasm, opening a new window for targeted drug delivery and imaging (Anilkumar et al., 2011). It has been observed that, though the fluorescent dyes are broadly used for imaging purpose, but are associated with blinking and lower photostability. On the other hand, all the carbon-based quantum dots are having non-blinking properties, but those derived by the top-down approach are having more resistance towards photobleaching compared to that prepared by the bottom-up processes (Liu et al., 2009; Liu et al., 2012; Yang et al., 2009). Whatever the synthesis method followed, the quantum dots thus formed having a large number of oxygen-containing functional groups (−COOH, −OH, −CHO, −CO−, etc.) on their surface. These functional entities make the particles more hydrophilic in nature and provide a huge platform to conjugate foreign molecules/nanoparticles for their multimodal applications. Furthermore, the optical behavior of these particles plays a dominant role for many of the biomedical applications. Generally, they shows absorption in the ultraviolet region (i.e., 230 nm, 300 nm, 340 nm, etc.) of the electromagnetic radiation attributed to the π-π* (from the aromatic sp^2 carbons) and n-π* (due to the presence of oxygen moieties) electronic transitions. It has been observed that with the absorbance of UV-light, the particles emit a blue to green luminescence with an average photoluminescence quantum yield (PLQY) of ~10%. But in some cases, higher value of PLQY has also been realized by improving the synthetic procedure or by passivating the quantum dot surface or conjugating with other nanoparticles (Anilkumar et al., 2011; J. Yang et al., 2009; Cao et al., 2007). The carbon-based quantum dot shows an upconversion photoluminescence behavior like the transition metal quantum dots. In a particular study, Shen et al. (2011) have observed an upconversion behavior of the graphene quantum dots when excited with a bunch of low energy photons. These photons excite the π-electron from the valence band to the conduction band (higher energy state, i.e., π*) and then the electrons from the transition state get returned to the lower energy state (σ orbital) giving a luminescence of lower wavelength. They assume that this behavior is due to an anti-Stokes photoluminescence instead of the multiphoton active process. Actually, when the GQDs are excited with a femtosecond pulsed laser, the coherent photons of higher power density get up-converted to blue light (Gan et al., 2013). Though the actual mechanism behind the luminescence behavior of the carbon-based quantum dots are yet uncovered, but many of the possible processes has been reported. According to the work of Pan's group, the luminescence behavior of the GQDs arises from the zigzag sites with a carbine like triplet ground state (Pan et al., 2010). Whereas Zhu et al. (2012) assumes the luminescence of the GQDs derived from graphene oxide due to both the defect and intrinsic state emissions leading to green and blue color emissions. In some cases, it has been assumed that the luminescence arises from the π-π* electronic transitions (Liu et al., 2013). Instead of a narrow emission spectrum, these quantum dots exhibits a broad emission band (ranging from blue to near-infrared region) facilitating to be used as the multi-color imaging agent. This behavior may be due to the presence of different sized particles, existence of different emissive traps on the particle surface, the methods followed for their synthesis and purification (Baker et al., 2010; Liu et al., 2007). But the actual mechanism behind the luminescence behavior of the GQDs is still a debate for discussion and more theoretical and experimental work is needed to elucidate it.

1.4. FUNCTIONALIZATION

Carbon quantum dots attract towards itself due to its fascinating properties over the traditional fluorescent dyes and transition metal-based quantum dots. However, the low quantum yield restricted their applications in a boundary. In order to increase the photoluminescence quantum yields for diverse applications, the surface engineering is strictly required. As discussed above, the surface of the carbon-based quantum dots contains a number of functionalizable groups that were developed during their synthesis. These functional groups (carboxyl, hydroxyl, amino groups, etc.) provide reactive sites to further conjugate the CQDs surface with a variety of organic molecules, enzymes, inorganic ligands and nanoparticles through simple chemistry. It develops a coating on CQDs surface that enhances the luminescence property enabling the easy attachment of biomolecule onto them. In this part of the article, a brief discussion on the surface functionalization (includes in-situ and ex-situ functionalization) of CQDs and its importance has been presented.

1.4.1. In-Situ Functionalization

In this process, doping of the CQDs with heteroatoms are carried out. Generally, heteroatom like nitrogen (N), sulfur (S), boron (B) and phosphorous (P) are used for this type of doping process. The lone pair of electrons present on them and the defect sites created after doping plays a vital role to develop the active sites for enhanced electronic properties and diverse biological applications (Yan et al., 2012; Wang et al., 2010; Xiao et al., 2005). Mostly this can be carried out during the synthesis of CQDs, and the heteroatom comes from the carbon precursor or solvent used to carry out the synthesis. There are a lot of natural resources available to be taken as starting material for the synthesis of doped carbon dots. Soymilk, one of the natural sources (contain an abundant number of carbon, nitrogen, and oxygen in them) chosen as the precursor for the synthesis of N-doped carbon dots (Zhu et al., 2012). The blue fluorescent carbon dots were obtained by means of hydrothermal treatment and give the photoluminescent quantum yield of about 2.6%. Although there is no such prominent enhancement in the PLQY, but the presence of nitrogen in the carbon skeleton enables them to act as a bifunctional material, i.e., are not only gives excellent fluorescence behavior but also acts as a good electrocatalyst towards the reduction of oxygen. Also, the N-containing plant like grass was adopted as the starting material and the N-doped carbon dots were developed after a simple hydrothermal treatment (Liu et al., 2012). The physical characterizations support the presence of nitrogen in them (along with a plenty of carbon and oxygen), which plays a vital role for the QDs to be acting as a label-free fluorescent platform for the detection of Cu^{2+} ions. The PLQY of 6.2% was found to be some extent more than that obtained from soymilk. Afterward, so much effort had paid by the researchers for further enhancement of the quantum yield values. The protein-based bio-polymer obtained from Bombyx mori silk was used to produce N-doped carbon dots (Wu et al., 2013). These CQDs were highly fluorescent (PLQY=13.9%) having amphoteric in nature and employed for bio-imaging due to their high biocompatibility. Niu et al. (2013) have taken the Glutamic acid as the source material for both the carbon and nitrogen precursor and prepare the N-doped carbon quantum dots (NCQDs) via a one-step pyrolysis method. The NCQDs give blue color luminescence with comparatively high PLQY (28%) and employed for the detection of amoxicillin (an antibacterial drug). By taking thiomalic acid as the precursor, Chandra et al. (2013) have prepared the S-doped CQDs through a simple reflux reaction. Due to the presence of highly negative surface charge, they were easily interacting with the positively charged molecules like DNA-PEI to produce bright blue color luminescence and are applied for the cell imaging and aims to be used for drug and gene delivery purpose. Along with the single element doping, so many reports are also there regarding the enhancement of optical as well as the electronic property of CQDs by means of co-doping i.e., the multi-element doping. Dong et al. (2013) have successfully prepared the S, N-co doped CQDs by taking citric acid (source of carbon) and L-cysteine (source of nitrogen and sulfur) as the starting material through a one-step hydrothermal synthetic route. These S, N-CQDs shows very high photoluminescence property with a PLQY of 73%. Such enhanced PLQY was assumed to be due to the presence of heteroatoms (S, N) on the CQDs surface. (Figure 1.8) (Samantara et al., 2016)

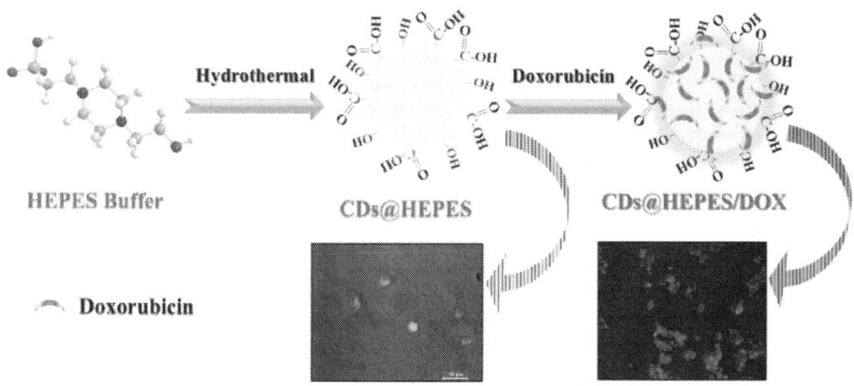

Figure 1.8: Schematic presentation of composite preparation of doxorubicin on the surface of CDs, (From Samantara et al., J. Mater. Chem. B, 2016, 4, 2412–2420).

Recently, Qu et al. (2013) developed the highly fluorescent nitrogen and sulfur co-doped GQDs from the hydrothermal treatment of citric acid, urea, and Thiourea. These N, S-GQDs are of highly photoluminescence (PLQY=78%) in nature and surprisingly shows a broad absorbance peak in the visible region. This absorbance in visible region makes them capable to be used as a visible light photocatalysts, and thus were employed for the preparation of the S, N-GQDs/TiO_2 composite and applied towards the degradation of Rhodamine B.

1.4.2. Ex-Situ Functionalization

The ex-situ surface modification process has been adopted by a number of researchers to obtain the functionalized CQDs. In other words, this is a post-treatment of the CQDs surface which makes them soluble in a broad range of solvents (from polar to non-polar solvents) depending upon the molecules/ligands chosen for modification. The main objective of such type of functionalization is to solubilize them in the organic solvents like toluene, chloroform, tetrahydrofuran (THF), hexane, etc. and to enhance the photoluminescence quantum yield. Sun et al. (2006) have prepared the carbon dots from laser ablation of the carbon target and were found to be less fluorescent in nature (having PLQY= 4%). After that, the surface passivation was carried out by treating with the PEG_{1500N} (diamine-terminated oligomeric polyethylene glycol) and PPEI-EI (poly (propionyl ethylene imine-*co*-ethylene imine). The surface passivated QDs have shown the enhanced fluorescence property with PLQY more than 10%. This enhancement in PLQY is due to the development of surface energy traps after the passivation. Also, the pi-bonds present on the surface of the CQDs helps to bind different biomolecules and therapeutic drugs onto their surface facilitating their targeted delivery.

In a particular work, heteroatom doped CQDs has been prepared from HEPES buffer molecule (CDs@HEPES) and prepare composite with a chemotherapeutic drug (Doxorubicin). This conjugate formation carried out by means of the π-π interaction/Van der Waal's interaction among the aromatic cluster present on the CQDs surface and the drug molecule. Not only the oxygen functionalities but also the pi-electron cloud assists to functionalize the CQDs surface.

1.5. APPLICATIONS

Though the carbon-based quantum dots are very new and are developed from last few years ago, but their ease of preparation/modification, tuneable optical properties, interesting physicochemical properties and negligible cytotoxicity make them suitable to employ in diverse fields ranging from energy to biomedical applications. These carbon-based tiny particles have potential to sustain their surface morphology and optical properties in different chemical environments and can be easily penetrate the cellular membrane. Furthermore, they show better hemocompatibility (dose up to 50 mg/kg of body weight) even when intravenously injected into the bloodstream without impairing the blood coagulation functions (Li et al., 2015). Therefore these bare CQDs and their conjugates are applied many more advanced biomedical applications. Out of others their application in the development of in vitro/in vivo bio-imaging agents, targeted drug delivery vehicle, fluorescence based biosensors, etc. has been exclusively studied. Gao et al. (2017) have synthesized the CQDs by hydrothermal process (from the 3, 4, 9, 10-perylenetetracarboxylic dianhydride) and functionalized to form the alkynylated CQDs (by means of amidation with propargylamine in the presence of 1, 1'-carbonyldiimidazole) using a simple click chemistry. Further they have conjugated these functionalized CQDs with DNA and gold nanoparticles and observed that these CQDs specifically target the nuclei of HEp-2 cell lines proving its potential for bio-imaging and sensing applications. Yang's group have prepared stabilizer free graphene quantum dots following electrochemical etching process from the graphite rods and observed a yellow luminescence with PLQY of 14% (Zhang et al., 2012).

These GQDs found to be easily enter into the stem cell without affecting their viability, proliferation and differentiation capability. Also their imaging performance has been successfully demonstrated with three types of stem cell lines (neurospheres cells, pancreas progenitor cells and cardiac progenitor cells). In another work, following a single step hydrothermal method highly luminescent (PLQY=47.39%) and heteroatom doped CQDs were derived from different HEPES buffer molecules (Samantara et al., 2016). These CQDs shows very negligible cytotoxicity (>95%) towards the Human Embryonic Kidney (HEK-293) cell lines (Figure 1.9). A composite of the carbon quantum dots (CDs@HEPES/DOX) with a model chemotherapeutic drug Doxorubicin has been prepared by a simple π-π/Van der Waal interaction. Very interestingly, a pH triggered targeted delivery of the doxorubicin in an in vitro analysis has been observed against the human oral squamous cell carcinoma (OSCC) cell line (H357).

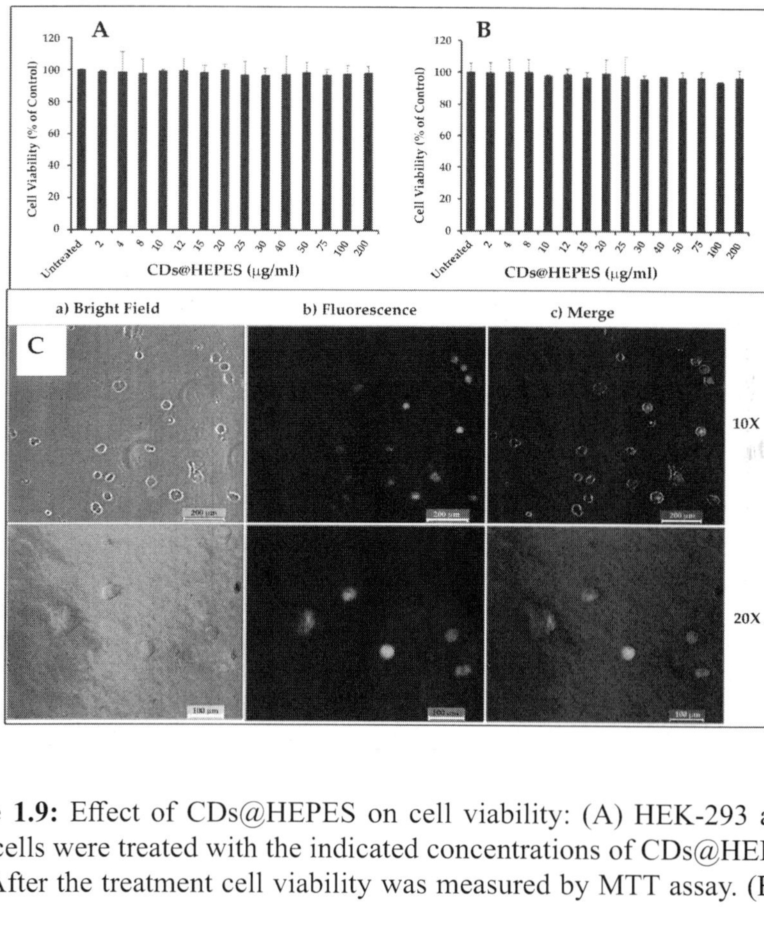

Figure 1.9: Effect of CDs@HEPES on cell viability: (A) HEK-293 and (B) H357 cells were treated with the indicated concentrations of CDs@HEPES for 48 h. After the treatment cell viability was measured by MTT assay. (Bars in-

dicate the standard deviation SD, n = 3.) (C) Cellular localization of CDs@HEPES: H357 cells were incubated with 20 mg/ml of CDs@HEPES for 6 h. Images were capture with a LEICA DMIL microscope and processed by LEICA LAS V4.2 software: (a) bright-field, (b) fluorescence and (c) merged images, (From Samantara et al., J. Mater. Chem. B, 2016, 4, 2412–2420).

Likewise graphene and fullerenes, the GQDs produce reactive oxygen species (ROS) on light irradiation. Therefore, they are broadly used as potential candidates in photodynamic therapy where the generated ROS kills the carcinogenic cells (Robertson et al., 2009). The photodynamic therapy also used to kill the microbial pathogens like bacteria, which plays a promising role to reduce the antibiotic resistances and applicable reduction of different therapies. Following the electrochemical method Ristic et al. (2014) have synthesized GQDs from graphite rod and studied the photodynamic antibacterial effect towards both the Gram-positive (Staphylococcus aureus) and Gram-negative (*Escherichia coli*) bacteria.

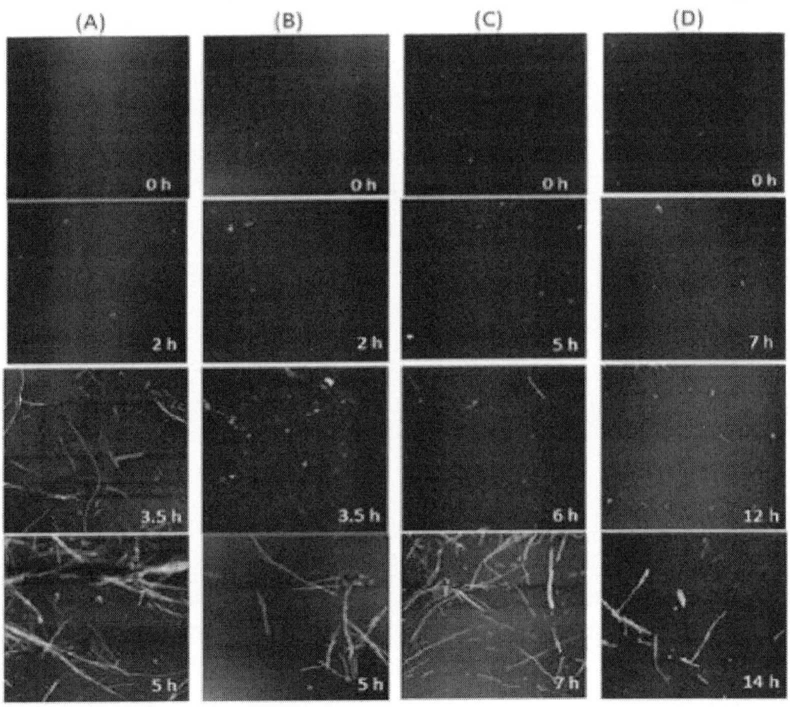

Figure 1.10: AFM images (size: 5×5 μm^2) of insulin incubated with different concentrations of C-Dots: (A) 0; (B) 0.2; (C) 2; and (D) 10 μg/mL, (From Li et al., ACS Appl. Mater. Interfaces, 2015, 7, 19153–19162).

Since the fibrillation of peptides/proteins in the extra cellular tissues have a potential effect in causing different diseases like Parkinson, Alzheimer's disease, diabetes (type-II), so its required to prevent the fibrillation process. To this end much more effort has been paid and nano particles like carbon nanotubes, surface capped transition metal quantum dots, organic molecules, polymers, etc. are employed for this purpose. Though the surface capped quantum dots shows better performances but their toxicity limits their application. Afterwards, Li et al. (2015) have synthesized the CQDs from the carbon powder by means of acid treatment and studied the anti-fibrillation effect of these carbon dots using the human insulin as the model protein. The inhibiting performance observed to be a dose dependent and it increases the lag period of human insulin (0.2 mg/ml) fibrillation from 2.5 hour to 5.5 hour on addition of 10 μg/ml of CQDs that has been clearly seen from Figure 1.10.

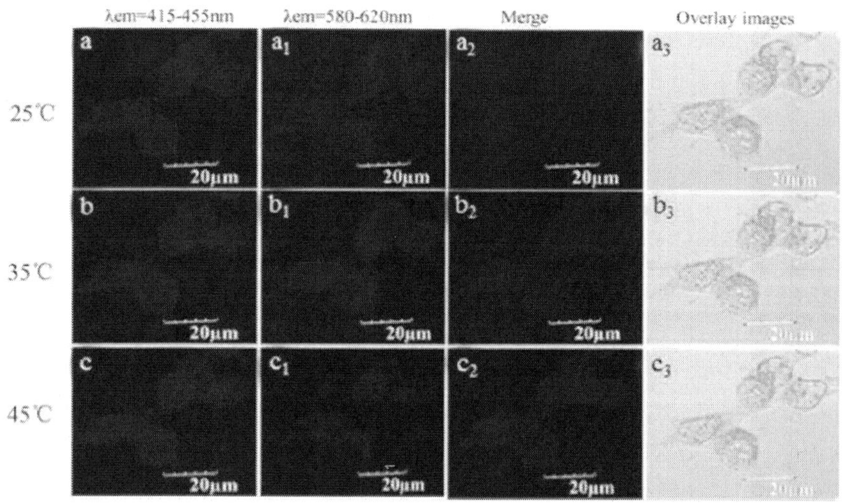

Figure 1.11: Confocal fluorescence microscopy images of 293T cells with internalized DEFNs at three different temperatures: a, 25°C; b, 35°C; c, 45°C. (1) Fluorescence images at the blue channel; (2) Fluorescence images at the red channel. (3) Overlap of the fluorescence images; (4) Corresponding bright field images, (From Li et al., Chem. Mater., 2015, 27, 1764–1771).

To monitor the biological reactions occurring inside the cell is strictly required to get more information regarding the biological processes and to facilitate the development of biomedicines. There are different parameters available to measure, but the measurement of physiological temperature and its gradient inside the living cell by means of fluorescence based sensors proved to have more potential. In 2016, Zhang's group have developed a ratio metric thermo senor by conjugating the blue luminescent CQDs (amine functionalized CQDs) with red luminescent gold nano clusters (glutathione stabilized). These conjugates shows better biocompatibility facilitating the measurement of physiological temperature and its gradient in the range of 25–45°C inside the cell (293T Cell line). Since on increasing the temperature, an increase in the non-radiative recombination of electrons and holes on the GSH-Au NCs surface takes place, therefore these functionalized gold nanoclusters (GSH-Au NCs) were employed as the nano thermometric device (due to their temperature based tunable fluorescent emission intensity and lifetime). Therefore a reduced red emission intensity (that from the GSH-Au NCs) has been observed on increasing the physiological temperature, whereas the blue emission from CQDs affected a little (Figure 1.11). This variation in the luminescence intensity with temperature make the conjugate not only to monitor the temperature/temperature gradient but also a temperature based cellular imaging has been recorded.

1.6. SUMMARY AND FUTURE PERSPECTIVES

As per the above discussion, the CQDs shows dominating behavior over the transition metal based quantum dots and fluorescent dyes. The easy synthetic route, negligible toxicity, interesting optical and chemical properties makes the CQDs to be used in broad range of scientific application. But some issues are still remain for further study, (a) Although there are so many reports available for the synthesis CQDs but still an easy route for bulk quantity synthesis required and also more work is needed to elucidate the mechanism for the fluorescence behavior of the CQDs, (b) all most all the reports involve the preparation of mixed size CQDs, but development of uniform sized particles is still desirable, (c) again the surface modification of the CQDs by organic molecules (like $-NH_2$ terminated molecules) for the ease of solubility in various organic solvents is well documented. But the attachment of important anticancer drugs onto the bare CQDs surface is still an ongoing challenge for the cancer therapy, (d) to investigate the actual mechanism of fluorescent enhancement through the surface functionalization of the CQDs, (d) the application of CQDs and their functionalized counterparts in biomedical to be broadly explored.

Un-doubtly, the quantum efficiency of the CQDs can be enhanced by means of surface functionalization. These highly fluorescent functionalized

CQDs must be used in various field of application i.e., from energy to biomedical applications due to their negligible toxicity, ease of synthesis, no blinking property, etc.

REFERENCES

1. Aldeek, F., Mustin, C., Balan, L., Roques-Carmes, T., Fontaine-Aupart, M. & Schneider, R. (2011). Surface-engineered quantum dots for the labeling of hydrophobic micro domains in bacterial bio films, *Biomaterials*, *32*, 5459.
2. Alivisatos, A., Gu, W., Larabell, C. (2005). Quantum dots as cellular probes, *Annu. Rev. Biomed. Eng*, 7, 55.
3. Anilkumar, P., Wang, X., Cao, L., Sahu, S., Liu, J., Wang, P., Korch, K., Tackett II, K., Parenzana, A., & Sun, Y. P. (2011). Toward quantitatively fluorescent carbon-based "quantum" dots, *Nanoscale*, *3*, 2023–2027.
4. Antonietti, M. (2001). Surfactants for novel templating applications. *Curr. Opin. Colloid Interface Sci., 6*, 244.
5. Baker, S., & Baker, G. A. (2010). Luminescent carbon nanodots: emergent nanolights. *Angew.Chem. Int. Ed. 49*, 6726.
6. Bera, D., Qian, L., Tseng, T., & Holloway P. H. (2010). Quantum dots and their multimodal applications: a review, *Materials*, *3*, 2260.
7. Bhunia, S., Pradhan, N., & Jana, N. R. (2014). Vitamin B1 derived blue and green fluorescent carbon nanoparticles for cell-imaging application, *ACS Appl. Mater. Interfaces*, *6*, 7672.
8. Bhunia, S., Saha, A., Maity, A., Ray, S., & Jana, N. R. (2013). Carbon nanoparticles-based fluorescent bioimaging probes, *Sci. Rep. 3,* 1473.
9. Bourlinos, A., Stassinopoulos, A., Anglos, D., Zboril, R., Karakassides, M., & Giannelis, E. P. (2008). surface functionalized carbogenic quantum dots. *Small*, *4*, 455.
10. Cao, L., Sahu, S., Anilkumar, P., Bunker, C., Xu, J., Fernando, K. A. S., Wang, P., Guliants, E., Tackett, K., & Sun, Y. P. (2011). Carbon nanoparticles as visible-light photocatalysts for efficient CO_2 conversion and beyond. *J Am Chem Soc*, *133*, 4754.
11. Cao, Li., Wang, Xin., Meziani, M., Lu, F., Wang, H., Luo, P., Lin, Y., Harruff, B., Veca, L., Murray, D., Xie, S., & Sun, Y. P. (2007). Carbon dots for multiphoton bioimaging. *J Am Chem Soc*, *129*, 11318.
12. Chandra, S., Patra, P., Pathan, S., Roy, S., Mitra, S., Layek, A., Bhar, R., Pramanik, P., & Goswami, A. (2013). Luminescent S-doped carbon dots: an emergent architecture for multimodal applications, *J. Mater. Chem. B*, *1*, 2375.
13. Demir, C., Baccile, N., Antonietti, M., & Titirici, M. M. (2009).

Carboxylate-rich carbonaceous materials via one-step hydrothermal carbonization of glucose in the presence of acrylic acid, *Chem. Mater.*, *21*, 484.

14. Ding, X. (2014). Direct synthesis of graphene quantum dots on hexagonal boron nitride substrate, *J. Mater. Chem. C*, *2*, 3717.
15. Dong, Y., Chen, C., Zheng, X., Gao, L., Cui, Z., Yang, H., Guo, C., Chi, Y., & Li, C. M. (2012). One-step and high yield simultaneous preparation of single- and multi-layer graphene quantum dots from CX-72 carbon black, *J. Mater. Chem.*, *22*, 18, 8764.
16. Dong, Y., Pang, H., Yang, H., Guo, C., Shao, J., Chi, Y., Li, C., & Yu, T. (2013). Carbon-based dots co-doped with nitrogen and sulfur for high quantum yield and excitation-independent emission, *Angew. Chem. Int. Ed.* 52, 7800.
17. Fang, Y., Guo, S., Li, D., Zhu, C., Ren, W., & Dong, S. (2012). Easy synthesis and imaging applications of cross-linked green fluorescent hollow carbon nanoparticles. *ACS Nano*, *6*, 400.
18. Gan, Z., Wu, X., Zhou, G., Shen, J., & Chu, P. K. (2013). Is there real up conversion photoluminescence from graphene quantum dots? *Adv. Opt. Mater*, *1,* 554–558.
19. Gao, J., & Xu, B. (2009). Applications of nanomaterials inside cells, *Nano Today*, *4*, 37.
20. Gao, M., Yang, L., Zheng, Y., Yang, X., Zou, H., Han, J., Liu, Z., Li, Y., & Huang, C. Z. (2017). "Click" on alkynylated carbon quantum dots: an efficient surface functionalization for specific biosensing and bioimaging, *Chem. Eur. J.*, *23*, 2171–2178.
21. Guo, X., Wang, C., Yu, Z., Chen, L., & Chen, S. (2012). Facile access to versatile fluorescent carbon dots toward light-emitting diodes. *Chem Commun*, *48*, 2692.
22. Hsu, P., Chang, H. T. (2012). Synthesis of high-quality carbon nanodots from hydrophilic compounds: role of functional groups, *Chem. Commun.*, *48*, 3984.
23. Jahan, S., Mansoor, F., Naz, S., Lei, J., & Kanwal, S. (2013). Oxidative synthesis of highly fluorescent boron/nitrogen co-doped carbon nanodots enabling detection of photosensitizer and carcinogenic dye, *Anal. Chem.* *85*, 10232.
24. Jaiswal, J., Mattoussi, H., Mauro, J., & Simon, S. (2013). *Nat. Biotechnol.*, *21*, 47–51.

25. Jie, G., Huang, H., Sun, X., Zhu, J. J. (2008). Electrochemiluminescence of CdSe quantum dots for immunosensing of human prealbumin, *Biosensors and Bioelectronics*, *23*, 1896.
26. Kim, S., Hwang, S., Kim, M., Shin, D., Shin, D., Kim, C., Yang, S., Park, J., Hwang, E., Choi, S., Ko, G., Sim, S., Sone, C., Choi, H., Bae, S., & Hong, B. H. (2012). Anomalous behaviors of visible luminescence from graphene quantum dots: interplay between size and shape. *ACS Nano*, *6*, 8203.
27. Kozák, O., Datta, K. K. R., Greplová, M., Ranc, V., Kašlík, J., & Zbořil, R. (2013). Surfactant-derived amphiphilic carbon dots with tunable photoluminescence, *J. Phys. Chem. C, 117*, 24991.
28. Kwon, W., Kim, Y., Lee, C., Lee, M., Choi, H., Lee, T., & Rhee, S. W. (2014). Electroluminescence from graphene quantum dots prepared by amidative cutting of tattered graphite, *Nano Lett., 14,* 3, 1306.
29. Lee, J., Kim, K., Park, W., Kim, B., Park, J., Kim, T., Bong, S., Kim, C., Chae, G., Jun, M., Hwang, Y., Jung, Y., & Jeon, S. (2012). Uniform graphene quantum dots patterned from self-assembled silica nanodots. *Nano Lett*, *12*, 6078.
30. Li, H., He, X., Kang, Z., Huang, H., Liu, Y., Liu, J., Lian, S., Tsang, C. H. A., Yang, X., & Lee, S. T. (2010). Water soluble fluorescent carbon quantum dots and photocatalyst design. *Angew Chem Int Ed*, *49*, 4430.
31. Li, L., & Yan, X. (2010). Colloidal Graphene Quantum Dots. *J Phys Chem Lett*, *1*, 2572.
32. Li, S., Guo, Z., Zhang, Y., Xue, W., & Liu, Z. (2015). Blood compatibility evaluations of fluorescent carbon dots, *ACS Appl. Mater. Interfaces*, *7*, 19153–19162.
33. Li, S., Wang, L., Chusuei, C., Suarez, V., Blackwelder, P., Micic, M., Orbulescu, J., & Leblanc, R. M. (2015). Nontoxic carbon dots potently inhibit human insulin fibrillation, *Chem. Mater.*, *27*, 1764–1771.
34. Li, X., Wang, H., Shimizu, Y., Pyatenko, A., Kawaguchi, K., & Koshizaki, N. (2011). Preparation of carbon quantum dots with tunable photoluminescence by rapid laser passivation in ordinary organic solvents. *Chem Comm*, *47*, 932.
35. Li, Y., Zhao, Y., Cheng, H., Hu, Y., Shi, G., Dai, L., & Qu, L. (2012). Nitrogen-doped graphene quantum dots with oxygen-rich functional groups, *J. Am. Chem. Soc. 134*, 15.
36. Liang, Q., Ma, W., Shi, Y., Li, Z., & Yang, X. (2013). Easy synthesis

of highly fluorescent carbon quantum dots from gelatin and their luminescent properties and applications, *Carbon*, *60*, 421.

37. Lin, L., & Zhang, S. (2012). Creating high yield water soluble luminescent graphene quantum dots via exfoliating and disintegrating carbon nanotubes and graphite flakes. *Chem Comm*, *48*, 10177.
38. Liu, C., Zhang, P., Zhai, X., Tian, F., Li, W., Yang, J., Liu, Y., Wang, H., Wang, W., & Liu, W. (2012). Nano-carrier for gene delivery and bioimaging based on carbon dots with PEI-passivation enhanced fluorescence, *Biomaterials*, *33*, 3604–3613.
39. Liu, H., Ye, T., & Mao, C. (2007). Fluorescent carbon nanoparticles derived from candle soot. *Angewandte Chemie International Edition*, *46*, 6473–6475.
40. Liu, Q., Guo, B., Rao, Z., Zhang, B., & Gong, J. R. (2013). Strong two-photon-induced fluorescence from photostable, biocompatible nitrogen-doped graphene quantum dots for cellular and deep-tissue imaging, *Nano Lett.*, *13*, 2436–2441.
41. Liu, R., Wu, D., Liu, S., Koynov, K., Knoll, W., Li, Q. (2009). An aqueous route to multicolor photo luminescent carbon dots, *Angew. Chem., Int. Ed.*, *48*, 4598.
42. Liu, S., Tian, J., Wang, L., Zhang, Y., Qin, X., Luo, Y., Asiri, A., Al-Youbi, A., & Sun, X. (2012). Hydrothermal treatment of grass: a low-cost, green route to nitrogen-doped, carbon-rich, photo luminescent polymer nanodots as an effective fluorescent sensing platform for label-free detection of cu (ii) ions, *Adv. Mater.*, *24*, 2037.
43. Liu, Y., Xiao, N., Gong, N., Wang, H., Shi, X., Gu, W., Ye, L. (2014). One-step microwave-assisted polyol synthesis of green luminescent carbon dots as optical nanoprobes, *Carbon*, *68*, 258.
44. Mitra, S., Chandra, S., Kundu, T., Banerjee, R., Pramanik, P., & Goswami, A. (2012). Rapid microwave synthesis of fluorescent hydrophobic carbon dots, *RSC Adv.*, *2*, 12129.
45. Niu, J., & Gao, H. (2014). Synthesis and drug detection performance of nitrogen-doped carbon dots, *Journal of Luminescence*, *149*, 159.
46. Pan, D., Zhang, J., Li, Z., & Wu, M. (2010). Hydrothermal route for cutting graphene sheets into blue-luminescent graphene quantum dots, *Adv. Mater.*, *22*, 6, 734.
47. Peng, H., Sejdic, J. T. (2009). Simple aqueous solution route to

luminescent carbogenic dots from carbohydrates, *Chem. Mater.*, *21*, 5563.

48. Peng, J., Gao, W., Gupta, B. K., Liu, Z., Aburto, R. R., Ge, L., Song, L., Alemany, L. B., Zhan, X., Gao, G., Vithayathil, S. A., Kaipparettu, B. A., Marti, A. A., Hayashi, T., Zhu, J. J., & Ajayan, P. M; (2012). Graphene quantum dots derived from carbon fibers, *Nano Lett.*, *12*, 2, 844.
49. Philpott, M., Kawazoe, Y. (2012). Graphene nanodots with intrinsically magnetic protrusions, *J Chem Phys*, *136*, 064706.
50. Pileni, M. P. (2007). Control of the size and shape of inorganic nanocrystals at various scales from Nano to macro domains. *J. Phys. Chem. C*, *111*, 9019.
51. Qiao, Z., Wang, Y., Gao, Y., Li, H., Dai, T., Liua, Y., & Huo, Q. (2010). Commercially activated carbon as the source for producing multicolor photo luminescent carbon dots by chemical oxidation, *Chem. Commun.*, *46*, 8812.
52. Qu, D., Zheng, M., Du, P., Zhou, Y., Zhang, L., Li, Di., Tan, H., Zhao, Z., Xie, Z., & Sun, Z. (2013). Highly luminescent S, N co-doped graphene quantum dots with broad visible absorption bands for visible light photo catalyst, *Nanoscale*, *5*, 12272.
53. Ristic, B., Milenkovic, M., Dakic, I., Markovic, B. M. T., Milosavljevic, M., Budimir, M., Paunovic, V., Dramicanin, M., Markovic, Z., & Trajkovic, V. S. (2014). Photodynamic antibacterial effect of graphene quantum dots, *Biomaterials*, *35*, 4428–4435.
54. Robertson, C., Evans, D., & Abrahamse, H. (2009). Photodynamic therapy (PDT): a short review on cellular mechanisms and cancer research applications for PDT. *J Photochem Photobiol B*, *96*, 1–8.
55. Samantara, A., Maji, S., Ghosh, A., Bag, B., Dash, R., & Jena, B. K. (2016). Good's buffer derived highly emissive carbon quantum dots: excellent biocompatible anticancer drug carrier, *J. Mater. Chem. B*, *4*, 2412–2420.
56. Shen, J., Zhu, Y., Chen, C., Yang, X., & Li, C. (2011). Facile preparation and up conversion luminescence of graphene quantum dots, *Chem. Commun.*, *47*, 2580–2582.
57. Shen, J., Zhu, Y., Yang, X., Zong, J., Zhang, J., & Li, C. (2012). One-pot hydrothermal synthesis of graphene quantum dots surface-passivated by polyethylene glycol and their photoelectric conversion under near-

infrared light. *New J Chem*, *36*, 97.

58. Shen, L., Zhang, L., Chen, M., Chen, X., & Wang, J. (2013). The production of pH-sensitive photoluminescent carbon nanoparticles by the carbonization of polyethyleneimine and their use for bioimaging, *Carbon*, *55*, 343–349.
59. Shinde, D., & Pillai, V. K. (2012). Electrochemical preparation of luminescent graphene quantum dots from multiwalled carbon nanotubes, *Chem. Eur. J. 18*, 12522.
60. Stassinopoulos, A., Bourlinos, A., Anglos, D., Zboril, R., Georgakilas, V., & Giannelis, E. P. (2008). Photo luminescent carbogenic dots, *Chem. Mater.*, *20*, 4539.
61. Sun, Y., Zhou, B., Lin, Y., Wang, W., Fernando, K. A. S., & Pathak, P. (2006). Quantum-sized carbon dots for bright and colorful photoluminescence, *J. Am. Chem. Soc.*, *128*, 7756.
62. Tang, L., Ji, R., Cao, X., Lin, J., Jiang, H., Li, X., Teng, K., Luk, C., Zeng, S., Hao, J., & Lau, S. P. (2012). deep ultraviolet photoluminescence of water-soluble self-passivated graphene quantum dots, *ACS Nano*, *6*, 5102.
63. Tang, Y., Yang, Q., Wu, T., Liu, L., Ding, Y., & Yu, B. (2014). Fluorescence enhancement of cadmium selenide quantum dots assembled on silver nanoparticles and its application to glucose detection, *Langmuir*, *30*, 6324.
64. Wang, C., Lin, H., Xu, Z., Huang, Y., Humphrey, M., & Zhang, C. (2016). Tunable carbon-dot-based dual-emission fluorescent nanohybrids for ratiometric optical thermometry in living cells, *ACS Appl. Mater. Interfaces*, *8*, 6621–6628.
65. Wang, F., Chen, Y., Liu, C., & Ma, D. (2011). White light-emitting devices based on carbon dots' electroluminescence. *Chem Commun*, *47*, 3502.
66. Wang, F., Pang, S., Wang, L., Li, Q., Kreiter, M., & Liu, C. Y. (2010). One-step synthesis of highly luminescent carbon dots in noncoordinating solvents. *Chem Mater*, *22*, 4528.
67. Wang, F., Xie, Z., Zhang, H., Liu, C., & Zhan, Y. G. (2011). Highly luminescent organosilane-functionalized carbon dots, *Adv. Funct. Mater. 21*, 1027.
68. Wang, J., Wang, C., & Chen, S. (2012). Amphiphilic egg-derived carbon dots: rapid plasma fabrication pyrolysis process, and multicolor

printing patterns, *Angew. Chem. Int. Ed.*, *51*, (9297)..

69. Wang, T., Wu, Y., & Zeng, A. J. (2010). Synthesis and characterization of amphiphilic pluronic (f68) –1,2-dipalmitoyl-sn-glycero-3-phosphoethanolamine copolymers and their micelles as a drug carrier, *Journal of Applied Polymer Science*, *117*, 604.
70. Wang, Y., Shao, Y., Matson, D., Li, J., & Lin, Y. (2010). Nitrogen-doped graphene and its application in electrochemical biosensing, *ACS Nano*, *4*, 1790.
71. Wei, W., Xu, C., Ren, J., Xu, B., & Qu, X. (2012). Sensing metal ions with ion selectivity of a crown ether and fluorescence resonance energy transfer between carbon dots and grapheme. *Chem Commun*, *48*, 1284.
72. Wu, Z., Zhang, P., Gao, M., Liu, C., Wang, W., Lenga, F., & Huang, C. Z. (2013). One-pot hydrothermal synthesis of highly luminescent nitrogen-doped amphoteric carbon dots for bio imaging from bombyx mori silk—natural proteins, *J. Mater. Chem. B*, *1*, 2868.
73. Xiao, K., Liu, Y., Hu, P., Yu, G., Sun, Y., & Zhu, D. (2005). N-Type field-effect transistors made of an individual nitrogen-doped multiwalled carbon nanotube, *J. Am. Chem. Soc. 127*, 8614.
74. Xu, X; Ray, R; Gu, Y; Ploehn, H.J; Gearheart, L; Raker, K; Scrivens, & Walter A. (2004). Electrophoretic analysis and purification of fluorescent single-walled carbon nanotube fragments, *J. Am. Chem. Soc.*, *126*, 12736–12737.
75. Yang, S., Cao, L., Luo, P., Lu, F., Wang, X. Wang, H., Meziani, M., Liu, Y., Qi, G., & Sun, Y., (2009). Carbon dots for optical imaging in vivo, *J. Am. Chem. Soc.*, *131*, 11308–11309.
76. Yang, Y., Cui, J., Zheng, M., Hu, C., Tan, S., & Xiao, Y. (2012). One step synthesis of amino-functionalized fluorescent carbon nanoparticles by hydrothermal carbonization of Chitosan, *Chem. Commun.*, *48*, 380.
77. Yang, Z., Wang, M., Yong, A., Wong, S. Yee, Xin-Hai Zhang, Happy Tan, Alex Yuangchi Chang, Xu Li & John Wang, (2011). Intrinsically fluorescent carbon dots with tunable emission derived from hydrothermal treatment of glucose in the presence of monopotassium phosphate, *Chem. Commun.*, *47*, 11615.
78. Yu, B., & Kwak, S. Y. (2012). Carbon quantum dots embedded with mesoporous hematite nanospheres as efficient visible light active photocatalysts. *J Mater Chem*, *22*, 8345.

79. Zhai, X., Zhang, P., Liu, C., Bai, T., Li, W., Daic, L., & Liu, W. (2012). Highly luminescent carbon nanodots by microwave-assisted Pyrolysis, *Chem. Commun.*, *48*, 7955.

80. Zhang, M., Bai, L., Shang, W., Xie, W., Ma, H., Fu, Y., Fang, D., Sun, H., Fan, L., Han, M., Liu, C., & Yang, S. (2012). Facile synthesis of water-soluble, highly fluorescent graphene quantum dots as a robust biological label for stem cells, *J. Mater. Chem.*, *22*, 7461.

81. Zhang, M., Bai, L., Shang, W., Xie, W., Ma, H., Fu, Y., Fang, D., Sun, H., Fan, L., Han, M., Liub, C., & Yang, S. (2012). Facile synthesis of water-soluble, highly fluorescent graphene quantum dots as a robust biological label for stem cells, *J. Mater. Chem.*, *22*, 7461–7467.

82. Zheng, L., Chi, Y., Dong, Y., Lin, J., & Wang, B. B. (2009). Electrochemiluminescence of water-soluble carbon nanocrystals released electrochemically from graphite, *J. Am. Chem. Soc.*, *131*, 4564.

83. Zhu, C., Zhai, J., & Dong, S. (2012). Bifunctional fluorescent carbon nanodots: green synthesis via soy milk and application as metal-free electrocatalysts for oxygen reduction, *Chem. Commun.*, *48*, 9367.

84. Zhu, S., Zhang, J., Qiao, C., Tang, S., Li, Y., Yuan, W., Li, B., Tian, L., Liu, F., Hu, R., Gao, H., Wei, H., Zhang, H., Sun, H., & Yang, B. (2011). Strongly green-photoluminescent graphene quantum dots for bioimaging applications, *Chem. Commun.*, *47*, 6858–6860.

85. Zhu, S., Zhang, J., Qiao, C., Tang, S., Li, Y., Yuan, W., Li, B., Tian, L., Liu, F., Hu, R., Gao, H., Wei, H., Zhang, H., Sun, H., & Yang, B., (2011). Strongly green-photo luminescent graphene quantum dots for bio imaging applications, *Chem. Commun.*, *47*, *24*, 6858–6860.

86. Zhu, S., Zhang, J., Tang, S., Qiao, C., Wang, L., Wang, H., Liu, X., Li, B., Li, Y., Yu, W., Wang, X., Sun, H., & Yang, B. (2012). Surface chemistry routes to modulate the photoluminescence of graphene quantum dots: from fluorescence mechanism to up-conversion bioimaging applications, *Adv. Funct. Mater*. *22*, 4732–4740.

2
CHAPTER

RECONNAISSANCE OF PROTEIN-NANOPARTICLE CONJUGATES: A MINI-REVIEW ON DEVELOPMENT AND APPLICATION

Lakkoji Satish[1] , Ayonbala Baral[2,3] , and Malay Kumar Ghosh[2,3]

[1] Department of Chemistry, National Institute of Technology, Rourkela – 769008, Odisha, India

[2] Hydro & Electrometallurgy Department, CSIR-Institute of Minerals and Materials Technology, Bhubaneswar – 751013, Odisha, India

[3] AcSIR CSIR-Institute of Minerals and Materials Technology, Bhubaneswar – 751013, Odisha, India

CONTENTS

ABSTRACT

The application of nanomaterials in wide range of areas such as food, chemical, pharmaceutical and biomedical sciences has gathered significant attention in recent years. Apart from their trending applications in tumor imaging, molecular profiling of cancer biomarkers, and targeted drug delivery, the immobilization of proteins on nanomaterials has been escalating in the last few decades (though it was introduced a century ago). The development of protein-nanoparticle conjugates is not only essential for improving biomedical research and applications but also for food and chemical industries. A major portion of the research has been focused on bio electronic systems and energy generation in biofuel cells as well as the protein/enzyme based biocatalytic reactions. For all such processes, the immobilization of protein/enzymes on these nanomaterials is required which can maintain and even enhance the performance. In this chapter, the current status and trends in protein immobilization applications in food, chemical, and pharmaceutical industries along with the available immobilization techniques have been presented. Also, a brief overview of the protein-nanoparticle interaction and the consequences of the physicochemical properties of nanoparticles such as size, composition, and surface charge, etc. on the structure and stability of proteins are provided. Hopefully the information given here will be helpful not only to develop novel nanomaterials but also in tuning the bioconjugates for a particular purpose.

Keywords: Nanoparticles, Protein corona, Immobilization, Spectroscopy, Surface analysis

2.1. INTRODUCTION

Rapid progress in nanotechnology has contributed immensely for the fabrication of nano sized materials and devices having various applications in medicine, agriculture and energy production (Gonzalez-Melendi et al., 2008; Engel et al., 2008; Heath and Davis, 2008; Navalakhe and Nandedkar, 2007; Chen et al., 2009; Gidaspow and Jiradilok, 2007). Recently, research on the synthesis and design of novel nanomaterials with different morphologies have been intensified. Nanomaterials (NMs) have been extensively studied owing to their distinctive structural, physicochemical characteristics, consider as ideal for different research areas (physics, chemistry, biology, and engineering) (Alivisatos, 1996; Chan and Nie, 1998; Dujardin and Mann,

2002; Nirmal and Brus, 1999). The unique properties these nanomaterials possess are due to their low dimensional shape (< 100nm, at least for a single dimension), however, there are only a few literature reports available in support of the upper limit (~100 nm) for the nanomaterials (Auffan et al., 2009; Kreyling et al., 2010). The nanomaterials cover a large number of organic and inorganic compounds, but there are few exceptions, e.g., naturally occurring biomolecules such as sugar or lipid, which cannot be considered as nanomaterial although the sizes of these molecules are in nanoscale. Herein, we assume that NMs are a class of material (whether biotic or abiotic), at the submicrometer level having at least a single dimension in the nanoscale and could exhibit altogether different properties as compared to their bulk counterpart. A nanomaterial must have distinct functional or structural parts on the surface or in the internal arrays possessing unique properties or compositions in nanoscale that might otherwise not be observed for the same in the macroscale. In this chapter, nanoparticles (NPs) produced from metals, noble metals, oxides, etc. along with a variety of variety of nonmetal NPs are discussed. Extensively studied polymeric NPs, metal oxide NPs, metallic NPs, and carbon-based nanocomposites, proteins, enzymes, and other assorted materials are also included.

Nanoparticles have drawn significant attention due to their diverse chemical, optoelectrical, mechanical, electrochemical, magnetic properties (Steigerwald and Brus, 2002; Wang, 2002; Weller, 1993). The dependence of optical, electrical, and magnetic properties on the size of nanoparticles make it an effective tool for biological applications such as bio sensing, bio imaging, therapeutics and in-vivo applications, cell targeting and drug delivery (Corot et al., 2006; de la Fuente et al., 2006; Medintz et al., 2005). The use of NPs towards biological and medical applications may originate from the basis that the size of NPs are very small and resemble with the cellular machinery, which simplify their interaction with NPs (Mahmoudi et al., 2011a; Ragnaill et al., 2011). In accordance with the reported theoretical data, it is well perceived that nanoparticles can be controlled in such a manner that they arrive at the inaccessible target machinery on the right time. This activity of NP could be helpful for the annihilation of pathogens like bacteria, viruses and also cancer cells. The targeted pathogenic agents which cause detrimental effect can be regulated and controlled without affecting healthy tissues. Apart from medical applications, nanoparticles can be used as carriers of therapeutic biomolecules (Alonso, 1996). The precise and versatile hybrid systems can be prepared by combining nanoparticles and biomolecules (proteins or specific polypeptides) which offer ample opportunities for the design of nano-bio conjugates that can be used for treating cancer and different immunological disorders (Drexler, 1981; Rajagopal and Schneider, 2004; Sinha et al., 2006).

In biological applications, NPs encounter a complex system consisting of cells and extracellular proteins. When nanoparticles are injected directly into the bloodstream, they come in contact with the red and white blood cells, clotting factors, and serum proteins. Several thousand proteins in the biological fluids of human body compete with each other for binding to the surface of nanoparticles, but only few hundred numbers of proteins show high affinity. These protein molecules adsorb onto the surface of the NPs and form a protein corona (Lynch et al., 2007; Treuel and Nienhaus, 2012; Walczyk et al., 2010; Walkey and Chan, 2012; Yang et al., 2013). The formation of nano-bio interface is one of the major biological responses towards NPs but the conjugation formed between biomolecules and NPs might be influenced by the fundamental features of NPs such as; shape, size, crystallinity, charge, hydrophobicity, and wettability, etc.), which control the process of the protein corona formation (Lindman et al., 2007; Nel et al., 2009). The biomolecule absorption to the surface of NPs determines the cellular/tissue response and deliberate an innovative "biological identity" in the biological environment (Lynch and Dawson, 2008; Norde and Lyklema, 1991a). Also, the surface area to volume ratio of these NPs are significantly high as compared to the bulk counterpart, hence, they offer a very active surface chemistry for the biological applications where the large surface energy of NPs drastically reduces due to the interaction with the dispersing media (Klein, 2007; Monopoli et al., 2011). Therefore, the dispersion of nanoparticles in a biological medium is a superficial phenomenon where NP surfaces are covered with a complex layer of biomolecules. The results of nanomaterials interaction with environmental biomolecules lead to the formation of a biological corona on the NP's surface which is relatively dissimilar to that forms by adsorption on a flat surface of the bulk material under the same set of parameters. When the biological individual such as organs, tissues and cells interact with NPs, they form nano-bio conjugates and the surfaces of the conjugates are found to be different from the native surface of NPs. The ability of proteins towards adsorption on the NPs surfaces may augment the probability of partial protein unfolding which can accelerate the protein association forming new protein clusters. The protein clusters may form harmful products (and can cause diseases through the processes like amyloidoses or could introduce new biological interactions at the biomolecules – NPs interface (Chiti and Dobson, 2006; Huff et al.,

2003; Koo et al., 1999; Dawson, 2002; Linse et al., 2007; Stradner et al., 2004). In order to produce safe NPs for biological applications, it is essential to explore NP-protein complex. The nano toxicology and safety assessment research are necessary for development for the vast use of nanoparticles in biological applications. The safety evaluation and biocompatibility of NPs for various biomedical applications can be established through protein-nanoparticle interactions. However, for an in-depth perceptive for the formation of protein corona and its consecutive biological interaction require dynamic tools to examine the nano-bio interfaces.

In this chapter, we present a broad discussion on the consequence of characteristics of NPs (such as shape, size, composition, surface chemistry, etc.) on the protein structure and function after the biomolecule-NP interaction. This would definitely provide an efficient perceptive on the protein-NP interactions which would be helpful to determine the fate of the NP and its behavior in biological medium. The evaluation of protein-NP interaction is not only important for the toxicology but also helpful for examining the application of NPs in protein/enzyme immobilization in immunoassay, biosensor, bio-separation, targeted drug delivery, and environmental analysis. Basically, enzyme-catalyzed reactions proceed through mild conditions and sometimes are considered as alternatives to diverse conventional chemical reactions, but free enzymes show reduced stability towards heat, pH or other factors and are difficult to recover and reuse (Kim et al., 2006; Tischer and Wedekind, 1999). This requires the development of various means to improve enzyme stability and its reusability. Immobilization techniques are used for the preparation of high-performing and less costly biocatalysts with enhanced stability and reusability. Hence, by improving the immobilization techniques and careful material selection, the immobilized enzymes can work within a wide range of pH and will have better thermal stability in comparison to native ones. Various immobilization technologies and new types of carriers have been examined for the enhancement of stability, loading and activity of enzymes which could increase the cost-effectiveness of enzyme biocatalysts in industrial application (Garcia-Galan et al., 2011; Romdhane et al., 2011; Stepankova et al., 2013).

2.2. NANOPARTICLE CLASSIFICATION AND APPLICATIONS

Nanotechnology can be defined by the designing of mater in an atomic, molecular and supramolecular scale. The technology involving in the fabrication, characterization, and application of different nano sized materials which has been enhanced the advancement of nanotechnology for the betterment of science and society. Different kind of nanoparticles, including metal, oxide, semiconductor, and even composite nanoparticles, have been extensively studied for different applications.(Luo et al., 2006) Herein, the classification of NPs are discussed which are categorized on the basis of different parameters such as; source (anthropogenic or natural), component forms (organic and inorganic), genesis (atmospheric, geogenic, biogenic, or anthropogenic), particle's shape, size, etc. (López-Serrano et al., 2014; L´opez-Serrano, 2014) (Figure 2.1).

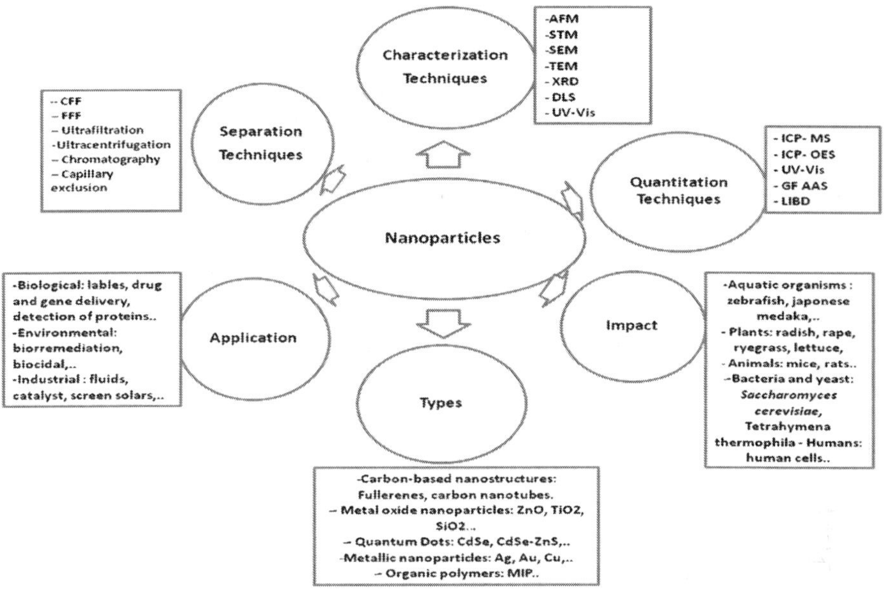

Figure 2.1: Nanoparticles types, characterization, application, and impact (From López-Serrano et al., Anal Methods, 6, 38, 2014).

2.2.1. Types of Nanoparticle Based on Chemical Composition

2.2.1.1. Polymer Nanoparticles

The use of polymer nanoparticles (PNP) is rapidly increasing and finding applications in wide areas such as electronics, photonics, sensors, medicine, environmental technology, etc. (Rao and Geckeler, 2011). Therefore, several approaches have been made to form multicomponent polymer structures with distinct and predictable phase-separation at the nanometer scale (Kietzke et al., 2003). The optically and electronically active semiconducting π-conjugated polymers (SPs) have been considered for many applications, ranging from electronic devices and sensors to tissue engineering. At present, semiconducting polymer nanoparticles (SPNs) are effectively used as nanoplatform that allow photons to generate ultrasound waves to assist in in-vivo photoacoustic molecular imaging (Pu et al., 2014).

2.2.1.2. Carbon Based Nanoparticles

Pure carbon or graphite is being potentially used as a starting material for constructing various types of carbon-based nanomaterials (CBNs), such as; nanotubes, graphene, quantum dots, fullerenes, and nano diamonds (Cha et al., 2013). The simplest form of carbon-based nanoparticle is fullerene (C60), is a spherical molecule made up of 60 carbon atoms. These carbon atoms are connected with each other forming icosahedron configuration and arranged at the corner points of the truncated isohedron. Fullerene possesses antiviral activity and is also used for enzyme interactions because it can potentially penetrate and establish links with the catalytic sites of some enzymes. The other form of carbon based nanoparticles, carbon nanotube has been popularized owing to its superior electrical, chemical and mechanical properties and it is quite useful in the energy sector, polymer industries and medicines (Mueller and Nowack, 2008).

2.2.1.3. Metal Oxide (MO) Nanoparticles

Several semiconductor metal oxide nanoparticles such as; MnO_2, Fe_2O_3, TiO_2, ZnO, and SiO_2 have been successfully used in different fields due to their unique physicochemical properties (Davoglio et al., 2018; Giri et al., 2007; Lassoued et al., 2017; Mueller and Nowack, 2008; Naveed Ul Haq et al., 2017). MO nanoparticles are extensively used in a variety of products like cosmetics, sunscreens, as catalysts, and in biomedicinc. As an example, MnO_2 NPs have been commonly used in catalysis, magnetic resonance imaging (MRI) contrast enhancement, drug delivery, immunoassay, biomolecular interactions, etc. (Baral et al., 2017).

2.2.1.4. Metallic Nano Particles

This group mainly consisting of noble metals like (Au and Ag) along with some transition metals such as (Zn and Fe) and Cu, Pt, Pd and Rh with several application: catalysis, biosensors, solar panels, bioremediation, etc.

(Doria et al., 2012). Besides the favorable functional properties, nanoscale metallic wires show excellent combinations of strength, stiffness, and ductility relative to bulk materials. Nanorods and nanowires have been explored at length in electronics, photonics, bio-chemical sensing, imaging, and drug delivery (Murphy et al., 2005).

2.3. IMPLICATION OF PROTEIN IMMOBILIZATION IN THERAPEUTIC/FOOD/ CHEMICAL INDUSTRIES AND ENVIRONMENTAL PROBLEMS

Immobilized proteins find wide applications in bioanalytical and biomedical field *viz*. drug delivery, tumor location analysis, biosensors, disease diagnosis, and rectification of congenital metabolic deficiencies, etc. (Padavattan et al., 2009). The usage of proteins to treat different diseases has attracted much attention from both industry and academia. Nonetheless, the advancement in therapeutic use is not very rapid because of the complex nature of human anatomy. Although, with the growth of nanotechnology in microelectronic devices biosensing evolved as a huge market with some million-dollars revenue across the globe. Numerous nanoparticles have been investigated to enhance the catalytic ability of proteins/enzymes. Immobilized proteins are found to be more effective as compared to the soluble ones due to the removal of various disadvantageous aspects of soluble proteins. They are of use in drug metabolism studies, particularly in metabolite identification, where multiple enzymes are immobilized. It is helpful in the investigation of sequential reactions in in-vivo systems; e.g., immobilization of luciferase, hexokinase, and pyruvate kinase was used to study substrate biotransformation in kidney, liver, etc. (Solanki et al., 2009). Research is still going on for the immobilization of several enzymes from various tissues and for large-scale production of several important metabolites from the immobilized enzymes. For instance, a few relevant studies are discussed here. In one study, immobilized cytochrome P-450 monooxygenase system was used for the oxidation of environmental pollutants, synthesis of drugs, neutralization of toxic intermediates produced in the metabolism of the human teratogen thalidomide and extracorporeal detoxification (Stuckler and Nestle, 2012). In another study, the immobilized UDP glucuronosyltransferase from rabbit liver was used for drugs conjugation as well as in extracorporeal glucuronidation of acetaminophen in rabbit blood in vivo and glucuronidation of endogenous phenols in human serum suffering from acute hepatic failure

(Ames et al., 1993; Olanoff et al., 1977). Likewise, immobilized Microsomal S-methyltransferase from pig liver was used in the metabolism of gentamicin; histone acetyltransferase immobilized onto nitrocellulose was used in the synthesis of acetylated nucleic acids (Gilissen et al., 1992). All these studies suggest that immobilized enzymes play key role in the biomedical field with promising applications in drug metabolism, metabolite synthesis, disease diagnosis, and extracorporeal detoxification, etc. The protein instability in extreme surrounding conditions including pH, temperature and solvents is a prime factor that puts a stop to its implementation in biomedical applications (Granados et al., 2014). During the usage of enzymes as therapeutics, they are exposed to various metabolites (glucose, urea, creatinine, ATP, etc.), drugs, chemicals (phosphates, ammonia, ethanol, acetic acid, etc.), proteins (viral antigens, antibodies, and serum proteins) and DNA, vitamins, hormones, etc. which sometimes denture the enzyme and reduce the catalytic activity (Cantor et al., 2012; Cantor et al., 2011). Protein immobilization is one of the best and cost-effective methods available, where the enzyme recovery is possible after their use. Currently, there are several biocompatible matrices that are non-immunogenic and can transport enzyme to the target cells successfully (Gurung et al., 2013). In addition, the simultaneous usage of several enzymes at a time is a new innovative step, especially when the disease is due to multiple enzyme deficiencies, where different enzymes can undergo multiple biochemical reactions. In a recent study, the immobilization of arginine deaminase on PEG has been found to successfully inhibit the growth of human melanoma and hepatocellular carcinomas by removing arginine from the surrounding tissues, demonstrating the achievement of enzyme immobilization in the cancer treatment (Jaenicke and Böhm, 1998). Furthermore, the advancement of biotechnology over the last decade leads to the production of safer more economical enzymes with superior potency and specificity by the pharmaceutical companies. Moreover, new initiatives by the Food and Drug Administration (FDA) are highly effective in the development of different enzyme drugs suitable for rare as well as common diseases. Though, the biomedical application of immobilized enzymes is still in its infancy, a lot of innovation in relevant area is expected in near future.

Immobilized proteins have also been used in various food and chemical industries such as paper products, fuels, agrochemicals, food, and beverages as well as production of organic acids, amino acids, antibiotics, etc. (Cheng and Zheng, 2014; Sumner, 1926). Enzymes from plant materials are cheaper and available in abundance, which play crucial role in starch, dairy and baking industries, textile, etc. (Li et al., 2012a; Rao et al., 1998). Proteases are one dominant plant-based enzymes which are widely used in detergent and dairy industries. Another one is carbohydrases (cellulases and amylases) mostly used in the starch, baking industries, textile, detergent industries, etc. (Datta et al., 2013; Tischer and Wedekind, 1999). Plants are the richest source of enzymes like carbohydrases, proteases, lipases, phytases, etc. Nowadays, almost all industries be it textile or bakery products, renewable biofuels or tires and adhesives use enzymes form plant resources (Jammer et al., 2015; Sergeeva and Vreugdenhil, 2002; Shnitka and Seligman, 1971). Enzyme immobilization opens a new path in enzyme technology as it allows quick scale-up, can be easily executed, products can be easily separated, lowers production cost up to several folds by enzyme recycling, and moreover, it is an environmentally friendly method. Therefore, industries are extensively using enzyme-based processes in lieu of chemical processes. Here, a few reports will be discussed where the stability and activity of enzymes are correlated during immobilization. From pea seeds β-galactosidase and from soybean seeds urease were immobilized on glutaraldehyde activated chitosan beads through covalent bonds. The efficiency of immobilization for β-galactosidase and urease were found to be 67.12 and 77 % with reusability of 14 washes and 10 washes, respectively (Dwevedi and Kayastha, 2009a). In addition, various innovative supports have been taken up to improve the physicochemical properties of the immobilized enzymes; for instance, immobilization of lipase from *Pachira aquatic* on polyvinyl alcohol coated calcium alginate beads, immobilization of β-amylase from fenugreek on chitosan beads coated with both PVC and PVP, immobilization of chlorophylase from *Brassica* sp. on magnetic alginate beads, immobilization of α-galactosidase from chickpea on graphene nanosheets, etc. (Dwevedi and Kayastha, 2009a, b; Dwevedi et al., 2009; Maheshwari et al., 2000; Rodríguez et al., 2010; Singh et al., 2014; Srivastava et al., 2015; Srivastava et al., 2014; Vieille and Zeikus, 2001).

Furthermore, protein immobilization in tackling numerous environmental issues has been intensified. Environmental pollution is one of the crucial problems in developing countries severely affecting the health of millions, annually, and is more severe than any other issues like malnutrition or fatal diseases. During the last few decades, significant focus has been put on the connection between environmental factors and health. Developing countries within South-east Asia and sub-Saharan Africa which account for about 25% of world's population are under major threat from environmental pollution. In those areas people are short of adequate shelter, safe drinking water, proper sanitation, and are frequently exposed to air pollution (Dwevedi, 2016). All these problems are caused by their low productivity, the high pace of damage to the ecosystems, unstable economy leading to insignificant financial plan towards public health. The main cause of pollution in water is due to several xenobiotics and recalcitrant pollutants. The aquatic pollution is largely effected by the effluents from waste disposals by the industries and household activities. The issue has further been aggravated due to the rapid development of technology in every sector in the last two decades. Various chemical or biological processes have been developed for wastewater treatment. However, their effectiveness is still debatable due to various reasons (Dwevedi, 2016). New alternate treatments have been investigated considering few key parameters; both time and cost-effectiveness and ease of installation. Treatment of water using enzymes has proved to be an efficient method abiding by all the above-said parameters. (Nakamura and Sueishi, 1991) In addition, immobilization of enzymes onto a suitable support provides additional advantages in the treatment of wastewater in terms of; hydrolyzing all types of recalcitrant and xenobiotic compounds, easy adaptability at wide range of contaminant concentrations, usability over wide range of pH, temperature and salinity, high selectivity and applicability even in dilute water, broad concentration range of diverse aromatic compounds, minor shock loading effects, least sludge volume owing to absence of biomass, lower treatment cost in comparison to other methods. Various enzymes being used for the remediation of water pollutants are given in Table 2.1.

Table 2.1: Remediation of Various Water Pollutants Using Different Enzymes*

Enzyme	**Source**	**Pollutants**
Peroxidase	Horseradish *Agromyces ramosus* Plant materials	Phenol, chlorophenol, anilines: degradation, phenol determination, Kraft effluent: decontamination Phenol, polyaromatics, herbicides: degradation, humic acid polymerization Water decontamination
Laccase	*Trametes hispida* *Pyricularia oryzae* *Trametes versicolor* Plant materials *Pycnoporus cinnabarinus*	Dyes decoloration Azo-dyes degradation Textile effluent, chlorophenol, urea derivatives: degradation Chlorophenol degradation, xenobiotic binding to humus Benzopyrenes degradation

Phenol oxidase	*Trametes versicolor Thermoascus aurantiacus*	Chlorinated compounds: degradation Kraft effluent decontamination
Tyrosinase	*Agaricus bisporus*	Catechol oxidation
*Remediation of various water pollutants using different enzymes: (Adapted from Dwevedi, 2016, Chapter 5: Enzyme Immobilization: Solution Towards Various Environmental Issues).		

Another major source of environmental pollution is the polluted air released from the chemical industries. The removal of air pollutants through biological route is an important task. Catalytic oxidation and adsorption onto activated carbon are the two major technologies used for controlling air pollutants (Graebin et al., 2012). But the associated cost and unavailability of any proper disposal methods make these technologies unaffordable for small-scale industries. Bio filters would be an alternative cost-effective technology for polluted air treatment due to the low energy consumption, which can be afforded by industries (Bayramoglu et al., 2011). Bio filters are the bed of immobilized microorganisms onto biofilms, which only require ambient temperature (30–40°C) and do not create any secondary pollutants (nitrogen oxides, spent activated carbon). Of late, biotrickling filters have been brought into treat very high concentration of pollutants in the air. It is made of pollutant-degrading microorganisms immobilized on a packed base consisted of inert material (structured plastic/poly urethanefoam) (Wang et al., 2009a). In recent years, considerable progress has taken place in optimizing and understanding the usage of biotrickling filters.

2.3.1. Role of Nanomaterials in Protein Immobilization

In the past decades, nanoparticles have received vast attention in the area of biotechnology and bioanalytical chemistry (Shan et al., 2007). They played an important role in protein immobilization, because of their ideal characteristics which balance the key factors that decide biocatalysts efficiency, including specific surface area, mass transfer resistance, and effective protein loading. Nanoparticles are considered as the ideal candidate as compared to macromolecules because NPs have high reactivity and less diffusion problem (Ahmad and Sardar, 2015). Numerous efforts have been performed for the immobilization of biomolecule with the support of nanostructures such as; polymer, carbon, oxide and metallic nanomaterials and composite materials (Yang et al., 2013). For enzyme immobilization, nanostructured carriers have been considered as key materials due to low dimension, high specific surface area, low mass transfer resistance, and excellent dispersibility. The protein stability and activity found to be increased when bound to nanoparticles as compared to unbound protein (Li et al., 2017). In addition, several studies revealed that immobilization of proteins to the nanoparticles has many advantageous effects are; improve protein stability and performance by minimizing protein unfolding (Du et al., 2015; Jiang et al., 2004; Kao et al., 2014; Li et al., 2017; Salimi et al., 2007). Few reports are available on enzyme immobilization with different kinds of nanoparticles (metal and metal oxide nanoparticles, magnetic nanoparticles, porous and polymeric nanoparticles) (Chen et al., 2017; Cipolatti et al., 2014; Ding et al., 2015; Husain, 2017; Vaghari et al., 2016). Nobel nanoparticles such as Au and Ag nanoparticles have been preferentially studied for the immobilization of proteins (Liu et al., 2003; Petkova et al., 2012). Here, few applications of nano-immobilized enzymes are given below (Table 2.2).

Table 2.2: Few Applications of Nano Immobilized Enzymes

Nano material	Enzyme	Application	Reference
Fe_3O_4	Holocellulase	Saccharification of paddy straw	(Kumar et al., 2017)
Carbon nanotubes	Lignin peroxidase	Biocatalysis	(Oliveira et al., 2016)
Graphene sheets	Cicer α-galactosidase	High immobilization efficiency and thermal stability	(Singh et al., 2014)
Phosphorylcholine (PC) group-covered nanoparticles	Peroxidase	Enzymatic activity	(Watanabe and Ishihara, 2008)
Zirconia Lipases from *C. rugosa* and Pseudomonas cepacia Resolution of (R,S)-ibuprofen and (R,S)-1-phenylethanol, respectively			(Chen et al., 2008)
Chitosan nanofibers	Lysozyme	Antibacterial	(Park et al., 2013)

2.4. METHODS OF IMMOBILIZATION

The most common immobilization methods include adsorption, covalent coupling, entrapment and cross-linking. Figure 2.2 shows the pictorial representation of immobilization methods. Various reports are available in the literature on the immobilization methods providing a detailed methodology, the protocol and advantages/disadvantages of each method. Herein, a brief discussion of each method has been provided in detail.

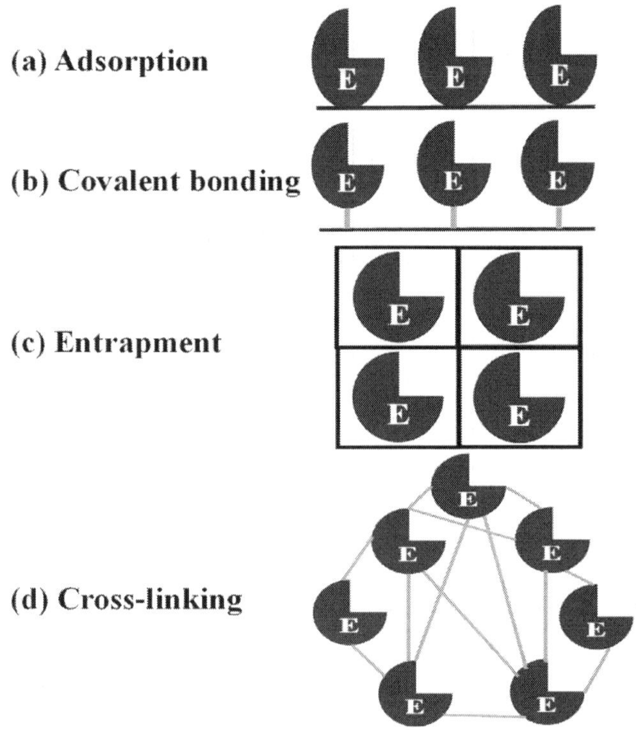

Figure 2.2: Pictorial representations of different enzyme immobilization methods (From Ahmad and Sardar, Biochemistry & Analytical Biochemistry, 4, 2015).

2.4.1. Adsorption

The physical adsorption of proteins on insoluble supports is a primitive, simple and straightforward method of reversible immobilization, which has wide applications and high loading capacity as compared to other immobilization methods. Here, proteins are physically attached to the support material by basically mixing them with a right adsorbent under suitable pH and ionic strength. After thorough washing of loosely bound and unbound enzyme, the immobilized enzyme is obtained in a directly usable form. Weak non-specific forces, for instance, van der Waals, hydrophobic interactions and hydrogen bonds are behind the adsorption process (Flickinger and Drew, 1999; Jegannathan et al., 2008). When the activity of the proteins degrades, they can be separated from the support under mild conditions and can be regenerated and reloaded with new substrate. However, leakage of the enzyme from the matrix is one of the drawbacks of those relatively weak nonspecific forces. Further, Immobilization is also possible through ionic forces between enzyme and support which is mostly observed in ion exchangers based on the principles used in chromatography (Bahulekar et al., 1991). The other physical method for immobilization is affinity binding that is based on the selection of complementary biomolecules (Gorecka and Jastrzebska, 2011; Guisan, 2006). The key benefits of this method are the selectivity of the interaction and minimal conformational changes induced which helps in retaining the catalytic activity. In addition, some reports revealed that enzymes can bind with the support surface through hydrophobic interactions (Hanefeld et al., 2009). The interaction strength between protein and adsorbent during immobilization is greatly influenced by the hydrophobicity of both and can be controlled through pH adjustment, temperature, and salt concentration (Brena and Batista-Viera, 2006; Porath, 1987).

2.4.2. Entrapment

Entrapment is an irreversible means of enzyme immobilization in which the movement of proteins is restricted in a porous gel. It is also expressed as physical restriction of the enzyme within a confined space or network (Won et al., 2005). This process could improve mechanical stability and curtail enzyme leaching and prevent denaturation as the concerned enzymes do not chemically interact with the support material (Shen et al., 2011). The method creates the most favorable microenvironment for the enzyme which is attained using wide ranges of materials like polymers, sol-gels, polymer/sol-gel composites and some inorganic materials (Gill, 2001; Jin and Brennan, 2002; Keeling-Tucker and Brennan, 2001; Tsai and Doong, 2007). The method can be achieved by using purely physical caging or a covalent binding. Studies have shown that entrapment of enzymes in natural polymers (agar/agarose/gelatin) through thermo reverse polymerization and alginate and carrageenan by ionotropic gelation (Datta et al., 2013). Several synthetic polymers like polyvinylalcohol hydrogel, polyacrylamide, etc. were also explored for this method (Deshpande et al., 1987; Grosova et al., 2008). Nonetheless, the main drawback of this method is the limited mass transfer of substrate or analytes to the enzyme active site (Gorecka and Jastrzebska, 2011). Other disadvantages are enzyme leakage (Sheldon, 2007) occurrence when the support matrix pores are too outsized, deactivation during immobilization and low loading capacity. Immobilized particle size to the pore size (of support material) ratio is an important factor, which must be considered before use (Gorecka and Jastrzebska, 2011).

2.4.3. Cross-Linking

This is an irreversible method of protein immobilization where a support is not required and is called as carrier-free immobilization (Sheldon, 2007). In this method, protein acts as its own carrier and the cross-linking can be developed by the connection of intermolecular cross-linkages between the enzymes through bi- or multifunctional reagents. Being an economical reagent, glutaraldehyde is most commonly used (Gorecka and Jastrzebska, 2011; Hanefeld et al., 2009; Sheldon, 2007). The cross-linking of proteins is achieved by the reaction of the free amino groups of lysine residues on the surface of protein molecules with oligomers or polymers of glutaraldehyde resulting from inter- and intramolecular aldol condensation via different pH dependent mechanisms. Since it has no support matrix for immobilization, the enzymes are no t firmly situated and are generally gelatinous, therefore this method is not usually preferred. Also, as there is direct covalent type linkage formed in the method leads to partial denaturation of the enzyme with loss of activity. Despite the disadvantages, still, it can be availed with the use of other immobilization techniques. For instance, nano diametric supports have been used for biocatalytic immobilizations (Datta et al., 2013; Huang and Cheng, 2008; Kim et al., 200; Ren et al., 2006; Wu et al., 2005). As well as, cross-linking of enzymes to electrospun nanofibers exhibit better residual activities. It was observed that lysozyme-immobilized electrospun Chitosan (CS) nanofibers have produced significant antibacterial applications (Park et al., 2013).

2.4.4. Covalent Binding

It is one of the most widely used methods for irreversible protein immobilization that involves the covalent bond formation between the protein and the support matrix. The functional groups of proteins are linked to support matrix. The functional groups that take part in the immobilization are usually the side chains of lysine (e-amino group), cysteine (thiol group), aspartic and glutamic acids (Guisan, 2006. ; Park et al., 2013; Sheldon, 2011). The activity of these immobilized proteins mostly depends on the size, shape, and composition of the carrier, nature of the coupling method, and specific conditions of the coupling reaction. For example, Candida rugosa lipase exhibited easy recovery and reusability in ester synthesis when it is covalently immobilized on the surface of silica nanoparticles and glutaraldehyde-activated aminopropyl glass beads. Likewise, covalently bound Candida rugosa lipase has been observed to be a robust biocatalyst for synthesis of aroma ester in a fluidized bed reactor (Damnjanović et al., 2012). This method provides a strong covalent bond between the lipase and its support matrix, which consequently enhances the reusability as compared to other methods (Flickinger and Drew, 1999; Guisan, 2006. ; Ovsejevi et al., 2013; Sheldon, 2011). It was also observed that thermal stability of enzymes increased when coupled with different supports like chitosan, mesoporous silica, etc. (Ispas et al., 2009). Moreover, the activity and stability of proteins enhanced significantly when peptide-modified surfaces are used (Fu et al., 2011). In some cases, the functional groups on the support material can be activated by certain reagents, which are linked to proteins through covalent binding either by direct linkage or via a spacer. Spacer molecules are very helpful to increase the activity of biocatalyst as they provide mobility to the coupled proteins. For instance, Cyanogen bromide (CNBr)-agarose and CNBr-activated-Sepharose containing carbohydrate moiety and glutaraldehyde as a spacer arm enhanced the thermal stability of covalently bound enzymes (Hartmann and Kostrov, 2013; Singh et al., 2013).

2.5. CHARACTERIZATION OF NANOMATERIAL BIOCONJUGATES

Various analytical means and approaches are available to study the interactions between protein and NPs in order to understand the underlying mechanism of biological activity of the NP and to make safe use of nanotechnology. The NP–protein complex formation can be determined by the binding affinity and binding ratio between NP and proteins. Also, the binding/adsorption of proteins onto NPs is generally accompanied by spectroscopic change, electron or energy transfer, size/shape alterations for both NPs and proteins. However, the alterations in above said parameters are helpful in exploring the NP–protein binding. In this section, the developments of different parameters for elucidation of the binding strategies between protein and NPs surface have been presented.

2.5.1. UV-Visible Spectroscopy

The binding of proteins onto NP results in alterations in the absorption of the NPs, which can be used to evaluate the nature of the bond (Aubin-Tam and Hamad-Schifferli, 2005; Chen et al., 1998; Delfino and Cannistraro, 2009; Edri and Regev, 2008; Jiang et al., 2005; Kathiravan et al., 2009; Sen et al., 2016). Nano particle size, aggregation, and local dielectric environments also affect the absorption spectra in terms of the shifting and broadening of the peaks. For example, the shifting and broadening in the absorption spectra of azurin (Az)–gold NP system completely depends on the Az concentration. However, this has not been consistent with all the NPs (Chen et al., 1998; Edri and Regev, 2008; Matsuura et al., 2006; Zorbas et al., 2005). In some cases, Beer-Lambert law can be used to calculate the concentration of one constituent present in the system when another constituent's concentration is known as observed in the SWCNT–BSA system (Edri and Regev, 2008). Not only the binding between NP and proteins can be monitored, but also UV-Vis spectroscopy is very helpful in exploring the conformational changes of proteins upon binding onto NP.

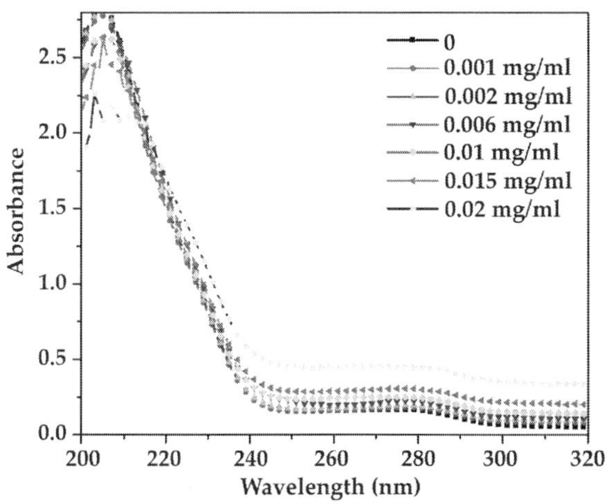

Figure 2.3: Absorption spectra of BSA in the presence of various concentrations of MnO2 (0, 0.001, 0.002, 0.006, 0.01, 0.015, and 0.02 mg ml^{-1}) at 25°C (From Baral et al., New J. Chem., 2017).

For example, BSA shows two characteristic peaks in the lower (200–230 nm) and higher (260–290 nm) wavelength regions (Figure 2.3). The one in the lower wavelength region is due to π-π* electronic transitions of the polypeptide backbone. The other one corresponds to the n-π* electronic transitions of fluorophore groups (Trp & Tyr residues). The alterations in the absorption peak at 280 nm (like red/blue shift, increase/decrease in intensity) indicates the conformational changes of protein particularly the microenvironment of the chromophores (Trp/Tyr residues) (Baral et al., 2017). Also, the absorption peaks in the UV-vis spectra can predict the formation of ground state complex between NP and proteins (Baral et al., 2017). There were studies where ground state complex formation between NP and proteins is confirmed by using this method (Jha and Venkatesu, 2016; Šimšíková, 2016; Wang et al., 2009b).

2.5.2. Fluorescence Spectroscopy

Fluorescence spectroscopy is very sensitive to protein dynamics because the excited fluorescence states exist for nanoseconds, which is in the time scale range of many important biological processes (rotational motion of protein side chains/ molecular binding/ protein conformational changes). Through fluorescence spectra valuable information on the polarity changes around the fluorophores can be gathered. When NPs are inherently luminescent or labeled with fluorescence probes, 36 fluorescence emission can also be detected from NPs. Steady-state or time-resolved fluorescence spectroscopy can be very helpful in monitoring NP– protein binding, fluorescence resonance energy transfer (FRET), etc. (Chen et al., 1999; Kathiravan et al., 2009; Lee et al., 2006; Macaroff et al., 2004; Mu et al.,2008; Pramanik et al., 2008; Sahoo et al., 2007; Shang et al., 2007a; Pompa et al., 2006; Pons et al., 2006; Xiao et al., 2008a).

2.5.2.1. Steady-State and Time-resolved Fluorescence Spectroscopy

Fluorescence intensity of a fluorophore is sensitive to the surrounding environment. This technique is very useful to investigate the microenvironment of chromophore molecules. Protein-NP interactions usually affect the local molecular environment of fluorophores and also quench the intrinsic fluorescence. Therefore, any change in the fluorescence intensity and shifting of the peaks suggest the interactions between NPs and proteins (Baral et al., 2017; Jha and Venkatesu, 2016). In addition, the lifetime of the fluorophores also changes if the fluorescence quenching is induced by collision effects. The fluorescence quenching happens due to many types of molecular interactions including molecular rearrangement, ground-state complex formation, excited state reactions, energy transfer, and collision quenching. The quenching can be static (ground-state complex formation) or dynamic (excited state). The mechanism of quenching can be elucidated through the Stern–Volmer equation (eq. 1) (Millan et al., 2016; Satish et al., 2017; Wang et al., 2009b).

$$\frac{I^0}{I} = 1 + K_{SV}[Q] \tag{1}$$

where, I^0 and I are the fluorescence intensity of protein in the absence and presence of quencher (NP), [Q] is the quencher concentration. The static and dynamic quenching mechanisms can be differentiated by studying the temperature dependent behavior of bimolecular quenching constants and binding constants. Usually, with an increase in temperature the binding constant decreases in the case of static quenching, whereas in dynamic quenching it increases. A linear Stern-Volmer plot always suggests a single mechanism of quenching, however, any deviations in the plot occur when both static and dynamic quenching take part in the quenching process. Deviations may also take place when high quencher concentration is used, which leads to a large degree of quenching (Keizer, 1983). In such a situation, a false static quenching is observed due to "sphere of action" mechanism, where the actual ground state complex formation is not observed. In those cases, absorption spectra of protein-NP complexes can give important information regarding the quenching mechanism. The relative strength and cooperativity of the NP–protein interaction can be achieved by quantifying several key parameters describing the association between proteins and NP nanomaterials, i.e., binding dissociation constant (K_D), binding association constant (K_A), and the Hill coefficient (n), assuming that the binding of the proteins to NP occurred under equilibrium conditions. Hill equation (Eqs. 2 & 3) can be used to calculate the necessary parameters (Ikeda et al., 2000; Kenry et al., 2016).

$$Q = \frac{I^0 - I}{I^0} \tag{2}$$

$$\frac{Q}{Q_{max}} = \frac{[NP]^n}{K_D^n + [NP]^n} \tag{3}$$

Where, Q_{max} is the saturation value of Q, K_D is the equilibrium binding dissociation constant which describes the relative strength of the NP–protein interaction, and n represents the Hill coefficient which defines the cooperativity of the NP– protein association.

2.5.2.2. Synchronous Fluorescence

It is a very useful technique to know the information about the local environment in the vicinity of fluorophore molecules. This method is more distinctive in comparison to the conventional excitation and emission spectra due to its several advantages such as simplified spectral information, sensitivity, reduction of bandwidth and avoidance of different perturbing effects (Millan et al., 2016). This technique provides the characteristics information of the amino acid environment (particularly Tyr or Trp residues) by fixing the $\Delta\lambda$ at 15 or 60 nm (Millan et al., 2017; Miller, 1979; Miller et al., 1982). The shifting in the maximum emission wavelength depends on the environmental polarity around fluorophore molecules. Žūkienė and Snitka demonstrated the effect of ZnO NPs on the microenvironment of Trp and Tyr residues of BSA. They observed that the hydrophobicity around Tyr residues was found to be decreased and it was increased around Trp residues. However, the covalent binding between BSA and ZnO NPs produced less hydrophobic Tyr environment than the non-covalent binding of BSA (Žūkienė and Snitka, 2015). Figure 2.4 shows the effect of Au nanoclusters on human serum albumin and it was observed that there was no perturbation in the microenvironment of Trp and Tyr residues (Yin et al., 2017).

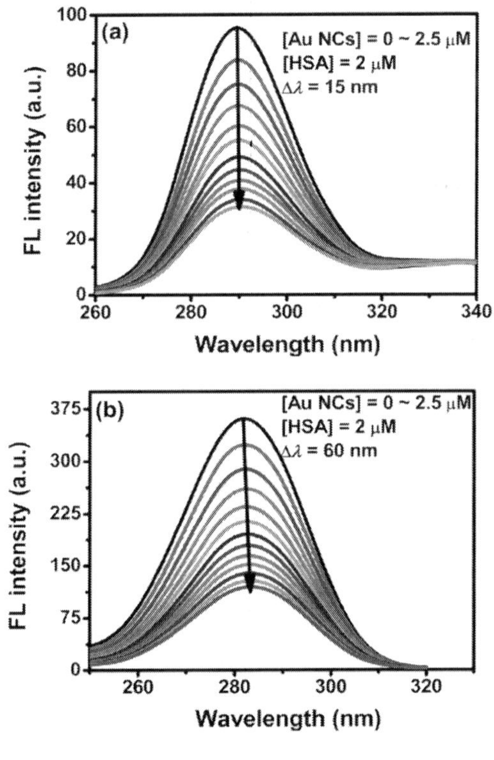

Figure 2.4: Synchronous fluorescence spectra of HSA in the absence and pres-

ence of Au NCs. [HSA] = 2 μM, [Au NCs] = 0–2.5 μM. (a) $\Delta\lambda = 15$ nm and (b) $\Delta\lambda = 60$ nm. (From Yin et al., Langmuir. 2017, 33, 5108–5116).

2.5.2.3. Forster Resonance Energy Transfer (FRET)

This phenomenon is based on non-radiative energy transfer between an excited molecular fluorophore (the donor) and an acceptor fluorophore through the intermolecular long-range dipole-dipole coupling. FRET is possible with any one of the two molecules being fluorescent and the efficiency is very high if the donor and acceptor are placed within the Forster radius so that energy transfer would be efficient. For the feasibility of measurements, the donor and acceptor molecules should be aligned for proper induction and there should be significantly overlap (>30%) between emission spectrum of a donor fluorophore and the absorption spectrum of an acceptor (Cardullo, 2013; Cardullo, 2002). The method has been highly used to monitor the conformational dynamics of proteins in solution as well as in single cells, cell organelles (Kenworthy, 2001; Xu et al., 2006). FRET is now a regular means to explore nanometer scale intermolecular and intramolecular distances. Sarkar et al., studied the conformational changes of human serum albumin (HSA) during thermal unfolding and refolding processes using FRET of intrinsic fluorophore (Trp214) to quantum dot covalently conjugated to HAS (Sarkar et al., 2007). Li et al., investigated the impact of blood proteins on the assembly structure and dynamics within nanoparticle using FRET measurements (Li et al., 2012b).

2.5.2.4. Fluorescence Correlation Spectroscopy (FCS)

Interaction study between proteins and NPs in aqueous medium can be carried out using FCS. It measures the fluorescence intensity of fluorophore particles in a small confocal volume. The fluorescence changes are observed with the time taken in the process when a fluorescent particles passes through the volume. The structural information thus observed can be extracted using Stokes-Einstein diffusion method. For instance, FCS has been used to find the human serum albumin adsorption on polymer-coated FePt and CdSe/ZnS NPs and it was found that a thickness of 3.3 nm protein corona formed on the NP surface (Rocker et al., 2009). In another study, the adsorption of human transferrin on small polymer-coated FePt NPs in a cooperative fashion was discussed and the formation of a monolayer with a thickness of 7 nm was measured using FCS (Jiang et al., 2010). FCS has the ability to provide significant quantitative information on the dynamics of the protein corona in a biological medium and is very helpful in exploring the dynamics of association/dissociation of proteins on NP surfaces.

2.5.3. Surface Plasmon Resonance

Protein adsorption on NP surfaces changes the oscillation of surface plasmon waves. SPR technology uses the alterations of these waves and it was mostly used to investigate the protein-NP binding kinetics (Cedervall et al., 2007a; Linse et al., 2007; Teichroeb et al., 2008). In a particular work, Cedervall et al., have used the SPR technique to study the binding/release of plasma proteins to/from NIPAM/ BAM copolymer NPs (Cedervall et al., 2007a). It was observed that significant amount of proteins bind to more hydrophobic NPs and the association and dissociation rates are clearly dependent on the hydrophobicity of NPs (Cedervall et al., 2007a). In another study by Zhao et al., specific cytochrome P450 enzymes are attached to the alternate gold surface, where SPR is able to distinguish between enzyme substrates and inhibitors based on the shifting in absorbance (Zhao et al., 2008). This feature is very useful in exploring safety profile for potential drug molecules.

2.5.4. Dynamic Light Scattering (DLS)

Generally, DLS has been used to investigate the size of NPs. It can also provide valuable information regarding the binding between protein and NPs. Various reports showed the application of DLS in bio conjugation and biomolecular binding studies (Jans et al., 2009; Liu et al., 2012; Piella et al., 2017; Sen et al., 2011; Tsai et al., 2011). DLS can determine small variations in the hydrodynamic size of particles. During the binding of proteins to NP surface, the NP size increases until a saturation point is reached. DLS can be used to monitor the binding ratio between protein and NP (Bhattacharya et al., 2006; Delfino and Cannistraro, 2009; Inomoto et al., 2009; Ipe et al., 2006; Tessier et al., 2008). For example, the molar ratio of cytochrome P450–Quantum dots (QDs: CdSe and CdS) was determined using the changes in the scattering intensity with gradual increase in the concentration of protein (Ipe et al., 2006). However, multiple factors like the hydration shell formation, the particle shape, and the counter ion binding significantly interfere with the DLS results.

2.5.5. Circular Dichroism (CD)

The secondary and tertiary structure of proteins in the presence of NPs can be monitored using far UV and near UV CD spectra. The secondary structures (α-helix, β-sheet, etc.) of proteins exhibit characteristic bands in the CD spectra in the far UV region. For example, the CD spectra of BSA produces strong bands at 208 and 222 nm in far UV region, which corresponds to the a-helical structure (Figure 2.5) (Chakraborti et al., 2012; Chatterjee and Mukherjee, 2014). The bands arise due to the peptide n-À* and À-À* transitions. Any changes in the spectra suggests the conformational changes of BSA. Generally, the band position depends on the transition energy between the ground state and the electronic excited states of polarizable electrons. The interaction of the electrons in the peptides with permanent or induced dipoles of solvent can able to stabilize/destabilize the energy levels of electronic states. The transition energies are supposed to be affected by the change in the solvent environment which ultimately leads to the peptide absorption band shifting of the peaks (Cascio and Wallace, 1995; Wallace et al., 2003).

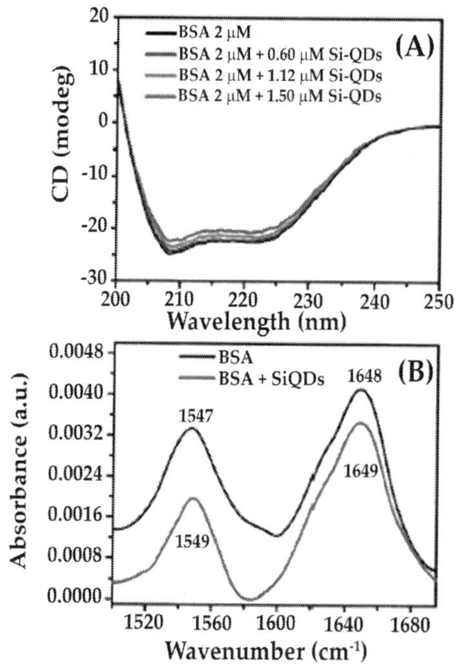

Figure 2.5: (A) The far-UV CD spectra of BSA (2 μM) in the absence and presence of different concentrations of Si QDs. (B) FTIR spectra of BSA (black line) and BSA in the presence of 1.50 μM allylamine-capped Si QDs (red line) (From Chatterjee and Mukherjee, Phys Chem Chem Phys. 16, 8400, 2014).

The near-UV CD spectral region is sensitive to the tertiary structure of the proteins. The chromophores sensitive to this region produce characteristic bands between 250–270 nm (phenylalanine residues), 270–290 nm (tyrosine residues) and 280–300 nm (tryptophan residues). Moreover, disulfide bonds also lead to broad weak signals all through the near-UV spectrum. The spectrum in this region is very sensitive to small variations in the tertiary structure caused by protein-protein interactions and/or alterations in solvent conditions. This technique has been extensively used for examining the conformational changes brought by protein–NP interactions (Aubin-Tam and Hamad-Schifferli, 2005; Chen et al., 1999; Jiang et al., 2005; Karlsson and Carlsson, 2005; Lundqvist et al., 2005; Matsuura et al., 2006). Though, NPs do not produce any CD spectra as they are not chiral, sometimes the scattering contribution of large size NPs cannot be ignored.

2.5.6. Fourier Transform Infrared (FTIR) and Raman Spectroscopy

FTIR and Raman spectroscopic techniques are helpful in analyzing the structural alterations in proteins after binding to the NP surfaces (Jiang et al., 2005; Karajanagi et al., 2004; Mandal and Kraatz, 2007; Peng et al., 2004). The secondary structure of protein can be estimated by the absorption of amide bonds using FTIR. The secondary structures exhibit characteristic peaks in the region of 1500−1600 cm^{-1} and1600−1700 cm^{-1} due to amide I and amide II bands and any changes in the spectra indicate the structural variations of proteins (Figure 2.5) (Chakraborti et al., 2013; Chakraborti et al., 2012; Chatterjee and Mukherjee, 2014). The amide I band is due to the C=O bond, which refers to the secondary structure of the protein and the amide II band appears due to the coupling of C−N bond stretching and N−H bond bending modes that refers to the adsorption of protein on solid surfaces. For example, the structure of the Leu-rich peptide Ac10L absorbed on different gold NPs) was studied using FTIR (Mandal and Kraatz, 2007). Likewise, the Raman spectra of proteins exhibit bands associated with the peptide main chain, aromatic and sulfur-containing side chains (Peng et al., 2004; Thomas, 1999). Though both methods provide the same information, they differ fundamentally in the mechanism involved. Both methods have their own advantages and disadvantages as far as the bio-applications are concerned. FTIR is simpler than Raman in terms of instrumentation and data collection, and also FITR intensity can be quantified by Beer's Law. However, Raman spectroscopy is more preferred over FTIR for studying protein-NP interactions because protein-NP aqueous solutions can be directly measured. Moreover, Raman spectroscopy provides simpler spectra than FTIR due to the localized vibrations of double/triple bond/electron-rich groups that produce high intense Raman bands in contrast to vibration of a single bond or electron-poor group (Thomas, 1999).

2.5.7. Isothermal Titration Calorimetry (ITC)

ITC has been used directly to evaluate the binding affinity constant (Ka), enthalpy changes (DH), and binding stoichiometry (n) between proteins and NPs in solution (De et al., 2007; Lindman et al., 2007; Linse et al., 2007; Rozhkov et al., 2003; Xiao et al., 2008b). In addition, temperature dependent measurements can provide the Gibbs energy changes (DG) and entropy changes (DS). Generally, proteins are directly titrated into NP solution. The recorded heat responses are then fitted to the isothermal function and the obtained thermodynamic parameters quantify protein binding as a function of the NP characteristics. Bouchemal has investigated the potential application of ITC for characterization of nanoparticles interaction with living systems (Bouchemal, 2008). Cedervall et al., used ITC in estimating the binding affinity and binding stoichiometry of proteinto nanoparticles (Cedervall et al., 2007a). Pilloni et al., also used ITC to establish the presence of PEG on the surface of nanocomposites during the biocompatibility studies of PEGylated magnetite–silica nanocomposites (Pilloni et al., 2010).

2.5.8. Nuclear Magnetic Resonance (NMR)

It is the most efficient technique available to analyze the conformational changes in proteins upon their interaction with NPs. In one of the studies, the interaction between copolymer NPs and different biomolecules; triglycerides, lipoproteins, cholesterol, and phospholipids from human plasma have been explored using NMR (Hellstrand et al., 2009). The lipid and protein binding patterns correspond closely to the composition of high-density lipoprotein (HDL) (Hellstrand et al., 2009). In addition, with the use of fractionated lipoprotein complexes, it was also shown that HDL binds to copolymer NPs with much higher specificity than other lipoprotein complexes (LDL, VHDL). In another study, the secondary structure of statherin peptides on biologically relevant hydroxyapatite (HAP) surfaces was determined using solid-state NMR (Stayton et al., 2003). Figure 2.6 demonstrate the specific direct interaction of Ubiquitin with Au NPs based on the chemical shifts of few peaks.(Calzolai et al., 2010) According to the principle, chemical shifts are very sensitive to the change in chemical environment and each peak represents a NH backbone of individual amino acid. In this case, only few peaks were altered after interaction with NP which describes that few amino acids are interacted with the Ubiquitin.

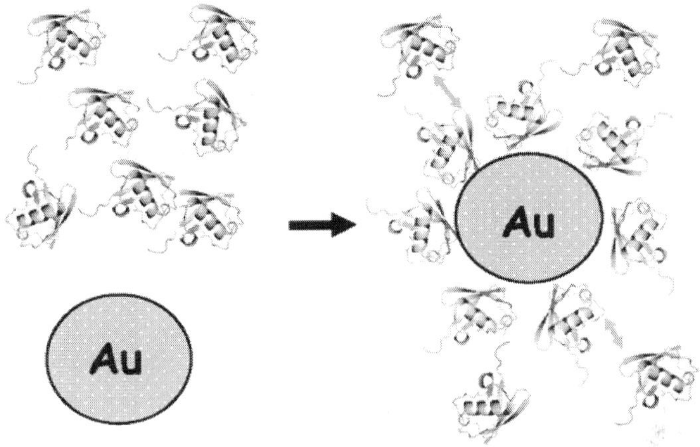

Figure 2.6: Section of 2D [^{15}N-^{1}H]-HSQC NMR spectra of free hUbq (red) and hUbq plus AuNP (yellow) samples, both at pH 7.7, showing chemical shift perturbation for some NH groups upon interaction with AuNP. Black arrows indicate NH backbone groups with the largest chemical shift perturbations (Q2, L15, and E18) in this spectral window (From Calzolai et al., Nano Lett. 2010, 10, 3101–3105).

2.5.9. Zeta Potential (ζ)

This method uses the ζ potential of NP surface to monitor the NP-protein binding. Most of the nanoparticles in solution possess an electrostatic charge on their surface, which holds back the NPs aggregation by electrostatic repulsion. In general, NPs possess large ζ potential and repel each other preventing flocculation, which helps in forming a stable dispersion. However, the protein adsorption on the NP surface would alter the overall surface charge on NPs. For example, the interaction between negatively charged BSA and positively charged lysozyme and a number of metal oxide NPs (such as Al_2O_3, SiO_2, TiO_2, or ZrO_2) was investigated with the help of ζ potential measurements (Rezwan et al., 2005). The ζ potentials and isoelectric points of the NPs were altered by the adsorption of proteins. In another study, the protein corona formation is observed from human plasma on several NPs with different surface properties and size (Lundqvist et al., 2008). Various polystyrene NPs (plain polystyrene, carboxyl-modified polystyrene, and amine-modified polystyrene) with different surface chemistries were systematically investigated to monitor the formation of protein corona on the basis of their surface properties and size.

2.5.10. X-ray Crystallography

This technique is now commonly used to establish how a drug molecule interacts with a targeted protein and the structural modifications needed in the drug molecule for better efficiency (Prakasham et al., 2010; Scapin, 2006). Recently, the immobilization of diastase enzyme on Ni-impregnated SiO_2 paramagnetic NPs was investigated using this technique (Prakasham et al., 2010). The results revealed that the binding and activities varied with the pH of the reaction mixture and the biocatalytic activity measurements showed higher activity in the case of an immobilized enzyme. Wei et al., demonstrated time-dependent, in situ growth of gold NPs within single crystals of lysozyme and revealed their three-dimensional distributions and the interaction of gold atoms with the lysozyme using X-ray crystallography as well as TEM (Wei et al., 2011). Jain et al., have employed this technique to characterize the "protein-capped" silver nanoparticles during the antibacterial studies against the Gram-positive and negative bacterial species (Jain et al., 2015).

2.5.11. Size Exclusion Chromatography (SEC)/Gel Filtration Chromatography (GFC)

This method is an old and versatile technique, used for the separation of chemicals, polymers and biological polymers including proteins, polysaccharides, and nucleic acids, on the basis of the sizes. A wide range of stationary phases (e.g., polyacrylamide, agarose, and dextran) have been used to separate biomolecules based on their size (Belgorodsky et al., 2005; Cedervall et al., 2007b; Lundqvist et al., 2004). This technique is useful for separating bound proteins and also for the detection of preferential binding and exchange rates during protein-NP interactions on the basis of protein's elution time. It is mainly helpful when the NPs have an electrostatic interaction with the stationary phase. Proteins or peptides can be separated on the basis of different retention times resulting from their hydrophobicity and their charge by reversed-phase chromatography (RPC) or ion exchange chromatography (IEC) respectively (Aubin-Tam and Hamad-Schifferli, 2005; Bhattacharya et al., 2006). The basic mechanism of RPC is that a polar stationary phase can adsorb a non-polar domain and proteins are continuously eluted by injecting a linear gradient of polar mobile phase. In IEC separation, the Coulombic attraction play the vital role between the biomolecule and the stationary phase and the bound substances are progressively eluted using a high ionic strength buffer gradient. The molecules are retained and afterward eluted by applying a linear gradient of high salt concentration on the basis of the charge (either positive or negative) on the IEC matrix. The protein charge depends on the pH of the solution (positive charge below their isoelectric point and a negative charge above it), which can be controlled delicately. Though the major work of this technique is for the purification of protein from a complex mixture, it has been currently integrated with mass spectrometry. IEC process is inevitable for both cation exchange and anion exchange chromatography. In cation exchange chromatography, the cations (positively charged proteins or NP protein complexes) are treated with a column of negatively charged beads whereas the anions using positively charged beads in anion exchange chromatography. Levy et al., have used IEC to separate gold NPs with various peptide cappings (Levy et al., 2004). Zheng et al., investigated the binding of proteins with Au nanoparticles using IEC (Zheng et al., 2003).

2.5.12. Electrophoresis

Electrophoresis is the most commonly employed technique for the separation and analysis of protein mixtures based on the dispersion of charged molecules in a fluid migrate under the influence of an applied electric field. Two commonly used methods for analyzing protein-NP conjugates are capillary electrophoresis (CE) and gel electrophoresis. CE separates proteins on the basis of charge and frictional forces and are detected by UV or fluorescence technique (Kim et al., 2007a; Kim et al., 2007b). It can be used to quantify adsorbed proteins without going through the desorption practice but the main drawback of the method is the easy adsorption of proteins on the inner surface of the capillary. CE was used to investigate the adsorption of plasma protein on poly (methoxy polyethyleneglycol cyanoacrylate-co-hexadecyl cyanoacrylate) (PEG-PHDCA) NPs (Kim et al., 2007b). Another method, SDS PAGE is the most common, cheap and reliable method which can be used to separate protein mixtures by molecular weight under the influence of an electric field (Cedervall et al., 2007b; Dutta et al., 2007; Lee et al., 2006; Lundqvist et al., 2008; Peng et al., 2004). The proteins migrate through a gel made of acrylamide cross-linked with bisacrylamide and are separated on the basis of size which possess different electrophoretic mobilities. Before making the proteins move through the gel, the proteins have to be denatured and must carry a negative charge. This is achieved by boiling the protein solution with a reducing agent (di thiothreitol or β-mercaptoethanol) and an anionic detergent (SDS). SDS makes the proteins negatively charged which in principle causes repulsion and induces detachment from the surface of NP. The gel resolves the proteins, which are stained with the available stains (Coomassie brilliant blue, silver nitrate staining, deep purple, etc.), and densitometry analysis is most commonly used to quantify protein abundance (Monopoli et al., 2011). Konduru et al., used gel electrophoresis to characterize the adsorbed proteins from the lung lining fluid on different NPs (CeO_2, silica-coated CeO_2, $BaSO_4$, and ZnO) and found that the amounts of proteins in the coronas formed on the four NPs varied significantly (Konduru et al., 2017). Deng et al., examined the interactions of poly(acrylic acid)-coated gold nanoparticles having hydrodynamic diameters of ~5, 10 and 20 nm with human plasma proteins using gel electrophoresis (Figure 2.7) (Deng et al., 2011).

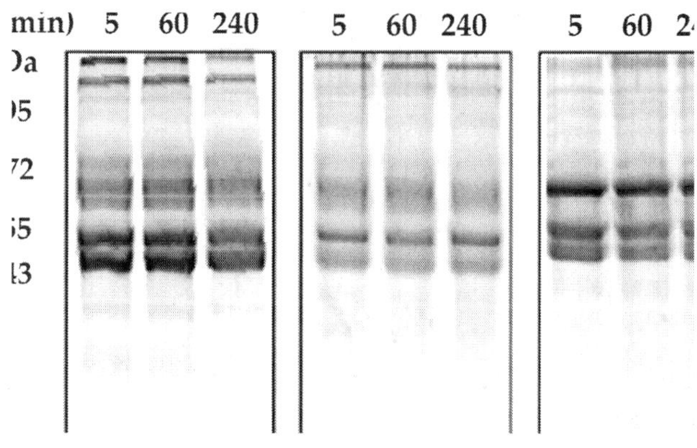

Figure 2.7: Human plasma protein adsorption on PAA-GNPs of 5 nm, 10 nm and 20 nm were examined. Sizes did not qualitatively affect the adsorption pattern. The adsorption patterns did not change qualitatively or quantitatively over 4 h of incubation, indicating the interaction saturated and reached equilibrium within the first few minutes of incubation (From Deng et al., Nat Nanotechnol, 6, 39, 2011).

2.5.13. Field Emission Scanning Electron Microscopy (FESEM) and Transmission Electron Microscopy (TEM)

The structural information from the topology of materials can be obtained through electron microscopy technique. If the materials size ranges in nanoscale region then high resolution techniques such as; FESEM, TEM, etc. can provide surface information. Generally, protein-NP interactions influence NP-NP interaction, NP aggregation as well as the conformational rearrangement of proteins. The particle aggregation decides the NP fate in NP-cell interactions (Lacerda et al., 2010). It is very important to investigate the morphological changes of both protein and NP in the NP-protein system as the toxicity of NPs are resulted from the formation of large aggregation in biological medium. The surface information of NPs generated from SEM/FESEM are limited because it cannot go to very high resolution. In SEM, an image is formed on a cathode ray tube whose raster is synchronized with the raster of a point beam of electrons scanned over the surface of a specimen. The electrons bounce off the surface of a sample to produce an image in this technique which is an improvement over the optical microscope because it extends the resolution limits up to 30,000X or as high as 60,000X and increase the depth of field by a factor of ~300. Further improvement of SEM technology was the FESEM which is capable of producing ultrahigh resolution over the entire accelerating voltage range due to the advancement in secondary electron detector technology. In analytical applications, both SEM and FESEM are the most employed methods to characterize the surface morphology and immobilization of enzymes. SEM technique is commonly used to examine the morphology of immobilized enzymes, whereas FESEM is used to examine very small topographic or morphology details (Chatterjee et al., 2014; Romdhane et al., 2010; Wan et al., 2007; Yudianti, 2011). The imaging of NP is a difficult task for SEM/FESEM, hence, for the improvement in imaging TEM technique was discovered which can provide high resolution two-dimensional projections of the materials (Koster et al., 2000). The working principle of TEM is based on the transmission of a beam of electrons through a specimen instead of scanning its surface morphology. The device helps to get much higher resolution image of NPs as compared to light microscope, making it possible to watch even a single column of atoms. In TEM technique, an electron beam generated from a tungsten filament is passed through the sample after which these electrons hit a phosphor screen, charge coupled device (CCD) or film producing an image. The advantage of this process is, sample of lower density allows more electrons to pass through resulting in brighter image. This method can provide images with as low as 0.2 nm resolution, i.e., smaller than the size of most atoms, hence reveals the real structural arrangement of atoms in the sample (Hayat, 1975). TEM has become a common technique for the investigation of surface topology as material science is rapidly approaching the nanometer scale. Apart from material science, TEM has also been used to visualize the protein/enzyme immobilization. The distribution of protein/enzyme on the support material is very important for the feasibility of bio catalysis (Noureddini and Gao, 2006). This technique provides the details of particle size and morphology of immobilized samples were mostly nano/micro size supports are adopted such as carbon nanotubes, nanoparticles and micron-size magnetic beads for the immobilization studies (Figure 2.8) (Liu et al., 2005; Shah et al., 2007; Verma et al., 2013).

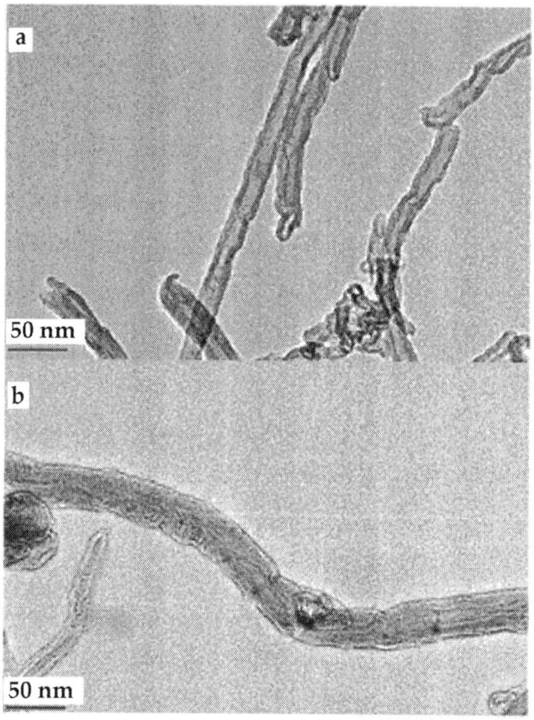

Figure 2.8: TEM images of functionalised MWNT (a) and MWNT bound lipase (b) (From Verma et al., PLoS One, 8, e73642, 2013).

2.6. PROTEIN-NANOPARTICLE INTERACTION STUDIES

It is well recognized that the application of NPs in bio environments is highly dependent on the nature and composition of protein corona as well as the possible disruption of the bound proteins (Gray, 2004; Lynch and Dawson,

2008; Lynch et al., 2006; Norde and Lyklema, 1991b). Therefore, it is the NP-protein complex and not the free NP which interacts with the living system (e.g., cells, tissues, organs, etc.). Research has been focused more on this mechanism as this is the key part which should be understood to develop smart nanomedicines and to ensure the safe realization of nanotechnology. Various parameters are responsible for the complex environment at the bio-nano interface, such as the NPs physiochemical properties, composition of the biological medium, and the possible conformational variations of the proteins (Nel et al., 2009). Herein, the influence of the most important physico-chemical properties of the NPs on the NP-protein interactions are discussed.

2.6.1. Effects of Nanoparticle Composition and Size

NPs composition, size, and its surface chemistry significantly influence the binding affinities and amounts of proteins adsorbed on the NP surfaces (Ehrenberg and McGrath, 2005; Gessner et al., 2001; Mahmoudi et al., 2008). Biomolecules adsorbed on the curved surface of NP undergo different conformational changes from the native structure in comparison with that adsorbed onto flat surfaces (Lundqvist et al., 2008; Lynch and Dawson, 2008; Walczyk et al., 2010). For example, NPs surface hydrophobicity and the available -CH_3 groups (reminiscent of fatty acid chains) can impact the protein corona formation (Cedervall et al., 2007a; Cedervall et al., 2007b). The results show that highest binding affinity exist between apolipoproteins and the most hydrophobic NPs. In another study, gold NPs of various sizes (i.e., 5 to 100 nm) were allowed to interact with blood proteins (albumin, fibrinogen, γ-globulin, histone, and insulin) (Lacerda et al., 2010). It was observed that the gold NPs had significant interactions with each protein and the degree of cooperativity of protein-NP binding depends on the NP size. It is reported that 12 nm sized negatively charged polyacrylic acid Au NPs bind fibrinogen with higher affinity compared to NPs of 7 nm size (Deng et al., 2012). In another study, 15 nm silica NPs brought a six-fold higher change in the secondary structure of human carbonic anhydrase I protein compared to its 6 nm counterpart (Lundqvist et al., 2004). Moreover, ribonuclease A was dramatically destabilized by silica NPs of increasing sizes Silica NPs with size of 100 nm induced greater structure loss in lysozyme as compared to 4 nm sized particle (Shang et al., 2007b; Vertegel et al., 2004).

2.6.2. Effect of Nanoparticle Shape and Surface Area

Protein adsorption on the NP surface is highly affected by the shape of NP, which consequently affect the way cells interact with it. *For example*, gold NPs interact with cell layers in a different manner when the shape is changed from spherical to rod geometry (Cox et al., 1999). As well as, it was observed that the protein concentration to NP surface area ratio has substantial effects on protein corona for some particular NPs. In one study, it was found that the protein binding pattern of sulfonated polystyrene NPs was very different for in vitro (3–10% proteins) and in vivo (55–80% proteins) studies (Monopoli et al., 2011). The shape and crystalline forms ZnO or TiO_2 nano-powders could influence the cytotoxicity and cytokine release by lung epithelial cells (Hsiao and Huang, 2011). Differential inflammatory potential and cellular association were observed for ZnO nano-spheres or nanosheets (Heng et al., 2011). Silver nanorods have been found to be toxic to the human lung epithelial cells, while nanospheres with the same mass concentration were nontoxic (Stoehr et al., 2011).

2.6.3. Effect of NP Surface Charge, Smoothness/Roughness, and Hydrophobicity/Hydrophilicity

The surface charge of NPs has a great impact on bio distribution of these particles because the compositions of protein corona are influenced by this parameter. *For example*, the application of some positive NPs is significantly diminished during in vivo studies as they are rapidly recognized by opsonins, which helps in removing the particles by the reticuloendothelial and mononuclear phagocytic systems (Jalilian et al., 2009; Mahmoudi et al., 2010; Mahmoudi et al., 2011b). The opsonization can be minimized by coating the surface of the NPs with proper polymers like polyethylene glycol (PEG), which reduces nonspecific binding and enable binding of certain surface groups at the NP surface specifically with the cell surface receptors (Gonzalo et al., 2006; Jalilian et al., 2011; Temming et al., 2006; Verel et al., 2003). The NP surface can also be chemically fabricated using polyethylene glycol (PEG), also referred to as "PEGylation," which is helpful in avoiding protein adsorption. This provides a "stealth character" to the NP surface, making it unrecognized by the immune cells (Gref et al., 2000). By controlling the PEG density on nanoparticles surface NPs can be made to remain in circulation for longer periods (Gref et al., 2000; Perry et al., 2012). Also, siliconate treatment of the NP surface has been reported to prevent protein adsorption. NP surfaces with no charge bind fewer proteins than their negatively charged (COOH functionalized) or positively charged (NH_2 functionalized) counterparts (Deng et al., 2012; Fertsch-Gapp et al., 2011). Polyethyleneimine-functionalization of ZnO NPs favorably interacted with albumin as compared to pristine ones (Cho et al., 2011). In a recent study, it was observed that structural modification of fibrinogen can be influenced by the surface properties of gold NPs. Negatively charged gold NPs bind fibrinogen in an orientation leading to cytokine release in human monocytic THP-1cells (Deng et al., 2013). Surface effects can also be varied by nanoscale surface roughness. Simulation studies of NP-synthetic membranes interactions reveal that nanoscale surface roughness significantly minimizes the repulsive interactions, thus promoting adhesion, which may render easier engulfment by cells (Nel et al., 2009). Moreover, the hydrophobicity/hydrophilicity of NP surface has much effect on protein corona. Proteins denature more on hydrophobic surfaces due to high affinities than when adsorbed on hydrophilic surfaces (Kondo et al., 1993). As per available literature, the major dominating forces are hydrophobic interactions in most cases till date,()however, the influence of electrostatic interactions cannot be disregarded (Cedervall et al., 2007a;Cedervall et al., 2007b; Ehrenberg et al., 2009; Chanteau et al., 2009). Cedervall et al., used ITC to show the effect of particle hydrophobicity on adsorption of the number of protein molecules (stoichiometry) and it was found that there was a significant difference in the protein affinities for polymeric particles of different hydrophobicities (Cedervall et al., 2007a).

All these physical-chemical properties of the NPs significantly influence the structural stability of proteins. Here, a few selective studies on NP induced structural modifications of proteins has been provided (Table 2.3).

Table 2.3: A Brief Summary of Literature on Protein-Np Interaction Studies

NP (type & size)	Model proteins investigated	Observations	Reference
Gold (45 nm)	Bovine serum protein	NP dosedependant conformational changes	(Wangoo et al., 2008)
Gold (5–100 nm)	Albumin Fibrionogen Insulin	Minor conformational changes	(Lacerda et al., 2010)
Gold (7–22 nm)	Human Fibrinogen	Unfolding induced immune response in THP-1 cells	(Deng et al., 2011)

SWCNTs	Horse radish peroxidase Subtilisin Carlsberg Chicken egg white lysozyme Soybean peroxidase Trypsin Proteinase K	No changes in conformation Higher stabilization	(Asuri et al., 2007) (Asuri et al., 2006)
Carbon black (14 nm)	Human transferrin, plasminogen, serum albumin, Apoliporpotein A1, Immunoglobulins	Higher protein adsorption due to hydrophobic interactions	(Ruh et al., 2012)
Graphene and Graphite	Dodecamer peptide	Distorted alpha helical	(Katoch et al., 2012)
Graphene oxide nanosheet (different size distributions)	Albumin Fibrinogen Globulin	GO size dependent protein adsorption and conformational changes	(Kenry et al., 2016)
SiO_2 NPs (6,9,15 nm)	Human Carbonic anhydrase	Minor changes but Protein activity was retained	(Lundqvist et al., 2004)
SiO_2 (~40 nm)	Bovine serum protein Hen egg lysozyme RNASe A Lactoperoxidase	Conformation changed No conformational changes No conformational changes Conformation changed	(Turci et al., 2010)
SiO_2 (20 nm)	Lysozyme	pH and lysozyme concentration-dependent aggregation of NP	(Bharti et al., 2011)
SiO_2 (7.5 –82 nm) (trimethylsilyl-capped)	Bovine serum albumin Bovine Fibrinogen	Structure stabilized Structure distorted	(Roach et al., 2006)
SiO_2 (4.2 nm)	β-casein	Protein-NP complex temporal changes during migration from interface into solution	(Ang et al., 2010)
ZnO (25 nm)	Vibrio cholera Tox r	NP-protein complex susceptible to denaturation	(Chatterjee et al., 2010)
Fe_3O_4 (8 nm) (PVA-coated)	Transferrin	Irreversible conformational changes	(Mahmoudi et al., 2011c)
Fe_3O_4 (7.3 nm) (TMAOH stabilized)	Bovine serum protein	Hard protein corona formationa	(Casals et al., 2010)
Fe_2O_3 (15nm)	Insulin	Selective binding and extraction of insulin fibrils	(Skaat et al., 2009)
TiO_2 (20 nm)	Aβ(1_42)	b-amyloid fibrillation	(Wu et al., 2008)
MnO_2	Bovine serum protein	No significant conformational changes	(Baral et al., 2017)
CdS (9 nm)	Hemoglobin	Marginal conformational changes	(Shen et al., 2007)

CdSe/ZnS (15–20 nm) (coated with the polymer with terminal carboxy groups)	Human serum albumin	Protein forms a monolayer with anti-cooperative binding isotherm	(Rocker et al., 2009)
Chitosan based Ag NP	Human serum albumin	Inhibition of fibril formation	(Sen et al., 2016)

2.7. CONCLUSIONS AND PERSPECTIVE

Nanoscience and nanoengineering are considered as two major fundamental building blocks of nanotechnology that regulate all physical entities. Recently, nanoscience has shown incredible potential in the field of biomedical research and applications. Several reports are available on the application of NPs towards biomolecular interactions but research on the use of NPs for medicine (as therapeutics) is still in an evolving state. The upshots of this review reveal that nanosystems could be the epitome of solution for the improved bioavailability of NP, biodistribution, and safety. The conjugates of nanoparticles and proteins could significantly regulate the potential of nanovectors for drug delivery. In a way, nanoparticles have gained special attention in biological fields as they could easily be engineered precisely to meet biological needs. To investigate the bioactivity of NPs, it is important to explore the nature of bond exists between nanoparticle-biomolecule during interactions. In the long run, hybrid nanosystems will be helpful for the implementation of new methodologies in the diagnostic and therapeutic field. Herein, the significance upshots of protein-nanoparticle interactions in toxicity and nanomedicine are demonstrated. In addition, this chapter outlines the protein or peptide interaction with NP and focused on the NP-protein complex formations along with the consequences such as the loss of the structural stability and activity of proteins. The absorption of the biomolecule on the surface of the nanoparticle can define the interaction of NP with living systems and help in carrying out the necessary modification of the cellular responses to the NPs. Here, we discussed on different aspects of protein adsorption and cellular responses to NPs such as the physicochemical properties of NPs, size, structure, morphology, and chemistry which determine their effects on biological systems.

Moreover, a detailed discussion of immobilization process has shown that the selection and designing of supporting matrix are important in protein immobilization for increasing the stability and activity. This study also suggests that when proteins are immobilized on nanomaterial surfaces, the stability and activity of proteins are found to be enhanced. Nanomaterials are the emerging constituents in the future market of advanced technology which could potentially control the mechanical properties such as stiffness and elasticity of the material and create biocompatibility environment for protein/enzyme immobilization. Recently, the immobilized enzymes have been applied in drug delivery, development of bioreactors, and biosensors that has attracted several industries across the world. The immobilized enzymes have been used for the identification and characterization of metabolites in the drug metabolism studies. It also helps for the study of consecutive reactions in vivo model systems; for instance, co-immobilization of luciferase, pyruvate kinase, and hexokinase was used to study substrate biotransformations in organs such as kidney, lungs, and liver. Further progress is focused on the immobilization of various enzymes from different tissues and make them useful in the large-scale production of different kind of metabolites.

Highlights of this review include the physicochemical characterizations and analysis of protein-NP conjugates, which is an important step to understand nature of interactions exist in the conjugates as well as the NP mediated biological effects. In view of literature, we noticed that protein adsorption promotes the translocation of NPs across cellular barriers and accumulation in crucial body organs. This bionanoconjugates are mostly applied for in vitro systems, whereas the extrapolation of NPs for in vivo systems still remains a challenging task and needs further investigation. Therefore, systematic analysis of binding characteristics of NPs with proteins of different structure, shape and functional properties will positively enhance the existing knowledge. Future studies of NP-protein interactions could lead to strategic manipulation of NP surfaces having high affinity to specific functional proteins or small drug molecules intended for delivery in vivo. Thus, the fundamental outcome of the dynamics of the complex interactions could provide critical insights into biological applications such as cytotoxicity, drug delivery, etc., and generate safer and value-added nanomaterials.

REFERENCES

1. Ahmad, R., & Sardar, M. (2015). Enzyme Immobilization: An overview on nanoparticles as immobilization matrix. *Biochemistry & Analytical Biochemistry. 04.*
2. Alivisatos, A. P. (1996). Semiconductor clusters, nanocrystals, and quantum dots. *Science. 271*, 933–937.
3. Alonso, M. J. (1996). Nanoparticulate drug carrier technology. Microparticulate systems for the delivery of proteins and vaccines, New York, NY: *Marcel Dekker*. 203–242.
4. Ames, B. N., K, S. M., & Hagen, T. M. (1993). Oxidants, antioxidants, and the degenerative diseases of aging. *Proc Natl Acad Sci USA. 90*, 7915–7922.
5. Ang, J. C., Lin, J. M., Yaron, P. N., & White, J. W. (2010). Protein trapping of silicananoparticles. *Soft Matter. 6,* 383–390.
6. Asuri, P., Bale, S. S., Pangule, R. C., Shah, D. A., Kane, R. S., & Dordick, J. S. (2007). Structure, function, and stability of enzymes covalently attached to single-walled carbon nanotubes. *Langmuir. 23*, 12318–12321.
7. Asuri, P., Karajanagi, S. S., Yang, H., Yim, T. J., Kane, R. S., & Dordick, J. S. (2006). Increasing protein stability through control of the nanoscale environment. *Langmuir. 22*, 5833–5836.
8. Aubin-Tam, M. E., & Hamad-Schifferli, K. (2005). Gold nanoparticle cytochrome c complexes: The effect of nanoparticle ligand charge on protein structure. *Langmuir. 21*, 12080–12084.
9. Auffan, M., Rose, J., Bottero, J. Y., Lowry, G. V., Jolivet, J. P., & Wiesner, M. R. (2009). Towards a definition of inorganic nanoparticles from an environmental, health and safety perspective. *Nat Nanotechnol. 4*, 634–641.
10. Bahulekar, R., Ayyangar, N. R., & Ponrathnam, S. (1991). Polyethyleneimine in immobilization of biocatalysts. *Enzyme Microb Technol. 13*, 858–868.
11. Baral, A., Satish, L., Das, D. P., Sahoo, H., & Ghosh, M. K. (2017). Construing the interactions between MnO2 nanoparticle and bovine serum albumin: insight into the structure and stability of a protein–nanoparticle complex. *New J Chem. 41*, 8130-8139
12. Bayramoglu, G., Karagoz, B., Altintas, B., Arica, M. Y., & Bicak, N. (2011). Poly(styrene-divinylbenzene) beads surface functionalized

with di-block polymer grafting and multi-modal ligand attachment: performance of reversibly immobilized lipase in ester synthesis. *Bioprocess Biosyst Eng. 34*, 735–746.

13. Belgorodsky, B., Fadeev, L., Ittah, V., Benyamini, H., Zelner, S., Huppert, D., Kotlyar, A. B., & Gozin, M. (2005). Formation and characterization of stable human serum albumin-tris-malonic acid C-60 fullerene complex. *Bioconjugate Chem. 16*, 1058–1062.
14. Bharti, B., Meissner, J., & Findenegg, G. H. (2011). Aggregation of silica nanoparticles directed by adsorption of lysozyme. *Langmuir. 27*, 9823–9833.
15. Bhattacharya, J., Choudhuri, U., Siwach, O., Sen, P., & Dasgupta, A. K. (2006). Interaction of hemoglobin and copper nanoparticles: implications in hemoglobinopathy. *Nanomedicine-Nanotechnology Biology and Medicine. 2*, 191–199.
16. Bouchemal, K. (2008). New challenges for pharmaceutical formulations and drug delivery systems characterization using isothermal titration calorimetry. *Drug Discov Today. 13*, 960–972.
17. Brena, B. M., & Batista-Viera, F. (2006). Immobilization of enzymes. Immobilization of enzymes and cells 2nd ed New Jersey (NJ): *Humana Press Inc.* 123–124.
18. Calzolai, L., Franchini, F., Gilliland, D., & Rossi, F. (2010). Protein-nanoparticle interaction: identification of the ubiquitin-gold nanoparticle interaction site, *Nano Lett. 10*, 3101–3105.
19. Cantor, J. R., Panayiotou, V., Agnello, G., Georgiou, G., & Stone, E. M. (2012). Engineering reduced-immunogenicity enzymes for amino acid depletion therapy in cancer. *Methods Enzymol. 502*, 291–319.
20. Cantor, J. R., Yoo, T. H., Dixit, A., Iverson, B. L., Forsthuber, T. G., & Georgiou, G. (2011). Therapeutic enzyme deimmunization by combinatorial T-cell epitope removal using neutral drift. *Proc Natl Acad Sci U S A. 108*, 1272–1277.
21. Cardullo, R. A. (2002). Principles of non-radiative FRET: the spectroscopic ruler.. *Microsc Anal. 15*, 19–21.
22. Cardullo, R. A. (2013). Theoretical principles and practical considerations for fluorescence resonance energy transfer microscopy. *Methods Cell Biol. 114*, 441–456.
23. Casals, E., Pfaller, T., Duschl, A., Oostingh, G. J., & Puntes, V. (2010). Time Evolution of the Nanoparticle Protein Corona. *ACS Nano. 4*,

3623–3632.

24. Cascio, M., & Wallace, B. A. (1995). Effects of local environment on the circular dichroism spectra of polypeptides. *Anal Biochem. 227*, 90–100.
25. Cedervall, T., Lynch, I., Foy, M., Berggard, T., Donnelly, S. C., Cagney, G., Linse, S., & Dawson, K. A. (2007b) Detailed identification of plasma proteins adsorbed on copolymer nanoparticles. *Angewandte Chemie-International Edition. 46*, 5754–5756.
26. Cedervall, T., Lynch, I., Lindman, S., Berggard, T., Thulin, E., Nilsson, H., Dawson, K. A., & Linse, S. (2007a) Understanding the nanoparticle-protein corona using methods to quantify exchange rates and affinities of proteins for nanoparticles. *Proc Natl Acad Sci U S A. 104*, 2050–2055.
27. Cha, C., Shin, S. R., Annabi, N., Dokmeci, M. R., & Khademhosseini, A. (2013). Carbon-based nanomaterials: multifunctional materials for biomedical engineering. *ACS Nano. 7*, 2891–2897.
28. Chakraborti, S., Joshi, P., Chakravarty, D., Shanker, V., Ansari, Z. A., Singh, S. P., & Chakrabarti, P. (2012). Interaction of polyethyleneimine-functionalized zno nanoparticles with bovine serum albumin. *Langmuir. 28*, 11142−11152.
29. Chakraborti, S., Sarwar, S., & Chakrabarti, P. (2013). The Effect of the Binding of ZnO Nanoparticle on the Structure and Stability of α-Lactalbumin: A Comparative Study. *J Phys Chem B. 117*, 13397−13408.
30. Chan, W. C. W., & Nie, S. M. (1998). Quantum dot bioconjugates for ultrasensitive nonisotopic detection. *Science. 281*, 2016–2018.
31. Chanteau, B., Fresnais, J., & Berret, J. F. (2009). Electrosteric enhanced stability of functional sub-10 nm cerium and iron oxide particles in cell culture medium. *Langmuir. 25*, 9064–9070.
32. Chatterjee, S., & Mukherjee, T. K. (2014). Spectroscopic investigation of interaction between bovine serum albumin and amine-functionalized silicon quantum dots. *Phys Chem Chem Phys. 16*, 8400–8408.
33. Chatterjee, S., Yadav, D., Barbora, L., Mahanta, P., & Goswami, P. (2014). Silk-cocoon matrix immobilized lipase catalyzed transesterification of sunflower oil for production of biodiesel. *Journal of Catalysts*. 1–7.
34. Chatterjee, T., Chakraborti, S., Joshi, P., Singh, S. P., & Gupta, V.,

Chakrabarti, P. (2010). The effect of zinc oxide nanoparticles on the structure of the periplasmic domain of the vibrio cholerae ToxR protein. *FEBS J. 277*, 4184–4194.

35. Chen, B. X., Wilson, S. R., Das, M., Coughlin, D. J., & Erlanger, B. F. (1998). Antigenicity of fullerenes: Antibodies specific for fullerenes and their characteristics. *Proc Natl Acad Sci U S A. 95*, 10809–10813.
36. Chen, D., Qiao, X., Qiu, X., & Chen, J. (2009). Synthesis and electrical properties of uniform silver nanoparticles for electronic applications. *Journal of Materials Science. 44*, 1076–1081.
37. Chen, M., Zeng, G., Xu, P., Lai, C., & Tang, L. (2017). How Do Enzymes 'Meet' Nanoparticles and Nanomaterials? *Trends Biochem Sci. 42*, 914–930.
38. Chen, Y. L., Zhang, X. F., Gong, Y. D., Zhao, N. M., Zeng, T. Y., & Song, X. Q. (1999). Conformational changes of fibrinogen adsorption onto hydroxyapatite and titanium oxide nanoparticles. *J Colloid Interface Sci. 214*, 38–45.
39. Chen, Y. Z., Yang, C. T., Ching, C. B., & Xu, R. (2008). Immobilization of lipases on hydrophobilized zirconia nanoparticles: highly enantioselective and reusable biocatalysts. *Langmuir. 24*, 8877–8884.
40. Cheng, G., & Zheng, S. Y. (2014). Construction of a high-performance magnetic enzyme nanosystem for rapid tryptic digestion. *Sci Rep. 4*, 6947.
41. Chiti, F., & Dobson, C. M. (2006). Protein misfolding, functional amyloid, and human disease. Vol. 75, *Annu Rev Biochem*. p. 333–366.
42. Cho, W. S., Duffin, R., Howie, S. E., Scotton, C. J., Wallace, W. A., Macnee, W., Bradley, M., Megson, I. L., & Donaldson, K. (2011). Progressive severe lung injury by zinc oxide nanoparticles; the role of Zn2+ dissolution inside lysosomes. *Part Fiber Toxicol. 8*, 27.
43. Cipolatti, E. P., Silva, M. J. A., Klein, M., Feddern, V., Feltes, M. M. C., Oliveira, J. V., Ninow, J. L., & de Oliveira, D. (2014). Current status and trends in enzymatic nanoimmobilization. *J Mol Catal B: Enzym. 99*, 56–67.
44. Corot, C., Robert, P., Idee, J. M., & Port, M. (2006). Recent advances in iron oxide nanocrystal technology for medical imaging. *Adv Drug Del Rev. 58*, 1471–1504.
45. Cox, M. C., Barnham, K. J., Frenkiel, T. A., Hoeschele, J. D., Mason, A. B., He, Q. Y., Woodworth, R. C., & Sadler, P. J. (1999). Identification

of platination sites on human serum transferrin using C-13 and N-15 NMR spectroscopy. *J Biol Inorg Chem. 4*, 621–631.

46. Damnjanović, J. J., Žuža, M. G., Savanović, J. K., Bezbradica, D. I., Mijin, D. Ž., Bošković-Vragolović, N., & Knežević-Jugović, Z. D. (2012). Covalently immobilized lipase catalyzing high-yielding optimized geranyl butyrate synthesis in a batch and fluidized bed reactor. *J Mol Catal B: Enzym. 75*, 50–59.
47. Datta, S., Christena, L. R., & Rajaram, Y. R. (2013). Enzyme immobilization: an overview on techniques and support materials. *Biotech*. 3, 1–9.
48. Davoglio, R. A., Cabello, G., Marco, J. F., & Biaggio, S. R. (2018). Synthesis and characterization of α-MnO 2 nanoneedles for electrochemical supercapacitors. *Electrochimica Acta. 261*, 428–435.
49. Dawson, K. A. (2002). The glass paradigm for colloidal glasses, gels, and other arrested states driven by attractive interactions. *Current Opinion in Colloid & Interface Science*. 7, 218–227.
50. de la Fuente, J. M., Berry, C. C., Riehle, M. O., & Curtis, A. S. G. (2006). Nanoparticle targeting at cells. *Langmuir. 22*, 3286–3293.
51. De, M., You, C. C., Srivastava, S., & Rotello, V. M. (2007). Biomimetic interactions of proteins with functionalized nanoparticles: A thermodynamic study. *J Am Chem Soc. 129*, 10747–10753.
52. Delfino, I., & Cannistraro, S. (2009). Optical investigation of the electron transfer protein azurin-gold nanoparticle system. *Biophys Chem. 139*, 1–7.
53. Deng, Z. J., Liang, M., Monteiro, M., Toth, I., & Minchin, R. F. (2011). Nanoparticle-induced unfolding of fibrinogen promotes Mac-1 receptor activation and inflammation. *Nat Nanotechnol. 6*, 39–44.
54. Deng, Z. J., Liang, M., Toth, I., Monteiro, M. J., & Minchin, R. F. (2012). Molecular interaction of poly(acrylic acid) gold nanoparticles with human fibrinogen. *ACS Nano. 6*, 8962–8969.
55. Deng, Z. J., Liang, M., Toth, I., Monteiro, M., & Minchin, R. F. (2013). Plasma protein binding of positively and negatively charged polymer-coated gold nanoparticles elicits different biological responses. *Nanotoxicology*. *7*, 314–322.
56. Deshpande, A., D'souza, S., & Nadkarni, G. (1987). Coimmobilization of Damino acid oxidase and catalase by entrapment ofTrigonopsis variabilis in radiation polymerised Polyacrylamide beads. *Journal of*

bioscience. *11*, 137–144.

57. Ding, S., Cargill, A. A., Medintz, I. L., & Claussen, J. C. (2015). Increasing the activity of immobilized enzymes with nanoparticle conjugation. *Curr Opin Biotechnol*. *34*, 242–250.
58. Doria, G., Conde, J., Veigas, B., Giestas, L., Almeida, C., Assuncao, M., Rosa, J., & Baptista, P. V. (2012). Noble metal nanoparticles for biosensing applications. *Sensors*. *12*, 1657–1687.
59. Drexler, K. E. (1981). Molecular engineering – an approach to the development of general capabilities for molecular manipulation. *Proceedings of the National Academy of Sciences of the United States of America-Physical Sciences*. *78*, 5275–5278.
60. Du, K., Sun, J., Zhou, X., Feng, W., Jiang, X., & Ji, P. (2015). A two-enzyme immobilization approach using carbon nanotubes/silica as support. *Biotechnol Prog*. *31*, 42–47.
61. Dujardin, E., & Mann, S. (2002). Bio-inspired materials chemistry. *Adv Mater*. *14*, 775–788.
62. Dutta, D., Sundaram, S. K., Teeguarden, J. G., Riley, B. J., Fifield, L. S., Jacobs, J. M., Addleman, S. R., Kaysen, G. A., Moudgil, B. M., & Weber, T. J. (2007). Adsorbed proteins influence the biological activity and molecular targeting of nanomaterials. *Toxicol Sci*. *100*, 303–315.
63. Dwevedi, A. (2016) Enzyme Immobilization: Advances in Industry, Agriculture, Medicine, and the Environment. *Springer International Publishing*. 1st Edition, X, 132.
64. Dwevedi, A., & Kayastha, A. M. (2009a). Optimal immobilization of beta-galactosidase from Pea (PsBGAL) onto Sephadex and chitosan beads using response surface methodology and its applications. *Bioresour Technol*. 2009a, 100, 2667–2675.
65. Dwevedi, A., & Kayastha, A. M. (2009b). Stabilization of beta-galactosidase (from peas) by immobilization onto amberlite MB-150 beads and its application in lactose hydrolysis. *J Agric Food Chem*. 2009b, 57, 682–688.
66. Dwevedi, A., Singh, A. K., Singh, D. P., Srivastava, O. N., & Kayastha, A. M. (2009). Lactose nano-probe optimized using response surface methodology. *Biosens Bioelectron*. *25*, 784–790.
67. Edri, E., Regev, O. (2008). pH effects on BSA-dispersed carbon nanotubes studied by spectroscopy-enhanced composition evaluation techniques. *Anal Chem*. *80*, 4049–4054.

68. Ehrenberg, M. S., Friedman, A. E., Finkelstein, J. N., Oberdoerster, G., & McGrath, J. L. (2009). The influence of protein adsorption on nanoparticle association with cultured endothelial cells. *Biomaterials*. *30*, 603–610.
69. Ehrenberg, M., & McGrath, J. L. (2005). Binding between particles and proteins in extracts: implications for microrheology and toxicity. *Acta Biomater*. *1*, 305–315.
70. Engel, E., Michiardi, A., Navarro, M., Lacroix, D., & Planell, J. A. (2008). Nanotechnology in regenerative medicine: the materials side. *Trends Biotechnol*. *26*, 39–47.
71. Fertsch-Gapp, S., Semmler-Behnke, M., Wenk, A., & Kreyling, W. G. (2011). Binding of polystyrene and carbon black nanoparticles to blood serum proteins. *Inhal Toxicol*. *23*, 468–475.
72. Flickinger, M. C., & Drew, S. W. (1999). Fermentation, biocatalysis and bioseparation. *Encyclopedia of Bioprocess Technology Vol 1,* 1st edn., New York (NY): Wiley.
73. Fu, J., Reinhold, J., & Woodbury, N. W. (2011). Peptide-modified surfaces for enzyme immobilization. *PLoS One*. *6*, e18692.
74. Garcia-Galan, C., Berenguer-Murcia, Á., Fernandez-Lafuente, R., & Rodrigues, R. C. (2011). Potential of different enzyme immobilization strategies to improve enzyme performance. *Adv Synth Catal*. *353*, 2885–2904.
75. Gessner, A., Olbrich, C., Schroder, W., Kayser, O., & Muller, R. H. (2001). The role of plasma proteins in brain targeting: species dependent protein adsorption patterns on brain-specific lipid drug conjugate (LDC) nanoparticles. *Int J Pharm*. *214*, 87–91.
76. Gidaspow, D., & Jiradilok, V. (2007). Nanoparticle gasifier fuel cell for sustainable energy future. *J Power Sources*. *166*, 400–410.
77. Gilissen, R. A. H. J., Meerman, J. H. N., & Mulder, G. J. (1992). Immobilization of solubilized UDP-glucuronosyltransferase from rat liver microsomes to Sepharose 4B. *Biochem Pharmacol*. *43*, 2661–2663.
78. Gill, I. (2001). Bio-doped Nanocomposite Polymers: Sol−Gel Bioencapsulates. Chem Mater. *13*, 3404–3421.
79. Giri, S., Trewyn, B. G., & Lin, V. S. (2007). Mesoporous silica nanomaterial-based biotechnological and biomedical delivery systems. *Nanomedicine (Lond)*. 2, 99–111.

80. Gonzalez-Melendi, P., Fernandez-Pacheco, R., Coronado, M. J., Corredor, E., Testillano, P. S., Risueno, M. C., Marquina, C., Ibarra, M. R., Rubiales, D., & Perez-de-Luque, A. (2008). Nanoparticles as smart treatment-delivery systems in plants: assessment of different techniques of microscopy for their visualization in plant tissues. *Ann Bot. 101*, 187–195.

81. Gonzalo, T., Talman, E. G., de Ven, A. V., Temming, K., Greupink, R., Beljaars, L., Reker-Smit, C., Meijer, D. K. F., Molema, G., Poelstra, K., & Kok, R. J. (2006). Selective targeting of pentoxifylline platinum-based to hepatic stellate cells using a novel linker technology. *J Control Release. 111*, 193–203.

82. Gorecka, E., & Jastrzebska, M. (2011). Immobilization techniques and biopolymer carriers. *Biotechnol Food Sci. 75*, 65–86.

83. Graebin, N. G., Martins, A. B., Lorenzoni, A. S., Garcia-Galan, C., Fernandez-Lafuente, R., Ayub, M. A., & Rodrigues, R. C. (2012). Immobilization of lipase B from Candida antarctica on porous styrene-divinylbenzene beads improves butyl acetate synthesis. *Biotechnol Prog. 28*, 406–412.

84. Granados, D. P., Sriranganadane, D., Daouda, T., Zieger, A., Laumont, C. M., Caron-Lizotte, O., Boucher, G., Hardy, M. P., Gendron, P., Cote, C., Lemieux, S., Thibault, P., & Perreault, C. (2014). Impact of genomic polymorphisms on the repertoire of human MHC class I-associated peptides. *Nature communications*. 5, 3600.

85. Gray, J. J. (2004). The interaction of proteins with solid surfaces. *Curr Opin Struct Biol. 14*, 110–115.

86. Gref, R., Luck, M., Quellec, P., Marchand, M., Dellacherie, E., Harnisch, S., Blunk, T., & Muller, R. H. (2000). 'Stealth' corona-core nanoparticles surface modified by polyethylene glycol (PEG): influences of the corona (PEG chain length and surface density) and of the core composition on phagocytic uptake and plasma protein adsorption. *Colloids and Surfaces B-Biointerfaces. 18*, 301–313.

87. Grosova, Z., Rosenberg, M., Rebros, M., Sipocz, M., & Sedlackova, B. (2008). Entrapment of beta-galactosidase in polyvinylalcohol hydrogel. *Biotechnol Lett. 30*, 763–767.

88. Guisan, J. M. (2006). Immobilization of enzymes as the 21st century begins. Immobilization of enzymes and cells, 2nd ed New Jersey (NJ): *Humana*. 1–13.

89. Gurung, N., Ray, S., Bose, S., & Rai, V. (2013). A broader view: Microbial enzymes and their relevance in industries, medicine, and beyond. *Bio Med Res Int.* 1–18.

90. Hanefeld, U., Gardossi, L., & Magner, E. (2009). Understanding enzyme immobilization. *Chem Soc Rev. 38*, 453–468.

91. Hartmann, M., & Kostrov, X. (2013). Immobilization of enzymes on porous silicas--benefits and challenges. *Chem Soc Rev. 42*, 6277–6289.

92. Hayat, M. A. (1975). Positive staining for electron microscopy.. New York (NY): Van Nostrand Reinhold Company.

93. Heath, J. R., & Davis, M. E. (2008). Nanotechnology and cancer. Vol. 59, *Annu Rev Med.* 251–265.

94. Hellstrand, E., Lynch, I., Andersson, A., Drakenberg, T., Dahlback, B., Dawson, K. A., Linse, S., & Cedervall, T. (2009). Complete high-density lipoproteins in nanoparticle corona. FEBS J. *276*, 3372–3381.

95. Heng, B. C., Zhao, X., Tan, E. C., Khamis, N., Assodani, A., Xiong, S., Ruedl, C., Ng, K. W., & Loo, J. S. (2011). Evaluation of the cytotoxic and inflammatory potential of differentially shaped zinc oxide nanoparticles. *Arch Toxicol. 85*, 1517–1528.

96. Hsiao, I. L., & Huang, Y. J. (2011). Effects of various physicochemical characteristics on the toxicities of ZnO and TiO nanoparticles toward human lung epithelial cells. *Sci Total Environ. 409*, 1219–1228.

97. Huang, L., & Cheng, Z. M. (2008). Immobilization of lipase on chemically modified bimodal ceramic foams for olive oil hydrolysis. *Chem Eng J. 144*, 103–109.

98. Huff, M. E., Balch, W. E., & Kelly, J. W. (2003). Pathological and functional amyloid formation orchestrated by the secretory pathway. *Curr Opin Struct Biol. 13*, 674–682.

99. Husain, Q. (2017). Nanomaterials Immobilized Cellulolytic Enzymes and their Industrial Applications: A Literature Review. *JSM Biochem Mol Biol*, 4, 1029.

100. Ikeda, Y., Taniguchi, N., & Noguchi, T. (2000). Dominant negative role of the glutamic acid residue conserved in the pyruvate kinase M-1 isozyme in the heterotropic allosteric effect involving fructose-1,6-bisphosphate. *J Biol Chem. 275*, 9150–9156.

101. Inomoto, N., Osaka, N., Suzuki, T., Hasegawa, U., Ozawa, Y., Endo, H., Akiyoshi, K., & Shibayama, M. (2009). Interaction of nanogel with cyclodextrin or protein: Study by dynamic light scattering and small-

angle neutron scattering. *Polymer. 50*, 541–546.

102. Ipe, B. I., Shukla, A., Lu, H. C., Zou, B., Rehage, H., & Niemeyer, C. M. (2006). Dynamic light-scattering analysis of the electrostatic interaction of hexahistidine-tagged cytochrome P450 enzyme with semiconductor quantum dots. *Chemphyschem. 7*, 1112–1118.
103. Ispas, C., Sokolov, I., & Andreescu, S. (2009). Enzyme-functionalized mesoporous silica for bioanalytical applications. *Anal Bioanal Chem. 393*, 543–554.
104. Jaenicke, R., & Böhm, G. (1998). The stability of proteins in extreme environments. *Curr Opin Struct Biol. 8*, 738–748.
105. Jain, N., Bhargava, A., Rathi, M., Dilip, R., & Panwar, J. (2015). Removal of Protein Capping Enhances the Antibacterial Efficiency of Biosynthesized Silver Nanoparticles. *PLoS One. 10*, e0134337.
106. Jalilian, A. R., Hosseini-Salekdeh, S. L., Mahmoudi, M., Yousefnia, H., Majdabadi, A., & Pouladian, M. (2011). Preparation and biological evaluation of radiolabeled-folate embedded superparamagnetic nanoparticles in wild-type rats. *J Radioanal Nucl Chem. 287*, 119–127.
107. Jalilian, A. R., Panahifar, A., Mahmoudi, M., Akhlaghi, M., & Simchi, A. (2009). Preparation and biological evaluation of Ga-67 -labeled-superparamagnetic nanoparticles in normal rats. *Radiochimica Acta. 97*, 51–56.
108. Jammer, A., Gasperl, A., Luschin-Ebengreuth, N., Heyneke, E., Chu, H., Cantero-Navarro, E., Grosskinsky, D. K., Albacete, A. A., Stabentheiner, E., Franzaring, J., Fangmeier, A., & van der Graaff, E., Roitsch, T. (2015). Simple and robust determination of the activity signature of key carbohydrate metabolism enzymes for physiological phenotyping in model and crop plants. *J Exp Bot. 66*, 5531–5542.
109. Jans, H., Liu, X., Austin, L., Maes, G., & Huo, Q. (2009). Dynamic Light Scattering as a Powerful Tool for Gold Nanoparticle Bioconjugation and Biomolecular Binding Studies. *Anal Chem. 81*, 9425–9432.
110. Jegannathan, K. R., Abang, S., Poncelet, D., Chan, E. S., & Ravindra, P. (2008). Production of biodiesel using immobilized lipase--a critical review. *Crit Rev Biotechnol. 28*, 253–264.
111. Jha, I., & Venkatesu, P. (2016). Deciphering the Interactions of Bromelain with Carbon Nanotubes: Role of Protein as Well as Carboxylated Multiwalled Carbon Nanotubes in a Complexation Mechanism. *Journal of Physical Chemistry C. 120*, 15436–15445.

112. Jiang, K., Schadler, L. S., Siegel, R. W., Zhang, X., Zhang, H., & Terrones, M. (2004). Protein immobilization on carbon nanotubes via a two-step process of diimide-activated amidation. *J Mater Chem. 14*, 37.

113. Jiang, X., Jiang, U. G., Jin, Y. D., Wang, E. K., & Dong, S. J. (2005). Effect of colloidal gold size on the conformational changes of adsorbed cytochrome c: Probing by circular dichroism, UV-visible, and infrared spectroscopy. *Biomacromolecules. 6*, 46–53.

114. Jiang, X., Weise, S., Hafner, M., Roecker, C., Zhang, F., Parak, W. J., & Nienhaus, G. U. (2010). Quantitative analysis of the protein corona on FePt nanoparticles formed by transferrin binding. *Journal of the Royal Society Interface. 7*, S5–S13.

115. Jin, W., & Brennan, J. D. (2002). Properties and applications of proteins encapsulated within sol–gel derived materials. *Anal Chim Acta. 461*, 1–36.

116. Kao, K. C., Lin, T. S., & Mou, C. Y. (2014). Enhanced activity and stability of lysozyme by immobilization in the matching nanochannels of mesoporous silica nanoparticles. *The Journal of Physical Chemistry C. 118*, 6734–6743.

117. Karajanagi, S. S., Vertegel, A. A., Kane, R. S., & Dordick, J. S. (2004). Structure and function of enzymes adsorbed onto single-walled carbon nanotubes. *Langmuir. 20*, 11594–11599.

118. Karlsson, M., & Carlsson, U. (2005). Protein adsorption orientation in the light of fluorescent probes: Mapping of the interaction between site-directly labeled human carbonic anhydrase II and silica nanoparticles. *Biophys J. 88*, 3536–3544.

119. Kathiravan, A., Renganathan, R., & Anandan, S. (2009). Interaction of colloidal AgTiO(2) nanoparticles with bovine serum albumin. *Polyhedron. 28*, 157–161.

120. Katoch, J., Kim, S. N., Kuang, Z., Farmer, B. L., Naik, R. R., Tatulian, S. A., & Ishigami, M. (2012). Structure of a peptide adsorbed on graphene and graphite. *Nano Lett. 12*, 2342–2346.

121. Keeling-Tucker, T., & Brennan, J. D. (2001). Fluorescent probes as reporters on the local structure and dynamics in sol-gel derived nanocomposite materials. *Chem Mater. 13*, 3331–3350.

122. Keizer, J. (1983). Nonlinear fluorescence quenching and the origin of positive curvature in Stern-Volmer Plots. *J Am Chem Soc. 105*, 1494–

1498.

123. Kenry; Loh, K. P., & Lim, C. T. (2016). Molecular interactions of graphene oxide with human blood plasma proteins. *Nanoscale*. 8, 9425–9441.

124. Kenworthy, A. K. (2001). Imaging protein-protein interactions using fluorescence resonance energy transfer microscopy. *Methods*. *24*, 289–296.

125. Kietzke, T., Neher, D., Landfester, K., Montenegro, R., Guntner, R., & Scherf, U. (2003). Novel approaches to polymer blends based on polymer nanoparticles. *Nat Mater*. 2, 408–412.

126. Kim, H. R., Andrieux, K., Delomenie, C., Chacun, H., Appel, M., Desmaele, D., Taran, F., Georgin, D., Couvreur, P., & Taverna, M. (2007b). Analysis of plasma protein adsorption onto PEGylated nanoparticles by complementary methods: 2-DE, CE and protein Lab-on-chip (R) system. *Electrophoresis*. *28*, 2252–2261.

127. Kim, H. R., Andrieux, K., Gil, S., Taverna, M., Chacun, H., Desmaele, D., Taran, F., Georgin, D., & Couvreur, P. (2007a). Translocation of poly(ethylene glycol-co-hexadecyl)cyanoacrylate nanoparticles into rat brain endothelial cells: Role of apolipoproteins in receptor-mediated endocytosis. *Biomacromolecules*. *8*, 793–799.

128. Kim, J., Grate, J. W., & Wang, P. (2006). Nanostructures for enzyme stabilization. *Chem Eng Sci*. *61*, 1017–1026.

129. Klein, J. (2007). Probing the interactions of proteins and nanoparticles. *Proc Natl Acad Sci U S A*. *104*, 2029–2030.

130. Kondo, A., Oku, S., Murakami, F., & Higashitani, K. (1993). Conformational changes in protein molecules upon adsorption on ultrafine particles. *Colloids Surf B Biointerfaces*. *1*, 197–201.

131. Konduru, N. V., Molina, R. M., Swami, A., Damiani, F., Pyrgiotakis, G., Lin, P., Andreozzi, P., Donaghey, T. C., Demokritou, P., Krol, S., Kreyling, W., & Brain, J. D. (2017). Protein corona: implications for nanoparticle interactions with pulmonary cells. *Part Fiber Toxicol*. *14*, 42.

132. Koo, E. H., Lansbury, P. T., & Kelly, J. W. (1999). Amyloid diseases: Abnormal protein aggregation in neurodegeneration. *Proc Natl Acad Sci U S A*. *96*, 9989–9990.

133. Koster, A. J., Ziese, U., Verkleij, A. J., Janssen, A. H., & de Jong, K. P. (2000). Three-dimensional transmission electron microscopy: a

novel imaging and characterization technique with nanometer scale resolution for materials science. *The Journal of Physical Chemistry B. 104*, 9368–9370.

134. Kreyling, W. G., Semmler-Behnke, M., & Chaudhry, Q. (2010). A complementary definition of nanomaterial. *Nano Today. 5*, 165–168.
135. Kumar, A., Singh, S., Tiwari, R., Goel, R., & Nain, L. (2017). Immobilization of indigenous holocellulase on iron oxide (Fe_2O_3) nanoparticles enhanced hydrolysis of alkali pretreated paddy straw. *Int J Biol Macromol. 96*, 538–549.
136. Lacerda, S. H. D. P., Park, J. J., Meuse, C., Pristinski, D., Becker, M. L., Karim, A., & Douglas, J. F. (2010). Interaction of gold nanoparticles with common human blood proteins. *ACS Nano. 4,* 365–379.
137. Lassoued, A., Dkhil, B., Gadri, A., & Ammar, S. (2017). Control of the shape and size of iron oxide (α-Fe_2O_3) nanoparticles synthesized through the chemical precipitation method. *Results in Physics. 7*, 3007–3015.
138. Lee, I. S., Lee, N., Park, J., Kim, B. H., Yi, Y. W., Kim, T., Kim, T. K., Lee, I. H., Paik, S. R., & Hyeon, T. (2006). Ni/NiO core/shell nanoparticles for selective binding and magnetic separation of histidine-tagged proteins. *J Am Chem Soc. 128*, 10658–10659.
139. Levy, R., Thanh, N. T. K., Doty, R. C., Hussain, I., Nichols, R. J., Schiffrin, D. J., Brust, M., & Fernig, D. G. (2004). Rational and combinatorial design of peptide capping Ligands for gold nanoparticles. *J Am Chem Soc. 126*, 10076–10084.
140. Li, C., Jiang, S., Zhao, X., & Liang, H. (2017). Co-Immobilization of enzymes and magnetic nanoparticles by metal-nucleotide hydrogelnanofibers for improving stability and recycling. *Molecules. 22*.
141. Li, S., Yang, X., Yang, S., Zhu, M., & Wang, X. (2012a). Technology prospecting on enzymes: application, marketing and engineering. *Computational and structural biotechnology journal*. 2, e201209017.
142. Li, Y., Budamagunta, M. S., Luo, J., Xiao, W., Voss, J. C., & Lam, K. S. (2012b) Probing of the Assembly Structure and Dynamics within Nanoparticles during Interaction with Blood Proteins. *ACS Nano*. 6, 9485–9495.
143. Lindman, S., Lynch, I., Thulin, E., Nilsson, H., Dawson, K. A., & Linse, S. (2007). Systematic investigation of the thermodynamics of

HSA adsorption to N-iso-propylacrylamide/N-tert-butylacrylamide copolymer nanoparticles. Effects of particle size and hydrophobicity. *Nano Lett.* 7, 914–920.

144. Linse, S., Cabaleiro-Lago, C., Xue, W. F., Lynch, I., Lindman, S., Thulin, E., Radford, S. E., & Dawson, K. A. (2007). Nucleation of protein fibrillation by nanoparticles. *Proc Natl Acad Sci U S A. 104*, 8691–8696.
145. Liu, H., Pierre-Pierre, N., & Huo, Q. (2012). Dynamic light scattering for gold nanorod size characterization and study of nanorod–protein interactions. *Gold Bulletin. 45*, 187–195.
146. Liu, S., Leech, D., & Ju, H. (2003). Application of Colloidal Gold in Protein Immobilization, Electron Transfer, and Biosensing. *Anal Lett. 36*, 1–19.
147. Liu, X., Guan, Y., Shen, R., & Liu, H. (2005). Immobilization of lipase onto micron-size magnetic beads. *J Chromatogr B Analyt Technol Biomed Life Sci. 822*, 91–97.
148. López-Serrano, A., Olivas, R. M., Landaluze, J. S., & Cámara, C. (2014). Nanoparticles: a global vision. Characterization, separation, and quantification methods. Potential environmental and health impact. *Anal Methods. 6*, 38–56.
149. Lundqvist, M., Sethson, I., & Jonsson, B. H. (2004). Protein adsorption onto silica nanoparticles: Conformational changes depend on the particles' curvature and the protein stability. *Langmuir. 20*, 10639–10647.
150. Lundqvist, M., Sethson, I., & Jonsson, B. H. (2005). High-resolution 2D H-1-N-15 NMR characterization of persistent structural alterations of proteins induced by interactions with silica nanoparticles. *Langmuir. 21*, 5974–5979.
151. Lundqvist, M., Stigler, J., Elia, G., Lynch, I., Cedervall, T., & Dawson, K. A. (2008). Nanoparticle size and surface properties determine the protein corona with possible implications for biological impacts. *Proc Natl Acad Sci USA. 105*, 14265–14270.
152. Luo, X., Morrin, A., Killard, A. J., & Smyth, M. R. (2006). Application of Nanoparticles in Electrochemical Sensors and Biosensors. *Electroanalysis. 18*, 319–326.
153. Lynch, I., & Dawson, K. A. (2008). Protein-nanoparticle interactions. *Nanotoday. 3*, 40–47.

154. Lynch, I., Cedervall, T., Lundqvist, M., Cabaleiro-Lago, C., Linse, S., & Dawson, K. A. (2007). The nanoparticle – protein complex as a biological entity; a complex fluids and surface science challenge for the 21st century. *Adv Colloid Interface Sci.* 134–35, 167–174.

155. Lynch, I., Dawson, K. A., Linse, S. (2006). Detecting cryptic epitopes created by nanoparticles. *Sci STKE*, 2006 pe14.

156. Macaroff, P. P., Oliveira, D. M., Lacava, Z. G. A., Azevedo, R. B., Lima, E. C. D., Morais, P. C., & Tedesco, A. C. (2004). The effect of bovine serum albumin on the binding constant and stoichiometry of biocompatible magnetic fluids. *IEEE Trans Magn. 40*, 3027–3029.

157. Maheshwari, R., Bharadwaj, G., & Bhat, M. K. (2000). Thermophilic fungi: Their physiology and enzymes.. *Microbiol Mol Biol Rev. 64*, 461–488.

158. Mahmoudi, M., Sahraian, M. A., Shokrgozar, M. A., Laurent, S. (2011a). Superparamagnetic Iron Oxide Nanoparticles: Promises for Diagnosis and Treatment of Multiple Sclerosis. *ACS Chem Neurosci. 2*, 118–140.

159. Mahmoudi, M., Sant, S., Wang, B., Laurent, S., Sen, T. (2011b). Superparamagnetic iron oxide nanoparticles (SPIONs): Development, surface modification and applications in chemotherapy. *Adv Drug Del Rev*. 2011b, 63, 24–46.

160. Mahmoudi, M., Shokrgozar, M. A., Sardari, S., Moghadam, M. K., Vali, H., Laurent, S., Stroeve, P. (2011c). Irreversible changes in protein conformation due to interaction with superparamagnetic iron oxide nanoparticles. *Nanoscale*. 2011c, 3, 1127–1138.

161. Mahmoudi, M., Simchi, A., Imani, M. (2010). Recent Advances in Surface Engineering of Superparamagnetic Iron Oxide Nanoparticles for Biomedical Applications. *Journal of the Iranian Chemical Society. 7*, S1–S27.

162. Mahmoudi, M., Simchi, A., Imani, M., Milani, A. S., Stroeve, P. (2008). Optimal design and characterization of superparamagnetic iron oxide nanoparticles coated with polyvinyl alcohol for targeted delivery and imaging. *J Phys Chem B. 112*, 14470–14481.

163. Mandal, H. S., & Kraatz, H. B. (2007). Effect of the surface curvature on the secondary structure of peptides adsorbed on nanoparticles. *J Am Chem Soc. 129*, 6356-+.

164. Matsuura, K., Saito, T., Okazaki, T., Ohshima, S., Yumura, M., Iijima,

S. (2006). Selectivity of water-soluble proteins in single-walled carbon nanotube dispersions. *Chem Phys Lett. 429*, 497–502.

165. Medintz, I. L., Uyeda, H. T., Goldman, E. R., Mattoussi, H 2005. Quantum dot bioconjugates for imaging, labeling and sensing. *Nature Materials. 4*, 435–446.

166. Millan, S., Satish, L., Bera, K., Susrisweta, B., Singh, D. V., Sahoo, H. (2017). A Spectroscopic and Molecular Simulation Approach toward the Binding Affinity between Lysozyme and Phenazinium Dyes: An Effect on Protein Conformation. *J Phys Chem B. 121*, 1475–1484.

167. Millan, S., Satish, L., Kesh, S., Chaudhary, Y. S., Sahoo, H. (2016). Interaction of Lysozyme with Rhodamine B: A combined analysis of spectroscopic & molecular docking. *J Photochem Photobiol B: Biol. 162*, 248–257.

168. Miller, J. N. (1979). Recent advances in molecular luminescence analysis. *Proceedings of the Analytical Division of the Chemical Society. 16*, 203–208.

169. Miller, J. N., Ahmad, T. A., & Fell, A. F. (1982). Derivative fluorescence spectroscopy. Anal Proc *19*, 37–41.

170. Monopoli, M. P., Walczyk, D., Campbell, A., Elia, G., Lynch, I., Bombelli, F. B., & Dawson, K. A. (2011). Physical-Chemical Aspects of Protein Corona: Relevance to in Vitro and in Vivo Biological Impacts of Nanoparticles. *J Am Chem Soc. 133*, 2525–2534.

171. Mu, Q., Liu, W., Xing, Y., Zhou, H., Li, Z., Zhang, Y., Ji, L., Wang, F., Si, Z., Zhang, B., Yan, B. (2008). Protein binding by functionalized multiwalled carbon nanotubes is governed by the surface chemistry of both parties and the nanotube diameter. *Journal of Physical Chemistry C. 112*, 3300–3307.

172. Mueller, N. C., Nowack, B. (2008). Exposure Modeling of Engineered Nanoparticles in the Environment. *Environ Sci Technol. 42*, 4447–4453.

173. Murphy, C. J., Sau, T. K., Gole, A. M., Orendorff, C. J., Gao, J., Gou, L., Hunyadi, S. E., Li, T. (2005). Anisotropic metal nanoparticles: Synthesis, assembly, and optical applications. *J Phys Chem B. 109*, 13857–13870.

174. Nakamura, M., Sueishi, T. (1991). Risk analysis issues in developing countries. Advances in risk analysis New York: Springer Science, *Business Media*. 555–564.

175. Navalakhe, R. M., & Nandedkar, T. D. (2007). Application of nanotechnology in biomedicine. *Indian J Exp Biol. 45*, 160–165.

176. Naveed Ul Haq, A., Nadhman, A., Ullah, I., Mustafa, G., Yasinzai, M., & Khan, I. (2017). Synthesis Approaches of Zinc Oxide Nanoparticles: The Dilemma of Ecotoxicity. *Journal of Nanomaterials*. 1–14.

177. Nel, A. E., Maedler, L., Velegol, D., Xia, T., Hoek, E. M. V., Somasundaran, P., Klaessig, F., Castranova, V., & Thompson, M. (2009). Understanding biophysicochemical interactions at the nano-bio interface. *Nature Materials. 8*, 543–557.

178. Nirmal, M., Brus, L. (1999). Luminescence photophysics in semiconductor nanocrystals. *Acc Chem Res. 32*, 407–414.

179. Norde, W., & Lyklema, J. (1991a). Why proteins prefer interfaces. *J Biomater Sci*, Polym Ed. *2*, 183–202.

180. Noureddini, H., Gao, X. (2006). Characterization of sol-gel immobilized lipases. *J Sol-Gel Sci Technol. 41*, 31–41.

181. Olanoff, L. S., Venkatasubramanian, K., Bernath, F. R., & Joyeuse, R. (1977). Perfusion trials with a collagen-immobilized enzyme in an extracorporeal reactor: Activity, stability, and biocompatibility. *J Biomed Mater Res. 11*, 125–136.

182. Oliveira, S. F., da Luz, J. M. R., Kasuya, M. C. M., Ladeira, L. O., & Correa Junior, A. (2016). Enzymatic extract containing lignin peroxidase immobilized on carbon nanotubes: Potential biocatalyst in dye decolorization. *Saudi J Biol Sci. 25*, 651-659.

183. Ovsejevi, K., Manta, C., & Batista-Viera, F. (2013). Reversible covalent immobilization of enzymes via disulfide bonds. *Methods Mol Biol. 1051*, 89–116.

184. Padavattan, S., Flicker, S., Schirmer, T., Madritsch, C., Randow, S., Reese, G., Vieths, S., Lupinek, C., Ebner, C., Valenta, R., & Markovic-Housley, Z. (2009). High-affinity IgE recognition of a conformational epitope of the major respiratory allergen Phl p 2 as revealed by X-ray crystallography. *J Immunol. 182*, 2141–2151.

185. Park, J. M., Kim, M., Park, H. S., Jang, A., Min, J., & Kim, Y. H. (2013). Immobilization of lysozyme-CLEA onto electrospun chitosan nanofiber for effective antibacterial applications. *Int J Biol Macromol. 54*, 37–43.

186. Peng, Z. G., Hidajat, K., & Uddin, M. S. (2004). Adsorption of bovine serum albumin on nanosized magnetic particles. *J Colloid Interface*

Sci. 271, 277–283.

187. Perry, J. L., Reuter, K. G., Kai, M. P., Herlihy, K. P., Jones, S. W., Luft, J. C., Napier, M., Bear, J. E., & DeSimone, J. M. (2012). PEGylated PRINT nanoparticles: the impact of PEG density on protein binding, macrophage association, biodistribution, and pharmacokinetics. *Nano Lett. 12*, 5304–5310.
188. Petkova, G. A., Zaruba, C. K., Zvatora, P., & Kral, V. (2012). Gold and silver nanoparticles for biomolecule immobilization and enzymatic catalysis. *Nanoscale research letters*. 7, 287.
189. Piella, J., Bastus, N. G., & Puntes, V. (2017). Size-Dependent Protein−Nanoparticle Interactions in CitrateStabilized Gold Nanoparticles: The Emergence of the Protein Corona. *Bioconjugate Chem. 28*, 88−97.
190. Pilloni, M., Nicolas, J., Marsaud, V., Bouchemal, K., Frongia, F., Scano, A., Ennas, G., & Dubernet, C. (2010). PEGylation and preliminary biocompatibility evaluation of magnetite-silica nanocomposites obtained by high energy ball milling. *Int J Pharm. 401*, 103–112.
191. Pompa, P. P., Chiuri, R., Manna, L., Pellegrino, T., del Mercato, L. L., Parak, W. J., Calabi, F., Cingolani, R., & Rinaldi, R. (2006). Fluorescence resonance energy transfer induced by conjugation of metalloproteins to nanoparticles. *Chem Phys Lett. 417*, 351–357.
192. Pons, T., Medintz, I. L., Wang, X., English, D. S., & Mattoussi, H. (2006). Solution-phase single quantum dot fluorescence resonance energy transfer. *J Am Chem Soc. 128*, 15324–15331.
193. Porath, J. Salting out adsorption techniques for protein purifications. *Biopolymers. 26*, S193–S204.
194. Prakasham, R. S., Devi, G. S., Rao, C. S., Sivakumar, V. S. S., Sathish, T., & Sarma, P. N. (2010). Nickel-Impregnated Silica Nanoparticle Synthesis and Their Evaluation for Biocatalyst Immobilization. *Appl Biochem Biotechnol. 160*, 1888–1895.
195. Pramanik, S., Banerjee, P., Sarkar, A., & Bhattacharya, S. C. (2008). Size-dependent interaction of gold nanoparticles with transport protein: A spectroscopic study. *J Lumin. 128*, 1969–1974.
196. Pu, K., Shuhendler, A. J., Jokerst, J. V., Mei, J., Gambhir, S. S., Bao, Z., & Rao, J. (2014). Semiconducting polymer nanoparticles as photoacoustic molecular imaging probes in living mice. *Nat Nanotechnol. 9*, 233–239.
197. Ragnaill, M. N., Brown, M., Ye, D., Bramini, M., Callanan, S., Lynch,

I., & Dawson, K. A. (2011). Internal benchmarking of a human blood-brain barrier cell model for screening of nanoparticle uptake and transcytosis. *Eur J Pharm Biopharm. 77*, 360–367.

198. Rajagopal, K., & Schneider, J. P. (2004). Self-assembling peptides and proteins for nanotechnological applications. *Curr Opin Struct Biol. 14*, 480–486.
199. Rao, J. P., & Geckeler, K. E. (2011). Polymer nanoparticles: Preparation techniques and size-control parameters. *Prog Polym Sci. 36*, 887–913.
200. Rao, M. B., Tanksale, A. M., Ghatge, M. S., & Deshpande, V. V. (1998). Molecular and biotechnological aspects of microbial proteases. *Microbiol Mol Biol Rev. 62*, 3597–3635.
201. Ren, G., Xu, X., Liu, Q., Cheng, J., Yuan, X., Wu, L., & Wan, Y. (2006). Electrospun poly(vinyl alcohol)/glucose oxidase biocomposite membranes for biosensor applications. *React Funct Polym. 66*, 1559–1564.
202. Rezwan, K., Studart, A. R., Voros, J., & Gauckler, L. J. (2005). Change of xi potential of biocompatible colloidal oxide particles upon adsorption of bovine serum albumin and lysozyme. *J Phys Chem B. 109*, 14469–14474.
203. Roach, P., Farrar, D., & Perry, C. C. (2006). Surface tailoring for controlled protein adsorption: effect of topography at the nanometer scale and chemistry. *J Am Chem Soc. 128*, 3939–3945.
204. Rocker, C., Potzl, M., Zhang, F., Parak, W. J., & Nienhaus, G. U. (2009). A quantitative fluorescence study of protein monolayer formation on colloidal nanoparticles. *Nat Nanotechnol. 4*, 577–580.
205. Rodríguez, L. A., Toro, M. E., Vazquez, F., Correa-Daneri, M. L., Gouiric, S. C., & Vallejo, M. D. (2010). Bioethanol production from grape and sugar beet pomaces by solid-state fermentation. *Int J Hydrogen Energy. 35*, 5914–5917.
206. Romdhane, I. B. B., Romdhane, Z. B., Gargouri, A., & Belghith, H. (2011). Esterification activity and stability of Talaromyces thermophilus lipase immobilized onto chitosan. *J Mol Catal B: Enzym. 68*, 230–239.
207. Romdhane, I. B.-B., Fendri, A., Gargouri, Y., Gargouri, A., & Belghith, H. (2010). A novel thermoactive and alkaline lipase from Talaromyces thermophilus fungus for use in laundry detergents. *Biochem Eng J. 53*, 112–120.
208. Rozhkov, S. P., Goryunov, A. S., Sukhanova, G. A., Borisova, A. G.,

Rozhkova, N. N., & Andrievsky, G. V. (2003). Protein interaction with hydrated C-60 fullerene in aqueous solutions. *Biochem Biophys Res Commun. 303*, 562–566.

209. Ruh, H., Kuhl, B., Brenner-Weiss, G., Hopf, C., Diabate, S., & Weiss, C. (2012). Identification of serum proteins bound to industrial nanomaterials. *Toxicol Lett. 208*, 41–50.
210. Sahoo, B., Goswami, M., Nag, S., & Maiti, S. (2007). Spontaneous formation of a protein corona prevents the loss of quantum dot fluorescence in physiological buffers. *Chem Phys Lett. 445*, 217–220.
211. Salimi, A., Hallaj, R., & Soltanian, S. (2007). Immobilization of hemoglobin on electrodeposited cobalt-oxide nanoparticles: direct voltammetry and electrocatalytic activity. *Biophys Chem. 130*, 122–131.
212. Sarkar, R., Narayanan, S. S., Palsson, L. O., Dias, F., Monkman, A., & Pal, S. K. (2007). Direct conjugation of semiconductor nanocrystals to a globular proteinto study protein-folding intermediates. *J Phys Chem B. 111*, 12294–12298.
213. Satish, L., Millan, S., & Sahoo, H. (2017). Spectroscopic insight into the interaction of bovine serumalbumin with imidazolium-based ionic liquids inaqueous solution. *Luminescence. 32*, 695–705.
214. Scapin, G. Structural biology and drug discovery. Curr Pharm Des. *12*, 2087–2097.
215. Sen, S., Konar, S., Das, B., Pathak, A., Dhara, S., Dasgupta, S., & DasGupta, S. (2016). Inhibition of fibrillation of human serum albumin through interaction with chitosan-based biocompatible silver nanoparticles. *RSC Adv. 6*, 43104–43115.
216. Sen, T., Mandal, S., Haldar, S., Chattopadhyay, K., & Patra, A. (2011). Interaction of Gold Nanoparticle with Human Serum Albumin (HSA) Protein Using Surface Energy Transfer. *Journal of Physical Chemistry C. 115*, 24037–24044.
217. Sergeeva, L. I., & Vreugdenhil, D. (2002). In situ staining of activities of enzymes involved in carbohydrate metabolism in plant tissues. *J Exp Bo*t. *53*, 361–370.
218. Shah, S., Solanki, K., & Gupta, M. N. (2007). Enhancement of lipase activity in non-aqueous media upon immobilization on multi-walled carbon nanotubes. *Chem Cent J. 1*, 30.
219. Shan, D., Zhu, M., Xue, H., & Cosnier, S. (2007). Development of

amperometric biosensor for glucose based on a novel attractive enzyme immobilization matrix: calcium carbonate nanoparticles. *Biosens Bioelectron. 22*, 1612–1617.

220. Shang, L., Wang, Y., Jiang, J., & Dong, S. (2007). pH-dependent protein conformational changes in albumin : gold nanoparticle bioconjugates: A spectroscopic study. *Langmuir. 23*, 2714–2721.
221. Shang, W., Nuffer, J. H., Dordick, J. S., Siegel, R. W. (2007b) Unfolding of Ribonuclease A on Silica Nanoparticle Surfaces. *Nano Lett. 7*, 1991–1995.
222. Sheldon, R. A. (2007). Cross-linked enzyme aggregates (CLEAs): stable and recyclable biocatalysts. *Biochem Soc. 35*, 1583–1587.
223. Sheldon, R. A. (2011). Characteristic features and biotechnological applications of cross-linked enzyme aggregates (CLEAs). *Appl Microbiol Biotechnol. 92*, 467–477.
224. Shen, Q., Yang, R., Hua, X., Ye, F., Zhang, W., & Zhao, W. (2011). Gelatin- templated biomimetic calcification for b-galactosidase immobilization. *Process Biochem. 46*, 1565–1571.
225. Shen, X.-C., Liou, X.-Y., Ye, L.-P., Liang, H., & Wang, Z. Y. (2007). Spectroscopic studies on the interaction between human hemoglobin and CdS quantum dots. *J Colloid Interface Sci. 311*, 400–406.
226. Shnitka, T. K., & Seligman, A. M. (1971). Ultrastructural localization of enzymes. *Annu Rev Biochem. 40*, 375–396.
227. Šimšíková, M. (2016). Interaction of graphene oxide with albumins: Effect of size, pH, and temperature. *Arch Biochem Biophys. 593*, 69–79.
228. Singh, N., Srivastava, G., Talat, M., Raghubanshi, H., Srivastava, O. N., & Kayastha, A. M. (2014). Cicer alpha-galactosidase immobilization onto functionalized graphene nanosheets using response surface method and its applications. *Food Chem. 142*, 430–438.
229. Singh, V., Sardar, M., & Gupta, M. N. (2013). Immobilization of Enzymes by Bioaffinity Layering. Immobilization of Enzymes and Cells (3rd edn), Springer, Humana Press, 129-137.
230. Sinha, R., Kim, G. J., Nie, S., & Shin, D. M. (2006). Nanotechnology in cancer therapeutics: bioconjugated nanoparticles for drug delivery. *Mol Cancer Ther. 5*, 1909–1917.
231. Skaat, H., Sorci, M., Belfort, G., & Margel, S. (2009). Effect of maghemite nanoparticles on insulin amyloid fibril formation: selective

labeling, kinetics, and fibril removal by a magnetic field. *J Biomed Mater Res A. 91*, 342–351.

232. Solanki, P. R., Dhand, C., Kaushik, A., Ansari, A. A., Sood, K. N., & Malhotra, B. D. (2009). Nanostructured cerium oxide film for triglyceride sensor. *Sens Actuators B: Chem, 141*, 551–556.

233. Srivastava, G., Roy, S., & Kayastha, A. M. (2015). Immobilization of Fenugreek beta-amylase on chitosan/PVP blend and chitosan coated PVC beads: a comparative study. *Food Chem. 172*, 844–851.

234. Srivastava, G., Singh, K., Talat, M., Srivastava, O. N., & Kayastha, A. M. (2014). Functionalized graphene sheets as immobilization matrix for Fenugreek beta-amylase: enzyme kinetics and stability studies. *PLoS One. 9*, e113408.

235. Stayton, P. S., Drobny, G. P., Shaw, W. J., Long, J. R., & Gilbert, M. (2003). Molecular recognition at the protein-hydroxyapatite interface. *Crit Rev Oral Biol Medicine. 14*, 370–376.

236. Steigerwald, M. L., & Brus, L. E. (2002). Semiconductor crystallites: a class of large molecules. *Acc Chem Res. 23*, 183–188.

237. Stepankova, V., Bidmanova, S., Koudelakova, T., Prokop, Z., Chaloupkova, R., & Damborsky, J. (2013). Strategies for stabilization of enzymes in organic solvents. *ACS Catalysis. 3*, 2823–2836.

238. Stoehr, L. C., Gonzalez, E., Stampfl, A., Casals, E., Duschl, A., Puntes, V., & Oostingh, G. J. (2011). Shape matters: effects of silver nanospheres and wires on human alveolar epithelial cells. *Part Fiber Toxicol. 8*, 36.

239. Stradner, A., Sedgwick, H., Cardinaux, F., Poon, W. C. K., Egelhaaf, S. U., & Schurtenberger, P. (2004). Equilibrium cluster formation in concentrated protein solutions and colloids. *Nature. 432*, 492–495.

240. Stuckler, D., & Nestle, M. (2012). Big food, food systems, and global health. *PLoS Med. 9*, e1001242.

241. Sumner, J. B. (1926). The isolation and crystallization of the enzyme urease. *J Biol Chem. 69*, 435–441.

242. Teichroeb, J. H., Forrest, J. A., & Jones, L. W. (2008). Size-dependent denaturing kinetics of bovine serum albumin adsorbed onto gold nanospheres. *European Physical Journal E. 26*, 411–415.

243. Temming, K., Lacombe, M., van der Hoeven, P., Prakash, J., Gonzalo, T., Dijkers, E. C. F., Orfi, L., Keri, G., Poelstra, K., Molema, G., & Kok, R. J. (2006). Delivery of the p38 MAPkinase inhibitor SB202190

to angiogenic endothelial cells: Development of novel RGD-equipped and PEGylated drug – Albumin conjugates using platinum(II)-based drug linker technology. *Bioconjugate Chem. 17*, 1246–1255.

244. Tessier, P. M., Jinkoji, J., Cheng, Y. C., Prentice, J. L., & Lenhoff, A. M. (2008). Self-interaction nanoparticle spectroscopy: A nanoparticle-based protein interaction assay. *J Am Chem Soc. 130*, 3106–3112.
245. Thomas, G. J. (1999). Raman spectroscopy of protein and nucleic acid assemblies. *Annu Rev Biophys Biomol Struct. 28*, 1.
246. Tischer, W., & Wedekind, F. (1999). Immobilized Enzymes: Methods and Applications. *Top Curr Chem. 200*, 95–126.
247. Treuel, L., & Nienhaus, G. U. (2012). Toward a molecular understanding of nanoparticle-protein interactions. *Biophysical Reviews. 4*, 137–147.
248. Tsai, D.-H., DelRio, F. W., Keene, A. M., Tyner, K. M., MacCuspie, R. I., Cho, T. J., Zachariah, M. R., & Hackley, V. A. (2011). Adsorption and Conformation of Serum Albumin Protein on Gold Nanoparticles Investigated Using Dimensional Measurements and in Situ Spectroscopic Methods. *Langmuir. 27*, 2464–2477.
249. Tsai, H. C., & Doong, R. A. (2007). Preparation and characterization of urease-encapsulated biosensors in poly(vinyl alcohol)-modified silica sol-gel materials. *Biosens Bioelectron. 23*, 66–73.
250. Turci, F., Ghibaudi, E., Colonna, M., Boscolo, B., Fenoglio, I., & Fubini, B. (2010). An integrated approach to the study of the interaction between proteins and nanoparticles. *Langmuir. 26*, 8336–8346.
251. Vaghari, H., Jafarizadeh-Malmiri, H., Mohammadlou, M., Berenjian, A., Anarjan, N., Jafari, N., & Nasiri, S. (2016). Application of magnetic nanoparticles in smart enzyme immobilization. *Biotechnol Lett. 38*, 223–233.
252. Verel, I., Visser, G. W. M., Boerman, O. C., van Eerd, J. E. M., Finn, R., Boellaard, R., Vosjan, M., Walsum, M. S. V., Snow, G. B., van & Dongen, G. (2003). Long-lived positron emitters zirconium-89 and iodine-124 for scouting of therapeutic radioimmunoconjugates with PET. *Cancer Biother Radiopharm. 18*, 655–661.
253. Verma, M. L., Naebe, M., Barrow, C. J., & Puri, M. (2013). Enzyme immobilization on amino-functionalised multi-walled carbon nanotubes: structural and biocatalytic characterisation. *PLoS One. 8*, e73642.
254. Vertegel, A. A., Siegel, R. W., & Dordick, J. S. (2004). Silica

Nanoparticle Size Influences the Structure and Enzymatic Activity of Adsorbed Lysozyme. *Langmuir. 20*, 6800–6807.

255. Vieille, C., & Zeikus, G. J. (2001). Hyperthermophilic enzymes: sources, uses, and molecular mechanisms for thermostability. *Microbiol Mol Biol Rev. 65*, 1–43.
256. Walczyk, D., Bombelli, F. B., Monopoli, M. P., Lynch, I., & Dawson, K. A. (2010). What the Cell "Sees" in Bionanoscience. *J Am Chem Soc. 132*, 5761–5768.
257. Walkey, C. D., & Chan, W. C. (2012). Understanding and controlling the interaction of nanomaterials with proteins in a physiological environment. *Chem Soc Rev. 41*, 2780–2799.
258. Wallace, A., Lees, J. G., Orry, A. J. W., Lobley, A., & Jane, R. W. (2003). Analyses of circular dichroism spectra of membrane proteins. *Protein Sci. 12*, 875–884.
259. Wan, L.-S., Ke, B.-B., Wu, J., Xu, Z.-K. (2007). Catalase Immobilization on Electrospun Nanofibers: Effects of Porphyrin Pendants and Carbon Nanotubes. *The Journal of Physical Chemistry C. 111*, 14091–14097.
260. Wang, X., Dou, P., Zhao, P., Zhao, C., Ding, Y., Xu, P. (2009a). Immobilization of lipases onto magnetic Fe(3)O(4) nanoparticles for application in biodiesel production. *Chem Sus Chem. 2*, 947–950.
261. Wang, Y. (2002). Nonlinear optical properties of nanometer-sized semiconductor clusters. *Acc Chem Res. 24*, 133–139.
262. Wang, Y. Q., Zhang, H.-M., Zhou, Q. H., Xu, H. L. (2009b). A study of the binding of colloidal Fe_3O_4 with bovine hemoglobin using optical spectroscopy. *Colloids and Surfaces A: Physicochem Eng Aspects. 337*, 102–108.
263. Wangoo, N., Suri, C. R., & Shekhawat, G. (2008). Interaction of gold nanoparticles with protein: A spectroscopic study to monitor protein conformational changes. *Appl Phys Lett. 92*.
264. Watanabe, J., & Ishihara, K. (2008). Multiple Protein-immobilized Phospholipid Polymer Nanoparticles: Effect of Spacer Length on Residual Enzymatic Activity and Molecular Diagnosis. *Nanobiotechnology*. 3, 76–82.
265. Wei, H., Wang, Z., Zhang, J., House, S., Gao, Y. G., Yang, L., Robinson, H., Tan, L. H., Xing, H., Hou, C., Robertson, I. M., Zuo, J. M., & Lu, Y. (2011). Time-dependent, protein-directed growth of gold nanoparticles within a single crystal of lysozyme. *Nat Nanotechnol. 6*, 93–97.

266. Weller, H. Quantized Semiconductor Particles: A novel state of matter for materials science. *Adv Mater. 5*, 88–95.

267. Won, K., Kim, S., Kim, K. J., Park, H. W., & Moon, S. J. (2005). Optimization of lipase entrapment in Ca-alginate gel beads. *Process Biochem. 40*, 2149–2154.

268. Wu, L., Yuan, X., & Sheng, J. (2005). Immobilization of cellulase in nanofibrous PVA membranes by electrospinning. *J Membr Sci. 250*, 167–173.

269. Wu, W. H., Sun, X., Yu, Y. P., Hu, J., Zhao, L., Liu, Q., Zhao, Y. F., & Li, Y. M. (2008). TiO2 nanoparticles promote beta-amyloid fibrillation in vitro. *Biochem Biophys Res Commun. 373*, 315–318.

270. Xiao, Q., Huang, S., Qi, Z. D., Zhou, B., He, Z. K., Liu, Y. (2008b). Conformation, thermodynamics and stoichiometry of HSA adsorbed to colloidal CdSe/ZnS quantum dots. *Biochim Biophys Acta. 1784*, 1020–1027.

271. Xiao, Q., Huang, S., Qi, Z.-D., Zhou, B., He, Z.-K., Liu, Y. (2008a). Conformation, thermodynamics and stoichiometry of HSA adsorbed to colloidal CdSe/ZnS quantum dots. *Biochimica Et Biophysica Acta-Proteins and Proteomics. 1784*, 1020–1027.

272. Xu, X., Brzostowski, J. A., & Jin, T. (2006). Using quantitative fluorescence microscopy and FRET imaging to measure spatiotemporal signaling events in single living cells. *Methods Mol Biol. 346*, 281–296.

273. Yang, S. T., Liu, Y., Wang, Y. W., & Cao, A. (2013). Biosafety and bioapplication of nanomaterials by designing protein-nanoparticle interactions. *Small. 9*, 1635–1653.

274. Yin, M. M., Dong, P., Chen, W. Q., Xu, S. P., Yang, L. Y., Jiang, F. L., & Liu, Y. (2017). Thermodynamics and Mechanisms of the Interactions between Ultrasmall Fluorescent Gold Nanoclusters and Human Serum Albumin, γ-Globulins, and Transferrin: A Spectroscopic Approach, *Langmuir*, *33*, 5108−5116.

275. Yudianti, R. (2011). Analysis of Functional Group Sited on Multi-Wall Carbon Nanotube Surface. *The Open Materials Science Journal. 5*, 242–247.

276. Zhao, J., Das, A., Schatz, G. C., Sligar, S. G., Van & Duyne, R. P. (2008). Resonance Localized Surface Plasmon Spectroscopy: Sensing Substrate and Inhibitor Binding to Cytochrome P450. *The Journal of Physical Chemistry C. 112*, 13084–13088.

277. Zheng, M., Davidson, F., & Huang, X. (2003). Ethylene Glycol Monolayer Protected Nanoparticles for Eliminating Nonspecific Binding with Biological Molecules. *J Am Chem Soc. 125*, 7790–7791.

278. Zorbas, V., Smith, A. L., Xie, H., Ortiz-Acevedo, A., Dalton, A. B., Dieckmann, G. R., Draper, R. K., Baughman, R. H., & Musselman, I. H. (2005). Importance of aromatic content for peptide/single-walled carbon nanotube interactions. *J Am Chem Soc. 127*, 12323–12328.

279. Žūkienė, R., & Snitka, V. (2005). Zinc oxide nanoparticle and bovine serum albumin interaction andnanoparticles influence on cytotoxicity in vitro. *Colloids Surf B Biointerfaces. 135*, 316–323.

3
CHAPTER

METAL NANOPARTICLES AS POTENTIAL ANTIMICROBIAL AGENTS

Diptipriya Sethi, Geetanjali Mishra, Barsha Dash, and Ramasamy Sakthivel

CSIR-Institute of Minerals & Materials Technology, Bhubaneswar-751013, India, E-mail: barsha.dash@gmail.com

CONTENTS

ABSTRACT

Despite of several existing effective antibiotics and other antimicrobial resources, contamination by microorganisms are still a major reason of many prolonged diseases. Consequently, there are much demand to develop supplementary resources because of the rising interest regarding multidrug-resistant bacterial strains and infections related to biofilms. Subsequently, attention has been paid towards new and emergent nanoparticle (NP) containing materials in the antimicrobial field. Now-a-days, metal nanoparticles are used extensively as antimicrobial agents as they are considered as an alternative source for antibiotics. In the present situation, metal NPs are treated as one of the most promising and innovative antimicrobial agents as they are beneficial in treating various bacterial infections. It is well-known that some of the metals and metallic compounds bears antimicrobial activity which is evident from the use of silver and copper in ancient times to treat burns, chronic wounds and to treat contaminated water to make it potable. Nowadays, the union of biology and nanotechnology has transformed the metals (in bulk form) to nanoparticles which may be implemented as potential antimicrobial agents. As NPs have distinctive physical and chemical properties, they can be easily employed for preferred uses. There has been a significant increase in the application of metal nanoparticles, i.e., aluminium, copper, silver, gold, magnesium, zinc and titanium as antimicrobials in industries, medicines, cosmetics, food packaging and textiles in the recent past. This chapter discusses the antimicrobial performances nanoparticles and its effect on drug-resistant bacteria, their mode of action and the possible risks that might occur on their use.

Keywords: Antimicrobial property, Metal nanoparticles, Oxidative stress, Antimicrobial agents, Biocompatibility

3.1. INTRODUCTION

Microbial contaminations are still a major problem of today's society. The increasing challenges considering the multidrug resistive bacterial strains and the biofilm-associated infections are some of the reasons that require urgent attention for the development of additional bactericidal means (Davey and Toole, 2000). Biofilm-associated infectious diseases are the reason for ~80 percent of microbial infections in human body, resulting in increased patient morbidity and medical expenses (Stoodley et al., 2004). Microbes like bacteria are naturally found in clinical, household and industrial surroundings in association with surfaces. Biofilms are group of microorganisms that attached to a substrate or surface. First, the bacteria gets attached to the surface which is a reversible process (Stoodley et al., 2004). Once settled, the bacteria multiply and form colonies inside peptidoglycan envelopes, which lead to the formation of a developed biofilm. Here, the bacteria not only turn to be resistant to antibacterial agents and the body's immune system, but also provide a reservoir of bacteria capable of triggering chronic infections throughout the body (Watnick and Kolter, 2000). This is why biofilms are a severe health threat. Moreover, biofilms respond slowly to conventional antibiotics and may get resistant towards it (Stoodley et al., 2004). Therefore, efforts have been devoted for the development of efficient antibacterial and biofilm preventing material for the serve of the society. The main sets of antibiotics used nowadays, generally affect three bacterial targets: cell wall synthesis, translational machinery, and DNA replication (Magiorakos et al., 2012). Unfortunately, bacterial resistance may also develop against each one of these modes of action. Mechanisms of bacterial resistance may occur in various ways, i.e., modification or degradation of antibiotic like β-lactamases and aminoglycosides, cell components such as cell wall modification as seen in vancomycin resistance and ribosomes in tetracycline's resistance and finally efflux pumps that gives multidrug resistance against numerous antibiotics (Magiorakos et al., 2012). The mechanism behind the mode of action of nanoparticle (NPs) is that, the straight contact of nanoparticle with the cell wall of bacteria therefore there is no need to penetrate into the cell. Now a days the majority of the bacteria are resistant to most of the antibiotics. This increases the hope that the NPs would be less vulnerable than antibiotics to endorse bacterial resistance. Subsequently, special consideration has been given for the development of antimicrobial materials instead of the nanoparticles. It is well-known that the antibacterial activity is the process by which growth of bacteria is destroyed/inhibited. It is also described because of the surface area interacting with the microorganisms (Seil et al., 2009). Antibacterial agents are drugs given at a selective concentration capable of damaging or inhibiting the growth of bacteria without affecting the host. Therefore these antibacterial agents are regarded as having bactericidal (if it kills bacteria) or as bacteriostatic (if it inhibits the growth of bacteria) properties. Various methods have been chosen for the in vitro study of antibacterial activity like broth dilution, disk diffusion, agar dilution, and the microtiter plate-based method. In this chapter, the prospective of various nanomaterials/ nanoparticles as antimicrobial agents has been described. The mechanism behind the antibacterial action of these nanoparticles and their interactions with microbial cells leading to cell death, including a detailed discussion of toxic and biocompatibility properties is also provided.

3.2. BIOCOMPATIBILITY OF NANOPARTICLES

The biocompatibility of nanomaterials must be considered prior to their use in the field of bio medicals like drug delivery, gene delivery, biosensors or the treatment of wound infections. In such applications, the tissues and cells come in direct contact with the NPs, where they can cause beneficial or destructive effects on the body. NPs as drugs can enter to the body by inhalation, oral ingestion, intravenous injection, and contact with the skin. The effect of NPs on various body tissues is not known, and the interaction of these NPs with cells and tissues is hardly understood. The toxicity or harmfulness of NPs can be evaluated by several in vitro and in vivo methods. The in vitro research is conducted on cell cultures. Cell culture studies are taken as a prescreening tool to understand the biological effects of NPs activity, their toxicity, and mechanism of action. A few inorganic and many synthesized polymeric NPs have been revealed to have different levels of biocompatibility.

3.3. ANTIMICROBIAL NANOPARTICLES

Nanoparticles, used as antibacterial agents, complementary to antibiotics are highly challenging and are creating interest as they might fill the gaps where antibiotics commonly fail. This includes combatting multidrug-resistant mutants and biofilms (Pelgrift and Friedman, 2013; Zhang et el., 2010). Antimicrobial nanoparticles that are currently in use (i.e., metal, metal oxide and organic nanoparticles) show a diversity of inherent and revised chemical composition properties. Thus, it is not surprising that they have numerous modes of action. Generally, the nanoparticles act in two major paths, which are correlated to each other and in most of the cases, may occur simultaneously;

a) Disruption of membrane potential and integrity.
b) Production of reactive oxygen species (ROS), also known as oxygen-free radicals where the NPs acts as nano catalysts (Pelgrift and Friedman, 2013; Huh and Kwon, 2011; Blecher et al., 2011).

When NPs bind electrostatically to the cell wall and cell membranes of bacteria, then membrane damage occurs, resulting to the change in the membrane potential/depolarization by losing its integrity. Which in turn results in an imbalanced transport, impaired respiration, interrupted energy transduction and/or cell lysis, and eventually cell death (Pelgrift and Friedman, 2013). ROS is considered as the most effective determining factor for both the in vitro and in vivo cytotoxicity of NPs (Nathan and Bussel, 2013). A burst of ROS generated via severe oxidative stress can potentially damage all the macromolecules of the cell leading to lipid peroxidation, alteration of proteins, and inhibition of enzymes and cripple the functioning of both RNA and DNA. At high concentrations, the ROS lead to cell death and at low doses cause severe DNA damage and mutations (Pan et al., 2010; Wang et al., 2011). In some cases, where the production of ROS is induced by visible or UV light (Matˇeejka and Tokarsk´y, 2014), the toxicity of NPs is photocatalytic. For instance, TiO_2 NPs, under near-UV irradiation, were found to induce lipid peroxidation and respiratory dysfunction leading death of *E. coli* cells (Maness et al., 1999).

Two types of nanoparticles having antimicrobial activity has been discussed in this chapter:

- Inorganic nanoparticles;
- Organic nanoparticles.

The detailed description about each of the nanoparticles has been discussed below.

3.3.1. Inorganic Nanoparticles

The antimicrobial activities of metals and metal oxides have been known since ancient days (Beyth et al., 2010). Metal and metal oxide nanoparticles, well known because of highly potent antibacterial effect, include metallic silver (Ag), Silver oxide (Ag_2O), titanium oxide (TiO_2), copper oxide (CuO), zinc oxide (ZnO), etc. Most of the metal oxide nanoparticles exhibit bactericidal properties through reactive oxygen *species generation as well as other properties.*

3.3.1.1. Silver Nanoparticles (Ag NPs)

Among all metal nanoparticles, silver nanoparticles (Ag NPs) are widely known as one of the promising antimicrobial agent against different microbes, i.e., bacteria, fungi, and viruses (Loomba and Scarabelli, 2013;

Rai et al., 2009; Mishra et al., 2013). Silver and its compounds have been used since long time to disinfect medical devices and for water purification. As medicine, these compounds are commonly used to treat burns, wounds, and a no. of infective diseases (Avalos et al., 2014; Elliott, 2010). The antimicrobial efficiency of Ag, as of other metals and metal oxide nanoparticles, was reported to be size-dependent (Aditya et al., 2013). Although the mechanism of Ag nanoparticle is still not clear, Ag nanoparticles with small diameters have a superior antimicrobial effect than those with larger diameters (Poulse et al., 2014). Moreover, antibacterial activity of Ag nanoparticle exceeds that of their bulk counterparts, because of their high surface to volume ratio and smaller particle size. Also silver has been reported as an efficient bactericidal agent against various pathogens *in vitro* and *in vivo* (Panacek et al., 2006). Moreover, it seems that bacteria are less susceptible to develop resistance against silver/silver based compounds compared to other conventional antibiotics (Simone et al., 2014; Leid et al., 2012). Though, the killing mechanism of Ag NPs are not fully understood, it is anticipated thatthe NPs interact physically with various bacterial cell surface (Chernousova and Epple, 2013). In *E. coli*, i.e., class of gram-negative bacteria, Ag nanoparticles were shown to cause "pits" in the cell wall there by increasing the membrane permeability and deactivating the respiratory chain (Beyth et al., 2010; Majdalawieh et al., 2014). Similarly, antimicrobial activity is also dependent on zeta potential. It has been observed that, the electrostatic forces developed during the interaction of nanoparticles having positive zeta potential with bacteria having negative zeta potential facilitates effective penetration of the NPs into the bacterial membrane (Aleksandar et al., 2012). Actually, the zeta potential and the size of the nanoparticles are two major factors which control the antimicrobial activity and thus nanoparticles having positive zeta potential and smaller size prove to be more effective. When Ag nanoparticles are dissolved in water or during penetration into the cells, some cationic silver is leached out from the nanoparticles. Thus, as compared to the ionic form of silver, NPs have a greater tendency towards antibacterial activity, whereas the properties are attributed to both the physical properties of nanoparticles and the leaching of silver ions (Aleksandar et al., 2012). The combined effect of the activity of nanoparticles and free ions contribute differently to produce a strong antibacterial impact on a broad spectrum of bacteria. Furthermore, the fact that bacterial resistance to elemental silver is extremely rare highlights the probability of synergistic action of multiple bactericidal mechanisms. The ions of silver bind to negatively charged protein and nucleic acid and thereby causing structural changes and deformations in the wall/membranes and in the nucleic acids of the bacterial cell (Gianluigi et al., 2015). In fact, silver ions interact with several functional groups containing electron donors like thiols, phosphates, hydroxyls, imidazole and indoles.

Other investigations revealed that the Ag ion, which has an affinity for sulfur and nitrogen, can bind to thiol and amino groups of proteins thereby causes inhibition and desorption of the protein structure (Sondi and Sondi, 2004). Furthermore, it has been suggested that silver NPs are photocatalytic and therefore is the reason for release of ROS forming free radicals which has bactericidal capacity, an observation that was contradicted by others showing that, at least in eukaryotic cells, this effect is cell-type dependent (Choi et al., 2008; Kumar et al., 2014; Ninganagouda et al., 2014; Luther et al., 2011). The intracellular structures of the microbial body got damaged because of silver ions or small Ag NPs as it can easily enter into the microbial body. Consequently because of binding of nanoparticles with the bacterial cell, protein synthesis will be blocked as denature of ribosomes occur along with blocking of translation/transcription process. Also the interaction of Ag NPs alters the protein synthesis and accumulates the immature precursors of the membrane proteins leading to the disruption of outer membrane. The mechanisms of toxicity of Ag NPs are summarized in Figure 3.1.

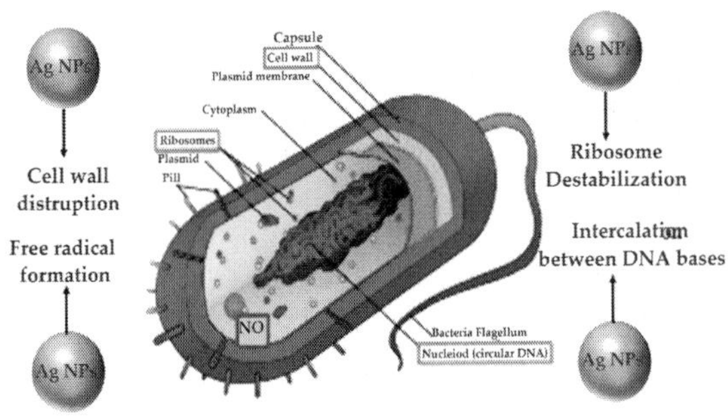

Figure 3.1: Mechanism of action of Ag NPs.

However, regardless of the controversies and discussions, Ag NPs are one of the most reliable antibacterial metal nanoparticles. Nowadays silver is mostly used as surface coating to inhibit biofilms. Also, nano silver has been used as anti-biofilms for coating catheters yielding positive results against both gram-positive and gram-negative bacteria. Moreover, after implantation of catheters into the bodies of the test animals, gathering of silver was not detected in the organs. Ag NPs can mixed with composites which having dental applications i.e., it can act against biofilms caused by *S. mutans* (Priscila et al., 2017). Also, Ag NPs are able to modify bone cements along with inhibiting the formation of biofilms on the surface of cements. Therefore, Ag NPs are being coated on medical devices as well as surgical masks and are already tried clinically. Likewise, Ag NPs are nowadays being used in dressings of wounds to reduce microbial growth and simultaneously improve healing capacity. Trials has been taken using Ag NPs embedded with chitosan hydrogel membrane which is bioactive in nature and the membrane revealed a synergistic activity of both chitosan and Ag NPs and consequently, the growth of different microorganisms like *E. coli, S. aureus, P. aeruginosa,* and *S. epidermidis* strains has been reduced (Sacco et al., 2015).

Ag NPs also having biomedical applications (Greulich et al., 2011). Compared to its ionic form, Ag NPs has high antimicrobial activity (Malarkodi et al., 2014). It is reported that the antibacterial activity of Ag NPs may be due to the damage caused to the bacterial outer membrane. It is also observed that, Ag NPs can form holes in the bacterial membrane which ultimately leads to fragmentation the cell. Another group reported that Ag ions interact with disulphide or sulfhydryl groups of enzymes leading to disruption of metabolic processes causing cell death (Jo et al., 2009). The influence of reduction of size on the antimicrobial and antifungal effect of Ag NPs has been studied. It was found that Ag NPs are non-toxic compared to the synthetic fungicides that are available (Malakodi et al., 2014). Mie et al., have successfully demonstrated the antibacterial activity of Ag NPs (19 nm) against microorganisms through disk diffusion method. These NPs have greater potential to be an antibacterial agent against the gram-negative bacteria, which opens many opportunities for their use in the pharmaceutical and biomedical industries (Egger et al., 2009). The bactericidal activity of Ag nanoparticles, ZnO, and Au on *Streptococcus mutans* has been compared and observed that the Ag NPs exhibited most of their activities towards controlling *S. mutans* (Mie et al., 2014). Likewise, the antibacterial effect of Ag NPs against *S. mutans* has been examined and revealed that the antibacterial effect of Ag NPs against *S. mutans* was more superior to that of chlorhexidine (Hernández-Sierraet al., 2008). Zarei et al., evaluated antibacterial effect of Ag NPs against different foodborne pathogens like *Listeria monocytogenes*, *Escherichia coli* (*E. coli*), *Salmonella typhimurium* (*S. typhimurium*) and *Vibrio parahaemolyticus* and observed that the Ag NPs have great antibacterial effect on all of the aforementioned pathogens

(Besinis et al., 2014). In another study, Bera et al., stated the size and shape-dependent antimicrobial activity of fluorescent Ag NPs (1–5 nm) against gram-positive (*Staphylococcus epidermidis* and *Bacillus megaterium*) and gram-negative bacteria (*Pseudomonas aeruginosa*) and found that the shape and size of the particles have significant control on their antibacterial activity. According to the above studies, the smaller particles can easily penetrates the cell wall and therefore have higher antimicrobial activity. Bahrami synthesized Ag–Au alloy nanoparticles and studied their antimicrobial effect against *Staphylococcus aureus* (*S. aureus*). Moreover, an enhanced antibacterial activity of the Ag-Au alloy has been observed when combined with penicillin G and piperacillin (Bera et al., 2014). Ag_2O nanoparticles have also antimicrobial activity (Bahrami et al., 2014). Sondi and Salopek-Sondi demonstrated antimicrobial efficacy of Ag_2O nanoparticles against *E. coli*. According to them, when *E. coli* were exposed to these nanoparticles, DNA replication has lost followed by DNA damage and therefore the cell cycle halted at the G2/M phase (Allahverdiyevet al., 2011). Furthermore, silver is reported to be less toxic than many other disinfectants. Antibacterial mechanisms of the Ag NPs and potential implications for human health and environment has been studied by many authors (Sondi and Sondi, 2004). Though many more work has already been carried out on the antimicrobial activity of the Ag NPs, but still more work is needed to develop silver-based compounds, composites and alloys with lower toxicity and having maximum antimicrobial activity.

3.3.1.2. Titanium Oxide Nanoparticles

Titanium dioxide (TiO_2) is another metal oxide which has been comprehensively considered for its antimicrobial activities. TiO_2 has long been known for its capacity to destroy both gram-positive and gram-negative bacteria and are also available in market as the antibacterial agent (Maness et al., 1999; Marambio-Jones and Hoek, 2010). More recent reports have revealed its efficiency against various viruses and parasites (Sethi et al., 2014). Similar to Au NPs, they are photocatalytic; their toxicity, induced by visible light, near-UV or UV, stimulates ROS burst which damage several macromolecules and functions of the bacterial cells (Marambio-Jones and Hoek, 2010; Blecher at al., 2011). TiO_2 is effective against the spores of *Bacillus* which is the most resistant organism known so far. Combinations of Ti or TiO_2 with other NPs have a synergistic effect showing an enhanced activity (Brady-Est´evez et al., 2008).

Photocatalysis based on TiO_2 nano catalysts is a very favourable technique for the treatment of contaminated water. Titanium dioxide is known as the best semiconductors used for photocatalytic reaction. TiO_2 has bandgap energy (Eg) = 3.2 eV, this means that it absorbs radiation below 390 nm, for photoexcitation. Moreover it is low cost, chemically and biologically inert and is stable towards photo/chemical corrosion. Also, when TiO_2 is suspended in water, its surface gets hydroxylated and the hydroxyl groups are the cause of the generation of hydroxyl radical which can oxidize maximum of the organic compounds to CO_2 and mineral acids (Reddy et al., 2007; Seil and Webster, 2012; Abibkia et al., 2012; Buzea et al., 2007). Different TiO_2 samples shows different photocatalytic activities towards the same organic substrate under similar experimental conditions (Adibkia et al., 2010). This is because of differences in their lateral size, surface morphology, crystal phase, specific surface area, surface charge, dopants/impurities and surface density of OH^- groups in the TiO_2 samples. Anatase and rutile are the two crystalline forms of TiO_2, but brookite form is also observed in some cases and these phases are temperature dependent. In most of the photocatalytic studies, due to large surface area, the anatase form is found to be more photoactive as compared to the rutile phase (Rai and Bai, 2011; Besinis et al., 2014). Therefore, much attention are paid towards the investigation of the properties of anatase TiO_2. The mechanism of photocatalysis is initiated by the absorption of the photon (hv_1). The energy of the photon absorption is equal to or greater than the band gap of TiO_2 (~3.3 eV for the anatase phase) which produces an electron-hole pair on the surface of TiO_2 nanoparticle as revealed in Figure 3.2. Thereafter a hole is generated by the promotion to an electron from the valence band (VB) to the conduction band (CB). Then the electrons present in the excited-state and holes can recombine and react with electron donors and electron acceptors adsorbed onto the semiconductor surface or within the surrounding electrical double layer of the charged particles. Then these holes can yield hydroxyl radicals after reaction with water having high redox potential. Depending upon the exact conditions, the holes, OH radicals, O^{2-}, H_2O_2 and O_2, it can play important roles in the photocatalytic reaction mechanism (Figure 3.2) (Samsudin et al., 2015).

Matsunaga et al., have investigated the photocatalytic reaction of TiO_2 NPs and presented the importance of photo catalysis in degradation of various types of microorganisms, such as *L. acidophilus* (gram-positive bacteria), *S. cerevisiae* (yeast), *E. coli* (gram-negative bacteria) and *C.vulgaris* (green algae) present in water (Mohammadi et al., 2010).

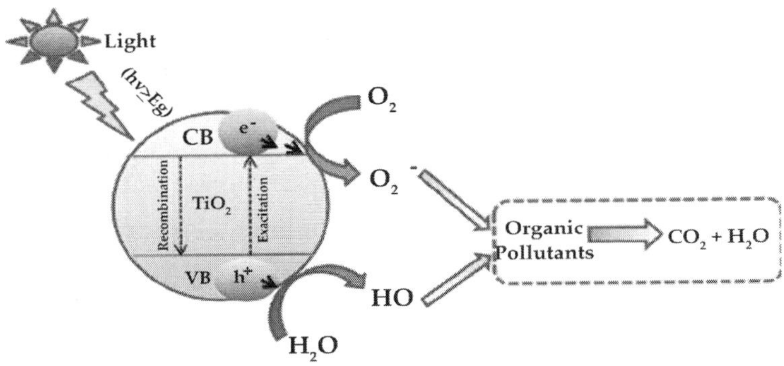

Figure 3.2: Mechanism of Photocatalysis (From Samsudin et al., Sains Malaysiana, 44, 1011, 2015).

Afterwards, a number of experiments have been conducted on the photocatalytic killing of various types of microorganisms, including viruses, fungi and many species of bacteria. However, major research works have been done on aqueous suspensions of TiO_2 to investigate its antibacterial properties (Zhang et al., 2008). Photocatalytic disinfection of *E. coli* by carbon modified TiO_2 photo catalyst was tested under the illumination of UV as well as visible light (Rao et al., 2013). It has been observed that by lowering the quantity of carbon in the catalyst antibacterial activity increases under visible light (4×18W, Philips, TL-D 18W/33–640) irradiation, whereas the catalysts having higher carbon content shows better antibacterial ability under UV light irradiation (6×20 W, Philips) (Figure 3.3) (M. Janus et al., 2012)

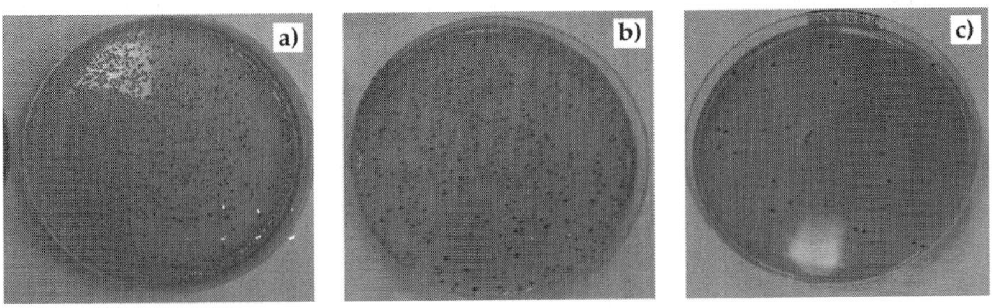

Figure 3.3: Plates with *E. coli* colony: a) in initial water sample, b) in water after 10 min of UV irradiation, c) in water with n-butanol TiO_2 photocatalyst after 10 min of UV irradiation. (From Janus et al., Environment Protection Engineering, 38, 89, 2012).

Similarly, the antibacterial action of TiO_2 against *E. coli* and two wild strains of the *Fusarium genus,* i.e., *F. solani,* and *F. anthophilum* has been studied and observed a complete disinfection in different seasons with different intensity of sunlight (Atkinson and Winge, 2009). Using the TiO_2, researchers successfully destroyed different types of bacteria and fungi, namely, *Escherichia coli, Staphylococcus aureus*, *Pseudomonas aeruginosa, Candidaalbicans*, *Saccharomyces cerevisiae* and *Aspergillus niger* (Cioffi and Rai, 2012). The efficiency of photocatalysis against microorganisms including bacteria cells, spores and biofilms, viruses, protozoa, fungi and algae has been investigated by a lot of researchers (Hosseinkhani et al., 2011; Karvani and Chehrazi, 2011). Tsuyoshi Ochiai used TiO_2-Modified Ti-Mesh Filter in synergistic Water-Treatment Reactors and successfully inhibit the growth of many waterborne pathogens like *Escherichia coli* NBRC13965 (*E. coli*), *Legionella pneumophila* ATCC49249 (*L. pneumophila*), Qβ phage NBRC20012 (Qβ), and *feline calicivirus* F-9 ATCC VR-782 (FCV), etc. (Figure 3.4) (Zaleska et al., 2008).

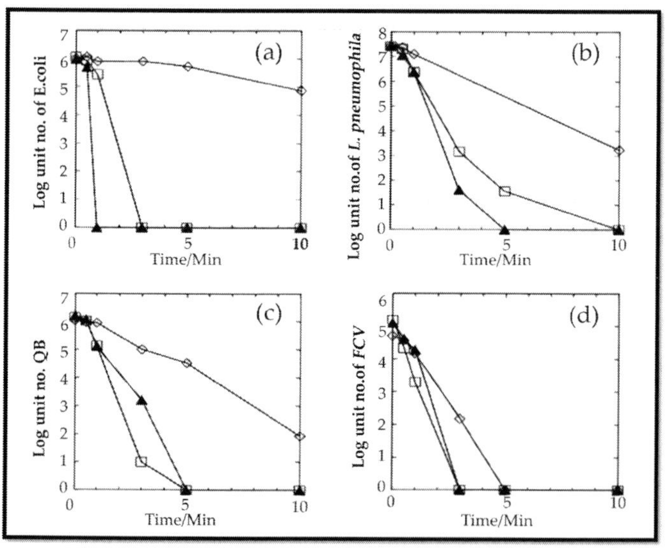

Figure 3.4: Time courses of log no. of a) *E. coli*, b) *L. pneumophila*, c) Q β and d) FCV in the water purification system open squares: O_3 bubbling alone (0.5–1.0 mg/L): solid triangles: BLB/TMiP/O_3 REACTOR (0.5–1.0 mg/L): open diamonds: excimer/TMiP reactor (From Ochiai et al., Water, 5, 1101, 2013).

Likewise, the solar photocatalytic disinfection with the use of novel cobalt and manganese-doped titania materials against reference strains of

E. coli and *K. pneumonia*. All the catalysts were very much effective for the removal of *K. pneumoniae* and *E. coli* under the solar light which is considered as opportunistic pathogens and are highly resistant in various water treatments (Sun et al., 2014). Many research has also been carried out for the inactivation of virus from the water source. Another group investigated the capability of TiO_2 NPs being doped with silver (nAg/TiO_2) for the inactivation of bacteriophage MS2 in aqueous solution under the irradiation of four 8 W UV-A (315–400 nm) lamps with peak emission at 350 nm (Hitachi) (Erich, 2011). The intensity of total light used in all these experiments was 2.5 mW/cm^2. It has been observed that the inactivation rate improved by more than 5 fold depending on the base TiO_2 material and the rate of deactivation improved by increasing the silver content (Gupta et al., 2013). Similarly, high surface area TiO_2 photocatalyst supported on a glass tube and a stainless steel plate was arranged and the antibacterial efficiency was tested with water spiked with *E. coli* in a quartz reactor with 350 nm light and solar light. It is observed that the number of *E. coli* cells diminished to a safe level from the initial concentration of 500–100,000 bacteria/mL during 4h of photolysis using 350 nm light and solar light (Rasmussen et al., 2010).

The antimicrobial action of the TiO_2 photocatalyst is because of the generation of OH• radicals, O^{2-} and H_2O_2 (as revealed in Figure 3.2). Many researchers from their study observed that the outer membranes of bacterial cell are the main sites for the attack of reactive photo generated oxygen species, leading to lipid peroxidation. Due to the cell membrane damage (Figure 3.5) and further oxidative degradation of intracellular components, the bacterial cell is eventually destroyed. Along with this, the photo oxidation of coenzyme A leads to the cell death by inhibiting the cell respiration.

Photocatalyst together with UV lights is capable of oxidizing toxic organic pollutants into nontoxic materials like CO_2 and water and can purify different types of microorganisms present in water (Figure 3.7). This method is so much efficient in eliminating the rest harmful organic compounds and also kills microorganisms from contaminated water. Nowadays Disinfection of microorganisms like *E. coli* in drinking water supply facilities is a great challenge. An annular type fluidized bed reactor has been constructed for disinfection of *E. coli* bacteria (Haghighi et al., 2013). Researchers have prepared the catalyst by mixing TiO_2 with glass bead and zeolite that shows extremely good bactericidal activity by killing 99.99% *E. coli* cells. It has been observed that the pilot scale is able to degrade 84% of *E. coli* within 80 minutes compared to 77.3% by using UV only and has the capacity of handling 46 liters/min. Degradation of *E. coli* has been verified and the result was compared with bare UV destruction and is very much promising. Due to these reasons, this method can be used on a continuous reactor to increase the efficiency for scalability. Peyton et al., screened three different semiconductors (TiO_2, ZnO, and Fe_2O_3) to degrade pollutant compounds in wastewater by using solar energy (Lonina and Narayanan, 2013). They observed that Fe_2O_3 was ineffective, whereas both ZnO and TiO_2 are found to be very much effective for the removal of all pollutants. Also the doping of different metal ions has been demonstrated to enhance the electronic properties of the TiO_2 photocatalyst and to realize the application of visible light. In this case, TiO_2 photocatalyst was studied for the decontamination of water. The UV-Visible absorption spectra of these metal ion-doped TiO_2 photo catalysts has a red-shift, i.e., a shift towards higher wavelengths of the electromagnetic spectrum depending on the amount and the kind of metal ions doped. Photocatalysis done by TiO_2 nanoparticles is used to remove the organic chemicals from water coming from various sources like wastewater, water coming from domestic and industrial sources.

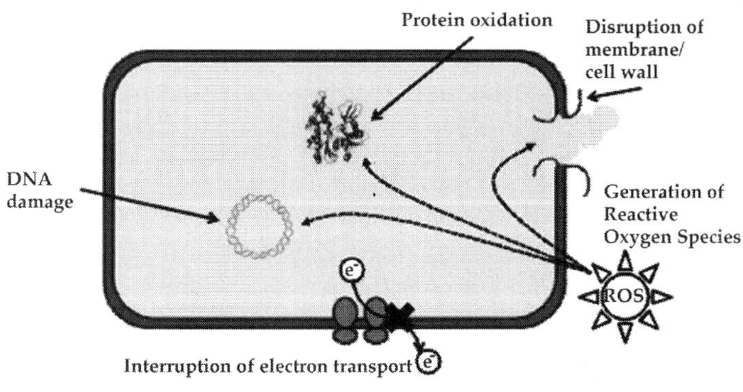

Figure 3.5: Mechanism of killing of bacterial cell by ROS (From Emamifar et al., 2011).

3.3.1.3. Zinc Oxide Nanoparticles

Additional wide spectrum bactericidal nanoparticles are ZnO-based materials (Cui et al., 2012). ZnO nanoparticles shows a comprehensive range of bactericidal activity against a diverse group of microorganisms, which depends on the concentration of nanoparticle and particle size (Cui et al., 2012) Moreover, ZnO nanoparticles have ability to destruct the growth of methicillin-sensitive *S. aureus* (MSSA), methicillin-resistant *S.*

aureus (MRSA), and methicillin-resistant *S. epidermidis* (MRSE) strains demonstrating as an efficient antibacterial agents that are not influenced by the drug-resistant mechanisms of MRSA and MRSE (Palanikumar et al., 2014; Ansari et al., 2012). Zinc oxide NPs are of relatively low cost and have size dependent effectiveness against a wide variety of bacteria. These include pathogens such as *Klebsiella pneumonia, Listeria monocytogenes, Salmonella enteritidis*, *Streptococcus mutans, Lactobacillus*, and *E. coli* also these nanoparticles are found to be not toxic to human cells (Huh and Kwon, 2011; Cui et al., 2012). Their various properties like white color, blocking of UV, and capability to inhibit formation of biofilm makes them appropriate for fabric and glass industries. Furthermore, treatment using zinc has been approved by the FDA and nowadays is available as a preservative of food Similar to TiO_2 NPs, the ZnO NPs have an effect on bacterial cells along the two pathways, described above, by disrupting the bacterial membrane after binding to the membrane and by inducing ROS production.

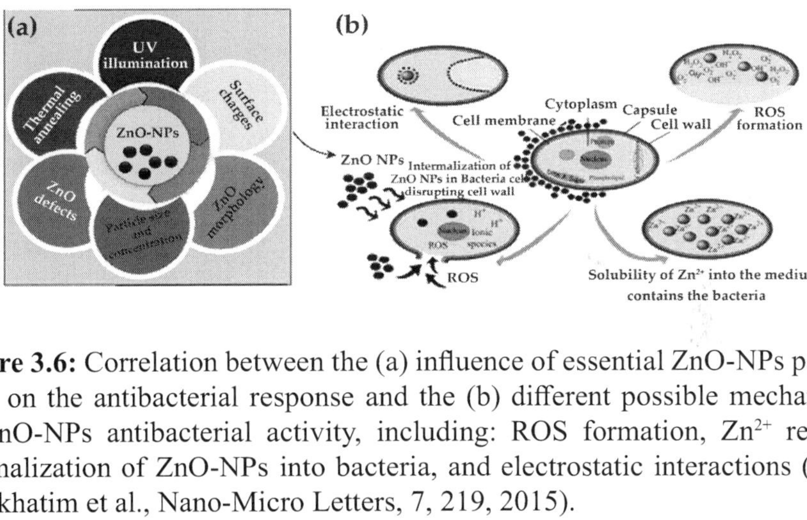

Figure 3.6: Correlation between the (a) influence of essential ZnO-NPs parameters on the antibacterial response and the (b) different possible mechanisms of ZnO-NPs antibacterial activity, including: ROS formation, Zn^{2+} release, internalization of ZnO-NPs into bacteria, and electrostatic interactions (From Sirelkhatim et al., Nano-Micro Letters, 7, 219, 2015).

It has been observed that, the synthesized nano ZnO shows different types of morphologies and also exhibit considerable bactericidal activity against a broad variety of microorganisms (Figure 3.6) (Sirelkhatim, 2015). The antibacterial activity of ZnO is currently being examined in both microstructure and nanostructure. But, the nanostructure ZnO nanoparticles shows enhanced antimicrobial activities compared to microstructure ZnO (Malka et al., 2013). Several studies reported that, ZnO NPs are non-toxic to human cells and hold good biocompatibility to human cells, this features of ZnO NPs makes their utilization in antibacterial compound

(Amna Sirelkhatim et al., 2015). The mechanism for antibacterial activity of nanomaterials is mainly due to their increased surface area-to-volume ratios, and their unique physicochemical properties (Seil and Webster, 2012). Nevertheless, the accurate mechanisms are yet under discuss, but numerous mechanisms proposed and adopted. Flow cytometry viability assays also used to determine the antibacterial activity of ZnO NPs against *E. coli* and *S. aureus* (Sawai, 2003). The most frequently used method to examine the antibacterial activity of ZnO NPs is the broth dilution method, in this method spread plating has been done after serial dilutionis of nanoparticle and bacterial suspension followed by incubation at 37°C for 24 h. and colony count. A number of researchers (Reddy et al., 2007) have studied the bactericidal activity of ZnO NPs to observe the growth of bacteria by tube dilution method after observing the turbidity of bacterial culture and the percentage of viable cells after the colony counts. But other researchers observed the enhancement in the antibacterial activity of ZnO NPs after modification of the procedure (Yamamoto, 2001). They found that there is enhancement in the antibacterial activity after decreasing the initial bacterial cell concentration from 10^6 to 10^2 colony forming unit (CFU). Nair et al., proposed that the initial bacterial cell concentration is very essential in the antibacterial activity estimation (Nair et al., 2009). The minimum inhibitory concentration (MIC) of an antimicrobial agent and Minimum bactericidal concentration (MBC) can be evaluated by using a method known as susceptibility test. But, nowadays variation in the established laboratory methods and protocols has been done to determine the bactericidal activity (Barry et al., 1999). The standardized and official method for the antibacterial activity test is the agar diffusion method (an indirect method). Through a direct test method, urease inhibition of inocula method, the bactericidal activity can be determined (Aslam et al., 2012). Another modified microdilution method utilizes multiple tests on a 96-well plate A large number of researchers uses modified procedures with the standard methods. In all of the above mentioned methods, the culture media, i.e., trypticase soy broth (TSB), Luria–Bertani broth (LB), nutrient agar (NA), tryptic soy agar (TSA), and blood agar (BA) were prepared accordingly and stored at 4–5°C. The stock solutions of ZnO NPs are prepared, diluted to desired concentrations, and then characterized by X-ray diffraction (XRD), field emission scanning electron microscope (FESEM), transmission electron microscope (TEM), energy dispersive-ray spectroscopy (EDX), electron spectroscopy imaging (ESI), etc., to compare the antibacterial response with ZnO properties. In some cases, the FESEM is used to reveal the complex mechanism behind the antibacterial activity of ZnO NPs by considering the modification in the bacterial cell morphology. Although, there are various issues linked with the antibacterial activity of ZnO NPs, but the accurate mechanism of toxicity is not entirely explored and still it is in controversy which required depth analysis and proper clarification. Distinguishing mechanisms for the bactericidal property that have been proposed in the literature are: direct contact of ZnO NPs with thc bacterial cell walls, resulting the destruction of bacteria cell through lipid peroxidation, release of bactericidal ions (mainly Zn^{2+} ions), and ROS generation (Brayner et al., 2006; Zhang et al., 2007; Adams et al., 2006; Kasemets et al., 2009; Jalal et al., 2010). However, the toxicity mechanism mainly depends upon the various media composition because in a variety of media the dissolved ZnO dissolute to give Zn^{2+} ion which further destructs the bacterial cell (Li et al., 2011).

The following points have an important effect on the antibacterial activity of the ZnO NPs:

- UV Illumination Effect;
- Impact of ZnO Morphology;
- Surface Modification by Thermal Annealing;
- Effect of ZnO Particle Size and Concentration;
- Minimum Inhibitory Concentration (MIC);
- Surface Defects;
- Reactive Oxygen Species Generation; and
- Zinc Ions (Zn^{2+}) Release.

3.3.1.4. Gold Nanoparticles (Au NPs)

The synthesis and biomedical application of noble metal nanoparticles, especially gold nanoparticles (Au NPs) is a promising field of research and has garnered the attention of vast groups of researchers. Bulk metallic gold (Au) is well-known to be chemically inert. Whereas, when the size is reduced to nanometer range, it starts to show diverse biological properties (Zheng et al., 2017). The Au NPs find application in many fields, ranging from electronics to agriculture, biomedicine, and pharmacy (Chowdhury et al., 2014). They can be prepared with ease and their surfaces can be customized through chemical modifications, allowing effective attachment of payloads through various means and shows higher gene transfection efficiency (Liu and Zhang, 2011; Dizaj et al., 2014). The tailored ability to ensure cellular uptake with controlled release of the drug to specific cell target, excellent stability, bioinertness and nontoxic like properties make the Au NPs a popular candidate for biomedical applications such as biosensors, drug delivery for cancer chemotherapy and radiotherapy (Paciotti et al., 2004; Podsiadlo et al., 2008). These NPs can be synthesized by a variety of processes like physical method, chemical synthesis and also from different biological agents. However, the biological synthesis method is more acceptable for being eco-friendly and simple (Ray et al., 2011). It allows the controlled synthesis of nanoparticles with definite size and shape (Bar et al., 2009). Due to their versatility in biological functionalization, biologically derived nanoparticles find a wide range of applications in the field of medicine and drug delivery (Chowdhury et al., 2014; Ray et al., 2011. Biosynthesis of nanoparticles using microorganisms like bacteria, actinomycetes, algae and fungi like yeast, *Aspergillus fumigatus, Phanerochaete chrysoporium*), *Macrophomina phaseolina* have also been investigated (Zhiruiguo et al., 2007; Sastry et al., 2003; Ramakrishn et al., 2016; Mourato et al., 2011; Bhainsa and D'Souza, 2006; Chhowdhury et al., 2014). Among them, fungi are drawing great attention as they have an excellent capacity of secreting extracellular enzymes and can go through a scalable production process easily. Furthermore, the large surface area of the mycelia and using fungi the downstream processing would be much easier (Chowdhury et al., 2014). Extracellular synthesis of Au NPs by green routes using edible mushrooms is a relatively new addition in this context ensuring less toxic chemicals adhering to the nanoparticles (Paciotti et al., 2004). Although Au NPs do not possess strong antimicrobial activity as silver nanoparticles, they can be exploited for having both antimicrobial and antifungal activities (Lima et al., 2013; Wani and Ahmed, 2013). The Au NPs with antibacterial properties have emerged as an alternative to high-dose administration of antibiotics and confirmed their efficiency against antibiotic-resistant microorganisms (Podsiadlo et al., 2008). As the nanoparticle toxicity and antibacterial mechanism depend on factors such as modification of surface, inherent properties, and the bacterial species, the smaller dimensions of Au NPs are more likely to penetrate the bacterial cells and cause cell damage followed by cell death (Bindhu and Umadevi, 2014). Among Au NPs with various morphologies, the triangular shaped Au NPs have been found to show improved antibacterial activity towards gram-positive and gram-negative bacteria (Smitha and Gopchandran, 2013). The sharp-faced triangular nanoparticles, irrespective of their surface chemistry, size or compositions, can penetrate through the membranes of endosomes and translocate to the cytoplasm where they can be retained. These features make them preferable to round shaped nanoparticles for drug delivery, gene delivery, subcellular targeting, and long-term tracking (Chu et al., 2014). In a recent report, the Au NPs of extremely small sizes (i.e., less than 2 nm) resulted in high antibacterial activity against gram-positive as well as gram-negative bacteria (Figure 3.7) (Zheng et al., 2017). Mode of action of antibacterial activity of Au NPs involves getting attached to the bacterial membrane by electrostatic interaction and disrupting its integrity (Tiwari and Soo, 2013). They can change membrane potential and decrease ATP levels within the cell, inhibiting the binding of tRNA with ribosomal subunit, affecting translation (Cui et al., 2012). These Au NPs can produce holes in the cell wall which causes leakage of cytoplasmic contents, and bind with the DNA, inhibiting transcription (Rai et al., 2010). Also the Au NPs aggregate within bacterial biofilms and bind to their surfaces causing cell wall distortions which can be utilized to reduce the duration of treatment and side effects of drugs (M. F. Zawrah et al., 2011). The interaction between ultra-small Au NPs (less than 2nm) and bacteria likely induce imbalance in metabolic pathway resulting in an enhancement of intracellular ROS species production that culminated in the destruction of the bacterial cell (Zheng et al., 2017). The possible use of Au NPs in biomedical areas especially for cancer therapy has been a subject of research, owing to its possible toxic effects related to human exposure with its biomolecules, cells, and tissues (Maurer et al., 2009). In humans, it is known that the primary site of Au NPs accumulation is the liver (Sonavane et al., 2008). Nowadays, gold nanoparticles used as vehicles of gene delivery. Modification of Au NPs surfaces with bimolecular is often an integral step providing high gene transfer efficacy with minimal cytotoxicity (Tiwari et al., 2013). The attractive features of Au NPs in this regard include their surface plasmon resonance, interaction with thiol groups, and their harmless nature (Dakrong et al., 2011). It is observed that the Au NPs are modified with basic amino acid and used as gene delivery vectors (Ghosh et al., 2008). Whereas their protein conjugates provide excellent biocompatibility towards normal cells reducing nonspecific toxicity (Shukla et al., 2005). They are extensively exploited in organisms owing to its biocompatibility, bioconjugation ability, strong absorbing and scattering properties as target drug delivery in various therapeutics and cancer treatment (Bhattacharyya and Mukherjee, 2008). The rate of permeability and retention of nanoparticles are enhanced in cancerous cells than the normal cells by virtue of the accumulation and entrapment process which is the result of the leaky nature of tumor blood vasculature (Bhattacharyya and Mukherjee, 2008). Moreover, ultra-small sizes of nanoparticles (smaller than 1 nanometer) ensure homogeneous distribution of the nanoparticles throughout the cells which is comparable to proteins such as bovine serum albumin alone (Matulionyte et al., 2017). Recently photo luminescent gold nanoclusters have acquired a considerable interest in biomedical research and applications due to their very small size and distinctive molecule-like optical properties Matulionyte et al., 2017). Capping of gold nanoparticles often ensures an increase in biocompatibility and reduction of toxicity Matulionyte et al., 2017). The capping of nanoparticles also ensures greater uptake and retention by cancer cells and other cells, which allowed greater time for integration of the delivered gene. Vectors can be Synthesized based on various cationic polymers such as polyethyleneimine (PEI), polypropylenimine (PPI) and polyamidoamine (PAMAM) and their derivatives are utilized for transfer of plasmid DNA into human cell lines and treatment of genetic diseases (Teimouri et al., 2016). The use of PEG (polyethylene glycol) is often used to functionalize Au NPs as non-viral vectors in the transfection of various cell lines (Encabo et al., 2017). GNPs attached with polyethylene Polyethylenimine (PEI) covalently can also be used as potent transfection reagents where expression levels of the transferred gene are similar to or greater than that obtained with commercially existing lipoplexes (Mar Encabo et al., 2017). In another study polyethyleneimine-g-bovine serum albumin (BSA-PEI) was used as a non-viral gene vector resulting in good colloid stability and high transfection efficiency (Zhang et al., 2017).

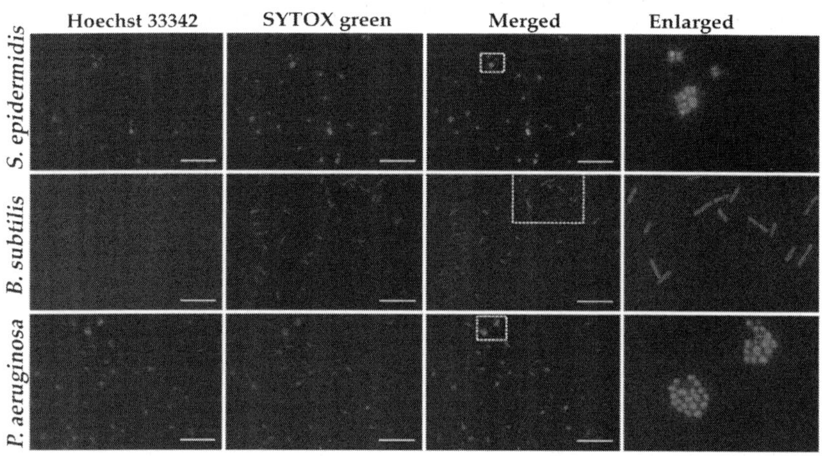

Figure 3.7: Au NCs showed high killing efficiency to S. aureus, which was absent when the cells were treated with Au NPs and Au (I)–MHA complexes

(Au(I) complex). Representative fluorescence images of the S. aureus after 2 h treatment. The dead cells were visualized by SYTOX green (false color: red), whereas the Hoechst 33342 (blue) helped to identify all cells. Scale bar is 25 μm (From Zheng et al., ACS Nano 11, 6904, 2017).

Au NPs have been reported to be bactericidal when photo thermally functionalized. In compared to Ag, Au NPs are less potent and have almost no bactericidal activity by themselves (Majdalawieh et al., 2014). Nevertheless, Au NPs attached with various antibiotics such as ampicillin, vancomycin, the antibacterial enzyme lysozyme, and even other NPs showed the bactericidal effect on many multidrug-resistant pathogens, including those which were penicillin and vancomycin resistant. The antibacterial activity of these Au NPs can also be enhanced by binding within antibiotic molecules like amino-substituted pyrimidines, citrate, etc., that induced ROS production and mutations in presence of light. The stability of Au-NP compared with that of other metal NPs, such as platinum (Pt). Most of the Pt NPs are applied for the cancer treatment. It has been observed that in mammalian cells, Pt NPs diffuse through the cell membranes and induce damage of DNA, cells are accumulated at the S-phase of the cell cycle leading apoptosis (Malka et al., 2013). Recently, the toxicity of Pt NPs to bacteria was also demonstrated and found to be size-dependent like that of Ag and Au NPs. Pt NPs having particle size of 1–3 nm are bactericidal to *P. aeruginosa* cells, whereas those of 4–21 nm size exhibited bacterio-compatible properties (Malka et al., 2013). Though either the Pt or Au NPs are nontoxic to bacteria, but the Au/Pt bimetallic NPs show a strong bactericidal effect. Interestingly, in contrast to other NPs, this effect was ROS-independent and the cell death takes place due to the membrane damage and a severe elevation of ATP. Au nanoparticles are considered as the most significant antibacterial agents because of their nontoxicity, easy modification of surface, polyvalent effects, easy detection and photothermal activity (Lima et al., 2013). Generally, the reason for the bactericidal activity of nanomaterials is a generation of ROS, but Au NPs do not generate ROS and the antimicrobial activity is due to some different mechanism (Cui et al., 2012). Cui et al., observed that bactericidal activity of the Au nanoparticles was due to the attack of these nanoparticles inside the membrane of bacteria and hamper the membrane potential and reduce the ATP level inhibiting the binding of t-RNA to the ribosome (Cui et al., 2012). In another work, Tiwari et al., have examined the antibacterial and antifungal activities of the Au NPs functionalized with 5-fluorouracil against *Micrococcus luteus, S. aureus, P. aeruginosa, E. coli, A. fumigates, and A. niger*. It shows better activity against Gram-negative bacteria than that of

Gram-positive bacteria. These nanoparticles also exhibited good antifungal activity against *A. fumigates* and *A. niger* as well (Tiwari et al., 2011). Zhou et al., have prepared the functionalized Au NPs with both the strongly bound (poly-allylamine hydrochloride) and weakly bound (citrate) capping agents and Ag NPs and investigated the antibacterial activity towards *E. coli* and *Bacillus Calmette-Guérin* (BCG). And these nanoparticles possess considerable bactericidal property against both Gram-negative (*E. coli)* and the Gram-positive bacteria (BCG) (Goodman et al., 2004). In another study, Lima et al., investigated the bactericidal activity of Au NPs (5 nm) against two bacteria such as, E. *coli* and *S. typhi.* These nanoparticles are able to reduce 90–95% colonies of *E. coli* and *S. typhi*. In comparison to other metal nanoparticles, the Au NPs are safer to be used as the antibacterial agent and their properties not only depend on their lateral dimension but also on their dispersion in the medium under analysis (Lima et al., 2013).

3.3.1.5. Copper Oxide Nanoparticles

Nowadays, Cu nanoparticles are gathering significant interest from the researchers because of their exclusive biological, chemical and physical properties, bactericidal property and also cost-effectiveness (Usman et al., 2013). Although copper oxide nanoparticles have been revealed to be effective against various bacterial pathogens, their antibacterial efficacy is somewhat inferior to that of Ag or ZnO based NPs. Hence, a comparatively higher concentration of nanoparticles is needed to achieve similar results. Moreover, the activity of CuO nanoparticle varies greatly depending on the bacterial species under investigation. Nonetheless, as Cu is much less expensive than other nanosized metal materials, it can be utilized for efficacy enhancement in the form of nanocomposites. Similar to other metallic nanoparticles, CuO NPs execute their antibacterial activity by disruption of bacterial membrane and generation of ROS. In general, Cu NPs are less effective than Ag NPs, while in some cases, the reverse is true. For example, *E. coli* and *S. aureus* are found to be more sensitive to silver, whereas *B. subtilis* and *B. anthracis* were more sensitive to Cu NPs (Wu et al., 2002). A possible explanation for this can be made in the context of the bacteria, such as *B.subtilis*, whose cell walls are rich in amine and carboxyl groups, more strongly adhere to CuO and thus are more susceptible to it (Huh and Kwon, 2011; Blecher et al., 2011). Thus it seems that in special cases it would be beneficial to use the CuO NPs instead of others, including silver. Usman et al., evaluated the antimicrobial and antifungal activities of Cu-chitosan nanoparticles (2–350 nm) against several microorganisms, such as methicillin-resistant *S.*

aureus, *B. subtilis*, *P. aeruginosa*, *Salmonella choleraesuis*, and *C. albicans* and these nanoparticles showed high antimicrobial property (Mahapatra et al., 2008). Mahapatra et al., observed experienced the high antibacterial activity of CuO nanoparticles against *Klebsiellapneumoniae, P. aeruginosa, Salmonella paratyphi* and *Shigella strains*. The authors proposed the mechanism that, invasion of nanoparticles in the bacterial cell membrane and then destruct the necessary enzymes of bacteria were the main factors that cause cell death. But they also observed that these nanoparticles do not any harmful effect on HeLa cell line (Azam et al., 2012). According to Azam et al., the antibacterial activity of CuO NPs mostly depends on their size, stability, and their concentration in the bacterial growth medium. They examined the antibacterial activities of CuO nanoparticles against two Gram-positive bacteria (*S. aureus and B. subtilis*) and two Gram-negative bacteria (*Pseudomonas aeruginosa* and *E. coli*) and observed an inhibitory effect of the NPs against both the groups of bacteria. Again another group of researchers stated that, the metal nanoparticles able to reduce the growth of bacterial by passing through the bacterial cell membranes (Dhapte et al., 2014). This study showed that CuO nanoparticles (23 nm) had significant bactericidal activity against a diverse range of bacterial strains (*E. coli, P. aeruginosa, K. pneumoniae, Enterococcus faecalis, Shigella flexneri, S. typhimurium, Proteus vulgaris, and S. aureus*). It has been observed that the CuO NPs possesses high antibacterial property than antifungal property (Mahapatra et al., 2008).

3.3.1.6. Silicon Nanoparticles

Antimicrobial activity of SiO_2 is higher at the nano dimension because of their increased surface area (Cousins et al., 2007). Cousins et al., investigated that Si nanoparticles are very effective in the inhibition of bacterial adherence to oral biofilms (Egger et al., 2009). Combination of Si nanoparticles with the other antibacterial metal like Ag has been studied extensively in the current years. Egger et al., observed an enhanced antimicrobial activity of novel Ag–Si nanocomposite against a diverse range of microorganisms compared to silver nitrate and silver zeolite (Mukha et al., 2010). In another study, Mukha et al., have synthesized Ag/SiO_2 and Au/SiO_2 nanocomposites and studied their antibacterial activities. They observed an improved antimicrobial performance of the Ag/SiO_2 nanocomposites against *E. coli*, *S. aureus*, and *C. Albicans* compared to that of Au/SiO_2 nanocomposites. Therefore, Ag/SiO_2 nanoparticles are potentially suitable material for water disinfection, medical and pharmaceutical applications (Lee et al., 2010). They have developed Ag NPs/Si nanowire composite and observed an enhanced antibacterial activity. Also, these nanocomposites are biocompatible towards the human lung adenocarcinoma epithelial cell line A549 (Yamamoto et al., 2010). All of the above studies conclude that Si compounds along with their nanocomposites with other antibacterial metal (i.e., Ag) demonstrate vast potential on the making of antimicrobial agents. So these nanocomposites could be used as antimicrobial agents in biomedical applications.

3.3.1.7. MgO and CaO Nanoparticles

Likewise, other metal and metal oxide nanoparticles, e.g., CaO and MgO show high antibacterial property because of their alkali nature and production of active oxygen species. It is confirmed that the mechanism for the antibacterial action of CaO and MgO nanoparticles is the creation of superoxides on their surface and also pH value increase due to hydration (Jin and He, 2011). According to various literature, MgO nanoparticles destruct the membrane of the bacterial cell and the subsequent leakage of intracellular contents is the main cause for the destruction of the bacterial cells (Leung et al., 2014). Hewitt et al., observed that MgO nanoparticles exhibit antibacterial activity against *E. coli* because of the production of active oxygen species. However, Leung et al., proposed that MgO nanoparticles possess strong antibacterial activity by damaging the cell membrane without forming any type of ROS species (Sawai et al., 2000). Whereas, in another study, Sawai et al., investigated the antibacterial activity of MgO against *E. coli* and *S. aureus* showing the bactericidal property is due to the generation of active oxygen, such as superoxide (Jeong et al., 2007). According to Jin, MgO nanoparticles alone and also with nisin (antibacterial peptide) and ZnO showed strong antibacterial activity (higher than 7 log reductions in Colony forming unit counts) against *E. coli* and *Salmonella Stanley*. Further enhancement in antibacterial activity of MgO NPs has been observed on increasing their concentrations (Leung et al., 2014) Jeong et al., investigated that CaO nanoparticles exhibit antimicrobial efficiency against *E. coli, S. typhimurium, S. aureus* and *B. subtilis* (Vidic et al., 2013). Yamamoto et al., observed that $CaCO_3$/MgO nanocomposites exhibit better antibacterial activity against Gram-positive (*S. aureus*) and Gram-negative (*E. coli*) bacteria. Here the mechanism involved is that the production of superoxide ion on their surface and also with water the pH value of CaO and MgO increases due to hydration (Jin and He, 2011). Vidic et al., observed the superior antibacterial activity of ZnO nanocrystals against both Gram-positive (*B. subtilis*) and Gram-negative (*E. coli*). MgO nanoparticles showed less antibacterial activity but the ZnO–MgO composite nanoparticles exhibited higher antibacterial activity against Gram-positive (*B.subtilis*) bacteria. From their microscopic analysis, it is observed that the bacterial cells were destructed after making contact with ZnO–MgO nanoparticles. So they confirmed that nanostructured ZnO–MgO composite could be used as a protected new remedial for bacterial diseases (Sawai and Igarashi, 2002). Above results show that alone MgO and CaO nanoparticles or in mixing with other antibacterial compounds, exhibit outstanding antibacterial activity. These nanoparticles also are not expensive, biocompatible and easily obtainable materials. Due to these valuable properties, these nanoparticles could become a promising antibacterial agent (Sawai et al., 2000). So, researchers give the suggestion that these materials can be used in the preservation of environmental preservation, processing of food and medical application (Sawai et al., 2000).

3.3.2. Organic Nanoparticles

Organic nanoparticles are made-up of proteins, carbohydrates, lipids and other organic compounds. It is recommended that the killing mechanism of organic nanoparticles is caused by the releasing antibiotics, antimicrobial peptides, and antimicrobial agents. Another mechanism involves for the killing of a bacterial cell is several mechanisms of action have been anticipated for these cationic groups for the disruption of the cell membrane of bacteria, certain lengths of hydrophobic chains enter and rupture the bacterial membrane. It has been confirmed that more no. of positive charge are able to exhibit higher antimicrobial properties. These finding suggest the chance of synthesis of a variety of polymer-based positively charged surfaces to generate a diverse range of contact-killing materials. A brief summary of the antimicrobial polymers have been listed below.

3.3.2.1. Quaternary Ammonium Compounds

Disinfectants like benzalkonium chloride, stearalkoniumchloride, and cetrimonium chloride are eminent quaternary ammonium compounds. Their antimicrobial activity is a function of the N-alkyl chain length and therefore lipophilicity happens (Jenings et al., 2015). Compounds with alkyls group with 14–16 carbon chains show improved antibacterial activity against Gram-negative bacteria compared to alkyl chain length of 12–14 against Gram-positive bacteria. Initially these compounds interacts with bacterial cell wall due to from electrostatic interaction between positively charged moieties of the compound and negatively charged membranes of bacteria, also due to the invasion of the hydrophobic tail of the compound into the cell membrane of bacteria, which is hydrophobic in nature, inside the cell they denature structural proteins and enzyme. The synthesis of antimicrobial polymers which have only one biocide end group on backbone of polymeric compound by opening of cationic ring polymerization of 2-alkyl-1,3-oxazolines, which terminates the macromolecule by a cationic surfactant (Waschinski and Tiller, 2005). Quaternary pyridiniums are compounds consist of a heterocyclic ring and nitrogen atom. Polymer chain contains pyridinium group shows antibacterial activity against various microorganisms. Imidazole derivatives are also known as antimicrobial polymer with aromatic/ heterocyclic groups and they can form hydrogen bond with drugs and proteins while its alkylated form (imidazolium) has the ability to aggregate electrostatically in spite of losing the hydrogen bond-forming capability of free imidazole. Chemically these compounds are stable and biocompatible and show enhanced biodegradability (Anderson and Long, 2010). Synthesized Copolymers of N-vinyl imidazole and phenacyl methacrylate were shows high antimicrobial activity against various microorganisms such as bacteria, fungi, and yeast (Soykon et al., 2005) (Figure 3.8). These immobilized surfaces are able to destruct of both waterborne and airborne bacteria and fungi, including disease-causing bacteria and antibiotic-resistant bacteria. Degradation of cell membrane is a most important mechanism behind the antibacterial action. These surfaces are harmless to mammalian cells. N-alkylated PEIs immobilized over varied woven textiles (cotton, wool, and polyester) also show higher antibacterial activity against numerous airborne bacteria. Another significant class of polymers is polysiloxanes, the linear polymers of silicon oxide. Sauvet et al., have been synthesized statistical and block siloxane copolymers which contains quaternary ammonium salt groups as a lateral substituent and the compound shows strong antibacterial activity against both *Escherichia coli* and *Staphylococcus aureus* (Sauvet et al., 2003). Polymers possessing quaternary phosphonium or sulfonium groups exhibit mechanisms for antimicrobial action like compounds containing. quaternary ammonium group. Phosphonium containing polycationic biocides are found to be destruct bacterial cell more efficiently than quaternary ammonium salt polymers. Various experiments has been done on water-soluble thermo sensitive concludes that there is an enhancement in antimicrobial activity with an enhancement in length of the alkyl chain and phosphonium units in the polymer (Nonaka et al., 2003).

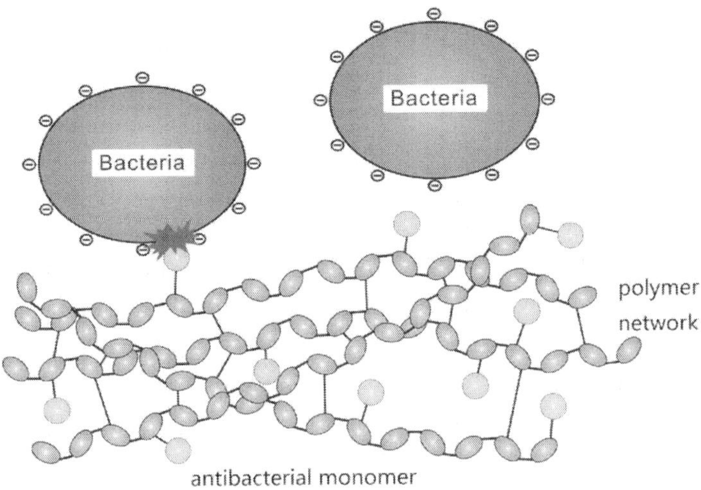

Figure 3.8: Schematic illustration of the antibacterial mechanism of quaternary ammonium monomers (From Ge et al., Materials, 8, 3532, 2015).

3.3.2.2. Cationic Quaternary Polyelectrolytes

Acrylic or methacrylic derivatives are the renowned cationic quaternary polyelectrolytes which are known as antimicrobial polymers and a number of methacrylic derivatives are synthesized from commercially available methacrylic monomers like 2-(dimethylamino)ethyl methacrylate. These polymers able to give large structural flexibility by the modification of hydrophobicity, molecular weight, surface charge, and other parameters (Muñoz-Bonilla and Fernández-García, 2012).

3.3.2.3. Benzoic Acid, Phenol, and p-Hydroxy Benzoate Esters

Benzoic acid, phenol, and p-hydroxybenzoate esters are mostly used as disinfectants and preservatives. These monomers compounds have set up their antimicrobial activities and efforts have been made to incorporate them with some polymer backbone for the synthesis of novel antimicrobial polymers which will enhance the biocidal activity. In compared to p-hydroxyphenyl acrylate, allyl phydroxyphenylacetate, and *p*-2-propen oxyphenol for their potentiality in antimicrobial action against both bacteria and fungi, the results shows that phydroxyphenylacrylate has been shown high antibacterial activity (Park et al., 2001). A factor which contributing more for antimicrobial action of p-hydroxyphenyl acrylate derivatives is the stereoelectronic effect of the phenyl group. After the binding of acryl or acryloxy groups to the phenyl moiety, they exhibit superior antimicrobial activities than aliphatic acrylates and hexyl acrylate (Jung et al., 2004). Another important compound of this class is "benzaldehyde," known for its antimicrobial activities. Benzaldehyde containing methyl methacrylate polymers have been prepared and tested against *Bacillus macroides, Pseudomonas aeruginosa*, and *Dunaliell atertiolecta* (Subramanyyam et al., 2009).

3.3.2.4. 5-Chloro-8-Hydroxy-Quinoline

To know the hydrolytic behavior of acrylate polymers containing 5-chloro-8-hydroxyquinoline, various parameter like physiological, acidic, and basic pH have been observed. Hydrolysis of these polymers happens due to the autocatalysis and these are potentiated by pH, temperature, and the content of hydrophilic polymers. Copolymerization of this polymer with N-vinyl pyrrolidone decreases the rate of hydrolysis because of steric hindrance (Bnkova et al., 1997). These polymers are also exhibit high antibacterial activity.

3.3.2.5. Polymeric Nanomaterials

Polymeric nanosized antimicrobial compounds possesses strong antimicrobial activity: they are not volatile in nature and chemically very much stable, able to attach to the surface of their interest, and barely infuse through biological membranes for example skin (Kenawy et al., 2007). Distinctively, polycationic bactericidal exhibit strong surface density of active groups which might result in enhancement in bactericidal activity. Polyamines regarded as extremely efficient antimicrobial nanoparticles which contain a wide range of bacterial targets when included in different polymeric matrixes (Shvero et al., 2015; Shvero et al., 2013) (Figure 3.9). Similarly, lipid nanoparticles are attractive for their biocompatibility, versatility, and their ability to target biofilm infections.

Organic antibacterial materials are believed to be less stable in nature, mainly at higher temperature when compared with inorganic materials. This may lead to difficulties that arise when designing products meant to be stable and able to withstand harsh process conditions. Therefore inorganic materials have been more often used as antimicrobial materials.

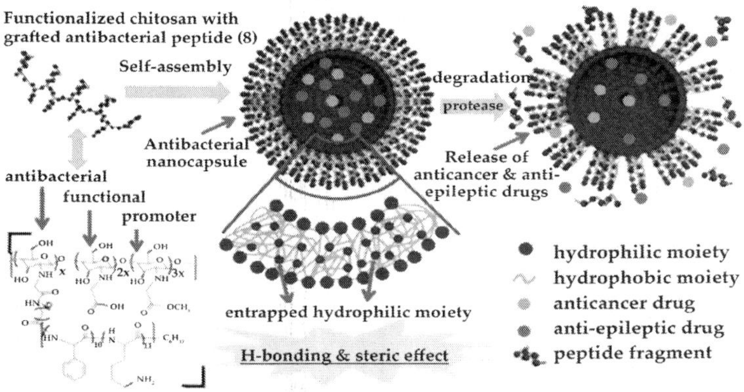

Figure 3.9: Schematic illustration of the self-assembly of polypeptide-grafted chitosan-based nanocapsules, and its ability for drug release upon enzymatic (protease) degradation (From Zhou et al., 2, 1021, 2013).

3.4. SUMMARY

Bacterial strains which are resistant to the antibiotics constantly possess a severe issue to the human health and thus intensifies the need to increase the preparation of new bactericidal materials. Therefore, there is a tough requirement for developing novel approaches and new materials that can deal with these serious issues. The development of nanotechnology has created many new antimicrobial options. The small size of the NPs is very suitable for carrying out antimicrobial biological operations. Metal, organic and other types of nanoparticle have exhibited fabulous potential as antibacterial and antifungal agents, which demonstrate their potential as proficient antibiotic reagents in wound care and related medical issues. The efficacy of these nanoparticles varies with their characteristics including size, shape, and concentration. Moreover, the atomic abundance on the particle's surface plays a major role in the antibacterial properties of such materials. As the particle size decreases, the percentage of atoms on the surface increases in relation to the total atoms of material, intensifying the activity. Various NPs exhibit antimicrobial activity against several pathogenic viral and bacterial species. Nowadays nanomaterials are a shows potential platform for alternative measures to reduce the bacterial infections. Antimicrobial NPs possesses broad range of classification and also applications. These antimicrobial NPs possess antimicrobial activity with insignificant toxicity, compared with other antimicrobial agents. Although there is availability of a no. of antibiotics but due to preserve antibiotic-resistant microbes there is a need to develop more antibacterial nanoparticle. Due to the exact targets of action of antibiotics less no.of bacteria can be destructed, but antimicrobial NPs able to destroy bacterial cell membranes and thus avert the development of drug-resistance microbes. Because of these advantages of antimicrobial NPs, various efforts have been put forward to coat these nanoparticles on the surfaces for medical devices, fibers, and textiles, rendering them antimicrobial. So due to improvement in quality of research, devoted efforts, successful application, and commercialization of antimicrobial NPs, the quality of life will be improved.

REFERENCES

1. Adams, L. K., Lyon, D. Y., & Alvarez, P. J. (2006). Comparative eco-toxicity of nanoscale TiO_2, SiO_2, and ZnO water suspensions. *Water Res.*, *40*, 3527–3532.
2. Adibkia, K., Alaei-Beirami, M., Barzegar-Jalali, M., Mohammadi, G., & Ardestani, M. S. (2012). Evaluation and optimization of factors affecting novel diclofenac sodium-eudragit RS100 nanoparticles. African *Journal of Pharmacy and Pharmacology*, *6*, 941–947.
3. Adibkia, K., Barzegar-Jalali, M., Nokhodchi, A., SiahiShadbad, M., Omidi, Y., Javadzadeh, Y., & Mohammadi, G. (2010). A review on the methods of preparation of pharmaceutical nanoparticles. *Journal of Pharmaceutical Science*, *15*, 303–314.
4. Aditya, N. P., Vathsala, P. G., Vieira, V., Murthy, R. S. R., & Souto, E. B. (2013). Advances in nanomedicines for malaria treatment. *Advances in Colloid and Interface Science*. 201–202, 1–17.
5. Aleksandar. F., Moreno, R., Lu, T. K., Puscasu, V. A., Yoon, C. J., Langer, R., & Farokhzad, O. C. (2012). Surface Charge-Switching Polymeric Nanoparticles for Bacterial Cell Wall-Targeted Delivery of Antibiotics. *ACS Nano*, *22*, 4279–4287.
6. Allahverdiyev, A. M., Abamor, E. S., Bagirova, M., & Rafailovich, M. (2011). Antimicrobial effects of TiO_2 and Ag_2O nanoparticles against drug-resistant bacteria and leishmanial parasites. *Future Microbiology*. *6*, 933–940.
7. Anderson, E. B., & Long, T. E. (2010). Imidazole- and imidazolium containing polymers for biology and material science applications. *Polymer*, *51*, 2447–2454.
8. Ansari, M. A., Khan, H. M., Khan, A. A., Cameotra, S. S., & Alzohairy, M. A. (2015). Anti-biofilm efficacy of silver nanoparticles against MRSA and MRSE isolated from wounds in a tertiary care hospital. *Indian J. Med. Microbiol*, *33*, 101–109.
9. Ansari, M. A., Khan, H. M., Khan, A. A., Sultan, A., & Azam, A. (2012). Characterization of clinical strains of MSSA, MRSA, and MRSE isolated from skin and soft tissue infections and the antibacterial activity of ZnO nanoparticles. *World Journal of Microbiology & Biotechnology*, *28*, 1605–1613.
10. Aslam, M., Anis, I., Afza, N., Hussain, M. T., Iqbal, L., Hussain, A., Iqbal, S., Bokhari, T. H., & Khalid, M. (2012). Synthesis, antibacterial,

lipoxygenase and urease inhibitory activities of 2-aminophenol derivatives. *Med. Chem. Drug Discov*. *3*, 80–86.

11. Atkinson, A., & Winge, D. R. (2009). Metal acquisition and availability in the mitochondria. *Chemical Review*, *109*, 4708–4721.
12. Avalos, A., Haza, A. I., Mateo, D., & Morales, P. (2014). Interactions of manufactured silver nanoparticles of different sizes with normal human dermal fibroblasts. *International Wound Journal*, *13*, 101-109.
13. Azam, A., Ahmed, A. S., Oves, M., Khan, M., & Memic, A. (2012). Size-dependent antimicrobial properties of CuO nanoparticles against Gram-positive and-negative bacterial strains. *International Journal of Nanomedicine*. *7*, *20*12.
14. Bahrami, K., Nazari, P., Nabavi, M., Golkar, M., Almasirad, A., & Shahverdi, A. R. (2014). Hydroxyl capped silver-gold alloy nanoparticles: characterization and their combined effect with different antibiotics against Staphylococcus aureus. *Nanomedicine Journal*, *1*, 155–161.
15. Bankova, M., Manolova, N., Markova, N., Radoucheva, T., Dilova, K., & Rashkov, I. (1997). Hydrolysis and antibacterial activity of polymers containing 8-quinolinyl acrylate. *Journal of Bioactive and Compatible Polymers*, *12*, 294–307.
16. Bar, H., Bhui, D. K., Sahoo, G. P., Sarkar, P., & Pyne, S. (2008). Green synthesis of silver nanoparticles using seed extract of *Jatropha curcas*. *Colloids and Surfaces A: Physicochemical and Engineering Aspects*. *348*, 212–216.
17. Barry, A. L., Craig, W. A., Nadler, H., Reller, L. B., Sanders, C. C., & Swenson, J. M. (1999). Methods for determining bactericidal activity of antimicrobial agents; approved guideline, *vol. 19*, 18th edn. (National Committee for Clinical Laboratory Standards, CLSI, Wayne, 1999).
18. Belapurkar, A. D., Sherkhane, P., & Kale, S. P. (2006). Disinfection of drinking water using photocatalytic technique. *Current science*, *91*, 73–76.
19. Bera, R. K., Mandal, S. M., & Retna Raj, C. (2014). Antimicrobial Activity of Fluorescent Ag Nanoparticles. *Letters in Applied Microbiology*, *58*, 520–526.
20. Besinis, A., Peralta, T. D., & Handy, R. D. (2014). The antibacterial effects of silver, titanium dioxide, and silica dioxide nanoparticles compared to the dental disinfectant chlorhexidine on *Streptococcus*

mutans using a suite of bioassays. *Nanotoxicology*, *8*, 1–16.

21. Beyth, N., Yudovin-Farber, I., Perez-Davidi, M., Domb, A. J., & Weiss, E. I. (2010). Polyethyleneimine nanoparticles incorporated into resin composite cause cell death and trigger biofilm stress in vivo. *Proceedings of the National Academy of Sciences of the United States of America*, *107*, 22038–22043.
22. Bhainsa, K. C., & D'Souza, S. F. (2006). Extracellular biosynthesis of silver nanoparticles using the fungus Aspergillus fumigates. *Colloids and Surfaces B: Bio interfaces*, *47*, 160–164.
23. Bhattacharya, R., & Mukherjee, P. (2008). Biological properties of 'naked' metal nanoparticles. *Advanced Drug Delivery Reviews*, *60*, 1289–1306.
24. Bindhu, M. R., & Umadevi, M. (2014). Antibacterial activities of green synthesized gold nanoparticles. *Materials Letters*, *120*, 122–125.
25. Blecher, K., Nasir, A., & Friedman, A. (2011). The growing role of nanotechnology in combating infectious disease. *Virulence*, *2*, 395–401.
26. Bonilla, A. M., & García, M. F. (2012). Polymeric materials with antimicrobial activity. *Progress in Polymer Science*, *37*, 281–339.
27. Brady-Est´evez, A. S., Kang, S., & Elimelech, M. (2008). A single-walled- carbon-nanotube filter for removal of viral and bacterial pathogens. *Small*, *4*, 481–484.
28. Brayner, R., Ferrari-Iliou, R., Brivois, N., Djediat, S., Benedetti, M. F., Fie´vet, F. (2006). Toxicological impact studies based on *Escherichia coli* bacteria in ultrafine ZnO nanoparticles colloidal medium. *Nano Lett.*, *6*, 866–870.
29. Buzea, C., Pacheco, I. I., & Robbie, K. (2007). Nanomaterials and nanoparticles: sources and toxicity. *Biointerphases*, 2, MR17–MR71.
30. Chernousova, S., & Epple, M. (2013). Silver as antibacterial agent: ion, nanoparticle, and metal. *Angewandte Chemie International Edition*, *52*, 1636–1653.
31. Choi, O., Deng, K. K., Kim, N. J., Ross Jr., L., Surampalli, R. Y., & Hu, Z. (2008). The inhibitory effects of silver nanoparticles, silver ions, and silver chloride colloids on microbial growth. *Water Research*, *42*, 3066–3074.
32. Chowdhury, S., Basu, A., & Kundu, S. (2014). Green synthesis of

protein capped silver nanoparticles from phytopathogenic fungus Macrophominaphaseolina (tassi) goid with anti-microbial properties against multidrug-resistant bacteria. *Nanoscale Research Letters*, *9*, 365–373.

33. Chu, Z., Zhang, S., Chunyuan, B., & Fang, Z. C. (2014). Unambiguous observation of shape effects on cellular fate of nanoparticles. *Scientific Reports*, *4*, 4495.
34. Cioffi, N., & Rai, M. (2012). Nano-antimicrobials, in: Nicola Cioffi, M. Rai (Eds.), Synthesis and Characterization of Novel Nano Antimicrobials. *Springer. Berlin Heidelberg,* XVI, 556.
35. Cousins, B., Allison, H., Doherty, P., Edward, C. S., Garvey, M., Martin, D., & Williams, R. (2007). Effects of a nanoparticulate silica substrate on cell attachment of *Candida albicans. Journal of Applied Microbiology*, *102*, 757–765.
36. Cui, Y., Zhao, Y., Tian, Y., Zhang, W., Lü, X., & Jiang, X. (2012). The molecular mechanism of action of bactericidal gold nanoparticles on *Escherichia coli. Biomaterials*, *33*, 2327–2333.
37. Dakrong, P., Takuro, N., & Michael, B. (2011). The forthcoming applications of gold nanoparticles in drug and gene delivery systems. *Journal of Controlled Release*, *149*, 65–71.
38. Davey, M. E., & O'Toole, G. A. (2000). Microbial biofilms: from ecology to molecular genetics. *Microbiology and Molecular Biology Reviews*, *64*, 847–867.
39. De Simone, S., Gallo, A. L., Paladini, F., Sannino, A., & Pollini, M. (2014). Development of silver nano-coatings on silk sutures as a novel approach against surgical infections. *Journal of Materials Science: Materials in Medicine*, *25*, 2205–2214.
40. Dhapte, V., Kadam, S., Pokharkar, V., Khanna, P. K., & Dhapte, V. (2014). Versatile SiO_2NanoparticlesPolymer Composites with Pragmatic Properties. ISRN *Inorganic Chemistry*. 1–8.
41. Dizaj, S. M., Jafari, S., & Khosroushahi, A. Y. (2014). A sight on the current nanoparticle-based gene delivery vectors. *Nanoscale Research Letters*. *9*, 252–260.
42. Egger, S., Lehmann, R. P., Height, M. J., Loessner, M. J., & Schuppler, M. (2009). Antimicrobial properties of a novel silver–silica nanocomposite material. *Applied Environmental Microbiology*, *75*, 2973–2976.

43. Elliott, C. (2010). The effects of silver dressings on hronic and burns wound healing. *British Journal of Nursing, 19*, S32–S36.
44. Emami-Karvani, Z., Chehrazi, P 2011. Antibacterial activity of ZnO nanoparticle on Gram-positive and gram-negative bacteria. *African Journal of Microbiology Research, 5*, 1368–1373.
45. Emamifar, A. (2011). Application of antimicrobial polymer nanocomposite in food packaging. Book-nanotechnology and nanomaterials. *ISBN* 978–953–307–347–7, Published-July 27, 2011, 299-318.
46. Erich, P. (2011). Test method for antimicrobial activity using titanium dioxide nanoparticles. Brno, Czech Republic. EU, *Nanocon. 9*, 21–23.
47. Ge, Y., Wang, S., Zhou, X., Wang, H., Xu, H. H. K., & Cheng, L. (2015). The Use of Quaternary Ammonium to Combat Dental Caries. *Materials, 8*, 3532–3549.
48. Ghosh, P. S., Kim, C., Han, G., & Forbes, N. S. (2008). Vincent M. Efficient Gene Delivery Vectors by Tuning the Surface Charge Density of Amino Acid-Functionalized Gold Nanoparticles. *ACS Nano, 2*, 2213–2218.
49. Goodman, C., McCusker, C., Yilmaz, T., & Rotello, V. (2004). Toxicity of gold nanoparticles functionalized with cationic and anionic side chains. *Bioconjugate Chemistry. 15*, 897–900.
50. Greulich, C., Diendorf, J., & Gessmann, J. (2011). Cell type-specific responses of peripheral blood mononuclear cells to silver nanoparticles. *Acta Biomaterialia, 7*, 3505–3514.
51. Gupta, K., Singh, R., Pandey, A., & Pandey, A. (2013). Photocatalytic antibacterial performance of TiO_2 and Ag-doped TiO_2 against *S. aureus. P. aeruginosa* and *E. coli. Beilstein Journal of Nanotechnology, 4*, 345–351.
52. Haghighi, F., Mohammadi, S. R., Mohammadi, P., Hosseinkhani, S., & Shipour, R. (2013). Antifungal Activity of TiO_2 nanoparticles and EDTA on *Candida albicans* Biofilms. *Infection Epidemiology and Medicine, 1*, 33–38.
53. Hall-Stoodley, L., Costerton, J. W., & Stoodley, P. (2004). Bacterial biofilms: from the natural environment to infectious diseases. *Nature Reviews Microbiology, 2*, 95–108.
54. Hernández-Sierra, J. F., Ruiz, F., Cruz Pena, D. C., Martínez-Gutiérrez, F., Martínez, A. E., Guillén, A. de J. P., Tapia-Pérez, H., & Castañón,

G. M. (2008). The antimicrobial sensitivity of *Streptococcus mutans* to nanoparticles of silver, zinc oxide, and gold. *Nanomedicine NBM, 4*, 237–240.

55. Hosseinkhani, P., Zand, A., Imani, S., Rezayi, M., & RezaeiZarchi, S. (2011). Determining the antibacterial effect of ZnO nanoparticle against the pathogenic bacterium, *Shigelladysenteriae* (type 1). *International Journal of Nano Dimensions, 1*, 279–285.
56. Huh, A. J., & Kwon, Y. J. (2011). Nanoantibiotics: a new paradigm for treating infectious diseases using nanomaterials in the antibiotics resistant era. *Journal of Controlled Release, 156*, 128–145.
57. Jalal, R., Goharshadi, E. K., Abareshi, M., Moosavi, M., Yousefi, A., & Nancarrow, P. (2010). ZnO nanofluids: green synthesis, characterization, and antibacterial activity. *Mater. Chem. Phys., 121*, 198–201.
58. Janus, M., Markowska-Szczupa, A., Usiak-Nejman, E., & Morawski, A. W. (2012). Disinfection of *E. coli* by carbon modified TiO_2 photocatalysts. *Environment Protection Engineering, 38*, 89–97.
59. Jeong, M. S., Park, J. S., Song, S. H., & Jang, S. B. (2007). Characterization of antibacterial nanoparticles from the scallop, Ptino pectenyessoensis. *Bioscience Biotechnology and Biochemistry, 71*, 2242–2247.
60. Jin, T., & He, Y. (2011). Antibacterial activities of magnesium oxide (MgO) nanoparticles against foodborne pathogens. *Journal of Nanoparticle Research, 13*, 6877–6885.
61. Jo, Y. K., Kim, B. H., & Jung, G. (2009). Antifungal activity of silver ions and nanoparticles on phytopathogenic fungi. *Plant Diseases, 93*, 1037–1043.
62. Jung, H. K., Park, E. S., & Jae, H. S. (2009). Antimicrobial activity of hydroxyphenyl acrylate derivatives. *Journal of Agricultural and Food Chemistry, 52*, 7480–7483.
63. Kasemets, K., Ivask, A., Dubourguier, H. C., & Kahru, A. (2009). Toxicity of nanoparticles of ZnO, CuO, and TiO_2 to yeast *Saccharomyces Cerevisiae. Toxicol. In Vitro, 23*, 1116–1122.
64. Kenawy, E. R., Worley, S. D., & Broughton, R. (2007). The chemistry and applications of antimicrobial polymers: a state-of-the-art review. *Biomacromolecules, 8*, 1359–1384.
65. Kumar, D. A., Palanichamy, V., & Roopan, S. M. (2014). Photocatalytic

action of AgCl nanoparticles and its antibacterial activity. *Journal of Photochemistry and Photobiology B: Biology*, *138*, 302–306.

66. Leid, J. G., Ditto, A. J., & Knapp, A. (2012). In vitro antimicrobial studies of silver carbene complexes: activity of free and nanoparticle carbene formulations against clinical isolates of pathogenic bacteria. *The Journal of Antimicrobial Chemotherapy*, *67*, 138–148.
67. Leung, Y. H., Ng, A., Xu, X., Shen, Z., Gethings, L. A., Wong, M. T., Chan, C., Guo, M. Y., Ng, Y. H., & Djurišić, A. B. (2014). Mechanisms of Antibacterial Activity of MgO: Non-ROS Mediated Toxicity of MgO Nanoparticles Towards *Escherichia coli*. *Small*, *10*, 1171–1183.
68. Li, M., Zhu, L., & Lin, D. (2011). Toxicity of ZnO nanoparticles to *Escherichia coli*: mechanism and the influence of medium components. *Environ. Sci. Technol.*, *45*, 1977–1983.
69. Lima, E., Guerra, R., Lara, V., & Guzmán, A. (2013). Gold nanoparticles as efficient antimicrobial agents for *Escherichia coli* and *Salmonella typhi*. *Chemistry Central Journal*, 7, 11–17.
70. Liu, C., & Zhang, N. (2011). Nanoparticles in gene therapy: principles, prospects, and challenges. *Progress in Molecular Biology and Translational Science.104*, 509–562.
71. Lolina, S., & Narayanan, V. (2013). Antimicrobial and Anticancer Activity of Gold Nanoparticles Synthesized from Grapes Fruit Extract. *Chemical Science Transactions*, *2*, S105–S110.
72. Loomba, L., & Scarabelli, T. (2013). Metallic nanoparticles and their medicinal potential. Part I. Gold and silver colloids. *Therapeutic Delivery*, *4*, 859–873.
73. Luther, E. M., Koehler, Y., Diendorf, J., Epple, M., Dringen, R. (2011). Accumulation of silver nanoparticles by cultured primary brain astrocytes. *Nanotechnology*, *22*, Article ID 375101.
74. Lv, M., Su, S., He, Y., Huang, Q., Hu, W., Li, D., Fan, C., & Lee, S. T. (2010). Long-Term Antimicrobial Effect of Silicon Nanowires Decorated with Silver Nanoparticles. *Advanced Materials*, *22*, 5463–5467.
75. Magiorakos, A. P., Srinivasan, A., & Carey, R. B. (2012). Multidrug-resistant, extensively drug-resistant and pan-drug-resistant bacteria: an international expert proposal for interim standard definitions for acquired resistance. *Clinical Microbiology and Infection*, *18*, 268–281.
76. Mahapatra, O., Bhagat, M., Gopalakrishnan, C., & Arunachalam, K. D.

(2008). Ultrafine dispersed CuO nanoparticles and their antibacterial activity. *Journal of Experimental Nanoscience*, *3*, 185–193.

77. Majdalawieh, A., Kanan, M. C., El-Kadri, O., & Kanan, S. M. (2014). Recent advances in gold and silver nanoparticles: synthesis and applications. *Journal of Nanoscience and Nanotechnology*, *14*, 74757–4780.
78. Malarkodi, C., Rajeshkumar, S., Paulkumar, K., Vanaja, M., Gnanajobitha, G., & Annadurai, G. (2014). Biosynthesis and Antimicrobial Activity of Semiconductor Nanoparticles against Oral Pathogens. *Bioinorganic Chemistry and Applications*. 1–10.
79. Malka, E., Perelshtein, I., & Lipovsky, A. (2013). Eradication of multidrug-resistant bacteria by a novel Zn-doped CuO nanocomposite, *Small, 9*, 4069–4076.
80. Maness, P. C., Smolinski, S., Blake, D. M., Huang, Z., Wolfrum, E. J., & Jacoby, W. A. (1999). Bactericidal activity of photocatalytic TiO_2 reaction: toward an understanding of its killing mechanism. *Applied and Environmental Microbiology*, *65*, 4094–4098.
81. Mar Encabo-Berzosa, M., Sancho-Albero, M., Sebastian, V., Irusta, S., & Arruebo, M. (2017). Polymer functionalized gold nanoparticles as non-viral gene delivery reagents. *Journal of Gene Medicine*, *19*, 1–7.
82. Marambio-Jones, C., & Hoek, E. M. (2010). A review of the antibacterial effects of silver nanomaterials and potential implications for human health and the environment. *Journal of Nanoparticle Research*, *12*, 1531–1551.
83. Mat˘eejka, V., Tokarsk´y, J. (2014). Photocatalytic nanocomposites: a review. *Journal of Nanoscience and Nanotechnology*, *14*, 1597–1616.
84. Matulionyte, M., Dapkute, D., Budenaite, L., Jarockyte, G., & Rotomskis, R. (2017). Photoluminescent gold nanoclusters in cancer cells: cellular uptake, toxicity, and generation of reactive oxygen species. *International Journal of Molecular Science*, *18*, 378–385.
85. Maurer-Jones, M. A., Bantz, K. C., Love, S. A., Marquis, B. J., & Haynes, C. L. (2009). Toxicity of therapeutic nanoparticles. *Nanomedicine*, *4*, 219–241.
86. Mie, R., Samsudin, M. W., Din, L. B., Ahmad, A., Ibrahim, N., Adnan, S. N. A. (2014). Synthesis of silver nanoparticles with antibacterial activity using the lichen *Parmotremapraesorediosum*. *International*

Journal of Nanomedicine, *9*, 121–127.

87. Mishra, G., Dash, B., Pandey, S., & Mohanty, P. P. (2013). Antibacterial actions of silver nanoparticles incorporated Zn-Al layered double hydroxide and its spinel, *Journal of Environmental and Chemical Engineering*, *1*, 1124–1130.
88. Mohammadi, G., Valizadeh, H., Barzegar-Jalali, M., Lotfipour, F., Adibkia, K., Milani, M., Azhdarzadeh, M., Kiafar, F., & Nokhodchi, A. (2010). Development of azithromycin–PLGA nanoparticles: Physicochemical characterization and antibacterial effect against Salmonella typhi. *Colloids and Surfaces B*, *80*, 34–39.
89. Mourato, A., Gadanho, M., Lino, A. R., & Tenreiro, R. (2011). Biosynthesis of Crystalline Silver and Gold Nanoparticles by Extremophilic Yeasts. *Bioinorganic Chemistry and Applications*. Article ID 546074, 8 pages.
90. Mukha, I., Eremenko, A., Korchak, G., & Michienkova, A. (2010). Antibacterial Action and Physicochemical Properties of Stabilized Silver and Gold Nanostructures on the Surface of Disperse Silica. *Journal of Water Resources, Protection*. *2*, 131–136.
91. Nair, S., Sasidharan, A., Rani, V. D., Menon, D., Nair, S., Manzoor, K., & Raina, S. (2009). Role of size scale of ZnO nanoparticles and microparticles on toxicity toward bacteria and osteoblast cancer cells. *J. Mater. Sci. Mater. Med*. *20*, 235–241.
92. Nathan, C., & Cunningham-Bussel, A. (2013). Beyond oxidative stress: an immunologist's guide to reactive oxygen species. *Nature Reviews Immunology*, *13*, 349–361.
93. Ninganagouda, S., Rathod, V., & Singh, D. (2014). Growth kinetics and mechanistic action of reactive oxygen species released by silver nanoparticles from *Aspergillus nigeron Escherichia coli*. *Bio Med Research International*. Article ID 753419, 9 pages.
94. Nonaka, T., Hua, L., Ogata, T., & Kurihara, S. (2003). Synthesis of water-soluble thermosensitive polymers having phosphonium groups from methacryloyloxy ethyltrialkyl phosphonium chlorides-N-isopropylacrylamide copolymers and their functions. *Journal of Applied Polymer Science*, *87*, 386–393.
95. Ochiai, T., Masuko, K., Tago, S., Nakano, R., Nakata, K., Hara, M., Nojima, Y., Suzuki, T., Ikekita, M., Morito, Y., & Fujishima, A. (2013).

Synergistic Water-Treatment Reactors Using a TiO_2 -Modified Ti-Mesh Filter. *Water*, *5,* 1101–1115.

96. Paciotti, G. F., Myer, L., Weinreich, D., Goia, D., & Pavel, N. (2004). Colloidal gold: a novel nanoparticle vector for tumor directed drug delivery. *Drug Delivery*, *11*,169–183.
97. Palanikumar, L., Ramasamy, S. N., & Balachandran, C. (2014). Size dependent antimicrobial response of zinc oxide nanoparticles. *IET Nano Biotechnology*, *8*, 111–117.
98. Pan, X., Redding, J. E., Wiley, P. A., Wen, L., McConnell, J. S., & Zhang, B. (2010). Mutagenicity evaluation of metal oxide nanoparticles by the bacterial reverse mutation assay. *Chemosphere*, *79*, 113–116.
99. Panacek, A., Kv′ıtek, L., & Prucek, R. (2006). Silver colloid nanoparticles: synthesis, characterization, and their antibacterial activity. *The Journal of Physical Chemistry B*, *110*, 16248–16253.
100. Park, E. S., Moon, W. S., Song, M.-J., Kim, M. N., Chung, K. H., & Yoon, J. S. (2001). Antimicrobial activity of phenol and benzoic acid derivatives. *International Biodeterioration and Biodegradation*, *47*, 209–214.
101. Pelgrift, R. Y., & Friedman, A. J. (2013). Nanotechnology as a therapeutic tool to combat microbial resistance. *Advanced Drug Delivery Reviews*, *65*, 13–14, 1803–1815.
102. Pey, P., Packiyaraj, M. S., Nigam, H., Agarwal, G. S., Singh, B., & Patra, M. K. (2014). Antimicrobial properties of CuO nanorods and multi-armed nanoparticles against *B. anthracis* vegetative cells and endospores. *Beilstein Journal of Nanotechnology*, *5*, 789–800.
103. Podsiadlo, P., Sinani, V. A., Bahng, J. H., Kam, N. W. S., & Lee, J. (2008). Gold nanoparticles enhance the anti-leukemia action of a 6-mercaptopurine chemotherapeutic agent. *Langmuir*, *24*, 568–574.
104. Poulose, S., Panda, T., Nair, P. P., Theodore, T 2014. Biosynthesis of silver nanoparticles. *Journal of Nanoscience and Nanotechnology*, *14*, 2038–2049.
105. Premanathan, M., Karthikeyan, K., Jeyasubramanian, K., & Manivannan, G. (2011). Selective toxicity of ZnO nanoparticles toward Gram-positive bacteria and cancer cells by apoptosis through lipid peroxidation. Nanomed. Nanotechnol. *Biol. Med.*, *7*, 184–192.
106. Priscila, L., Freire, L., Allan, J., Albuquerque, R., Fabio, C., Sampaio, Galembeck, A., Miguel Flores, A. P., Thayza, C., Stamford, M., &

Rosenblatt, A. (2017). AgNPs: The New Allies Against *S. Mutans* Biofilm – A Pilot Clinical Trial and Microbiological Assay, *Brazilian Dental Journal, 28*, 417–422.

107. Raghupathi, K. R., Koodali, R. T., & Manna, A. C. (2011). Size-dependent bacterial growth inhibition and mechanism of antibacterial activity of zinc oxide nanoparticles. *Langmuir, 27*, 4020–4028.
108. Rai, A., Prabhune, A., & Perry, C. C. (2010). Antibiotic-mediated synthesis of gold nanoparticles with potent antimicrobial activity and their application in antimicrobial coatings. *Journal of Material Chemistry, 20*, 6789–6798.
109. Rai, M., Yadav, A., & Gade, A. (2009). Silver nanoparticles as a new generation of antimicrobials. *Biotechnology Advances, 27*, 76–83.
110. Rai, V. R., & Bai, A. J. (2011). Nanoparticles and their potential application as antimicrobials. *Science against Microbial Pathogens: Communicating Current Research and Technological Advances*. 197–209.
111. Ramakrishna, M., Babu, R., Gengan, R. M., Chandra, S., & Rao, N. (2016). Green synthesis of gold nanoparticles using marine algae and evaluation of their catalytic activity. *Journal of Nanostructure in Chemistry, 6*, 1–9.
112. Rao, M. C., Ravindranadha, K., & Mary, R. (2013). Development of ZnO Nanoparticles for Clinical Applications. *Journal of Chemical Biological and Physical Sciences, 4*, 469–473.
113. Rasmussen, J. W., Martinez, E., Louka, P., Wingett, D. G. (2010). Zinc oxide nanoparticles for selective destruction of tumor cells and potential for drug delivery applications. *Expert Opinion on Drug Delivery, 7*, 1063–1077.
114. Ray, S., Sarkar, S., & Kundu, S. (2011). Extracellular biosynthesis of silver nanoparticles using the mycorrhizal mushroom Tricholomacrassum (berk.) SACC: its antimicrobial activity against pathogenic bacteria and fungus, including multidrug-resistant plant and human bacteria. *Digest Journal of Nanomaterials and Biostructures, 6*, 1289–1299.
115. Reddy, K. M., Feris, K., Bell, J., Wingett, D. G., Hanley, C., & Punnoose, A. (2007). Selective toxicity of zinc oxide nanoparticles to prokaryotic and eukaryotic systems. *Appl. Phys. Lett., 90*, 213902.

116. Reddy, M. P., Venugopal, A., & Subrahmanyam, M. (2007). Hydroxyapatite-supported Ag-TiO2 as *Escherichia coli* disinfection photocatalyst. *Water Research, 41*, 379– 386.
117. Sacco, P., Travan, A., BorgognaM., Paoletti, S., Marsich, E. (2015). Silver-containing antimicrobial membrane based on chitosan-TPP hydrogel for the treatment of wounds. *J Mater Sci. Mater Med., 26*, 128–136.
118. Samsudin, E. M., Wu, S. Y., Ling, T., Hamid, S. B., Juan, J. C. (2015). Evaluation on the Photocatalytic Degradation Activity of Reactive Blue 4 using Pure Anatase Nano-TiO_2. *Sains Malaysiana, 44*, 1011–1019.
119. Sanghi, R., Verma, P., & Puri, S. (2011). Enzymatic Formation of Gold Nanoparticles Using Phanerochaetechrysosporium. *Advances in Chemical Engineering and Science, 1*,154–162.
120. Sastry, M., Ahmad, A., Islam, M. K., Kumar, R. Biosynthesis of metal nanoparticles using fungi and actinomycete. *Current Science, 85*, 162–170.
121. Sauvet, G., Fortuniak, W., Kazmierski, K., & Chojnowski, J. (2003). Amphiphilic block and statistical siloxane copolymers with antimicrobial activity. *Journal of Polymer Science Part A: Polymer Chemistry, 41*, 2939–2948.
122. Sawai, J., & Igarashi, H. (2002). Evaluation of Antibacterial Activity of Inorganic Materials and Application of Natural Inorganic Materials to Controlling Microorganisms. *Foods and food ingredients Journal of Japan, 203*, 47–57.
123. Sawai, J., Kojima, H., Igarashi, H., Hashimoto, A., Shoji, S., Sawaki, T., Hakoda, A., Kawada, E., Kokugan, T., & Shimizu, M. (2000). Antibacterial characteristics of magnesium oxide powder. *World Journal of Microbiology and Biotechnology, 16*, 187–194.
124. Sawai, J. (2003). Quantitative evaluation of antibacterial activities of metallic oxide powders (ZnO, MgO, and CaO) by conductimetric assay. *J. Microbiol. Methods, 54*, 177–182.
125. Seil, J. T., Taylor, E. N., Webster, T. J. (2009). Reduced activity of Staphylococcus epidermidis in the presence of sonicated piezoelectric zinc oxide nanoparticles, in (2009). IEEE 35th Annual Northeast Bioengineering Conference, Boston, MA, USA, 3–5 April (2009). (IEEE, 2009), 1–2.
126. Seil, J. T., Webster, T. J. (2012). Antimicrobial applications of

nanotechnology: methods and literature. *International Journal of Nanomedicine*, *7*, 2767–2781.

127. Sethi, D., Pal, A., Sakthivel, R., Pandey, S., Dash, T., Das, T., & Kumar, R. (2014). Water disinfection through photoactive modified titania. *Journal of Photochemistry and Photobiology B: Biology*, *130*, 310–317.
128. Shukla, R., Bansal, V., Chaudhary, M., Basu, A., & Bhonde, R. R. (2005). Biocompatibility of gold nanoparticles and their endocytotic fate inside the cellular compartments microscopic overview. *Langmuir*, *21*, 10644–10654.
129. Shvero, D. K., Abramovitz, I., Zaltsman, N., Davidi, M. P., Weiss, E. I., & Beyth, N. (2013). Towards antibacterial endodontic sealers using quaternary ammonium nanoparticles. *International Endodontic Journal*, *46*, 747–754.
130. Shvero, D. K., Zatlsman, N., Hazan, R., Weiss, E. I., & Beyth, N. (2015). Characterisation of the antibacterial effect of polyethyleneimine nanoparticles in relation to particle distribution in resin composite. *Journal of Dentistry*, *43*, 287–294.
131. Sirelkhatim, A., Mahmud, S., Seeni, A., Haida, N., Kaus, M., Ann, L. C., Khadijah, S., Bakhori, M., Hasan, H., & Mohama, D. (2015). Review on Zinc Oxide Nanoparticles: Antibacterial Activity and Toxicity Mechanism. *Nano-Micro Letters*, *7*, 219–242.
132. Smitha, S. L., & Gopchandran, K. G. (2013). Surface-enhanced Raman scattering, antibacterial and antifungal active triangular gold nanoparticles. *Spectrochimica Acta Part A: Molecular and Biomolecular Spectroscopy*, *102*, 114–119.
133. Sonavane, G., Tomodaa, K., & Makinoa, K. (2008). Biodistribution of colloidal gold nanoparticles after intravenous administration: Effect of particle size. *Colloids Surf B Bio interfaces*, *66*, 274–280.
134. Sondi I., & Salopek-Sondi, B. (2004). Silver nanoparticles as antimicrobial agent: a case study on E. coli as a model for Gram-negative bacteria. *Journal of Colloid and Interface Science*, *275*, 177–182.
135. Soykan, C., Kun, R. C., & Delibas, A. (2005). Microbial screening of copolymers of N-vinyl imidazole with phenacyl methacrylate: synthesis and monomer reactivity ratios. *Journal of Macromolecular Science Part A: Pure and Applied Chemistry*, *42*, 1603–1619.

136. Subramanyam, E., Mohandoss, S., & Shin, H. W. (2009). Synthesis, characterization, and evaluation of antifouling polymers of 4-acryloyloxybenzaldehyde with methyl methacrylate. *Journal of Applied Polymer Science*, *112*, 2741–2749.
137. Sun, T., Hao, H., Hao, W. T., Yi, S. M., Li, X. P., & Li, J. R. (2014). Preparation and antibacterial properties of titanium-doped ZnO from different zinc salts. *Nanoscale Research Letters*, *9*, 98.
138. Teimouri, M., Nia, A. H., Abnous, K., Eshghi, H., & Ramezani, M. (2016). Graphene oxide-cationic polymer conjugates: synthesis and application as gene delivery vectors. *Plasmid*, *84*, 51–60.
139. Tiwari, P. K., & Soo Lee, Y. (2013). Gene delivery in conjunction with gold nanoparticle and tumor treating electric field. *Journal of Applied Physics*, *114*, 5–13.
140. Tiwari, P. M., Vig, K., Dennis V. A., & Singh, S. R. (2011). Functionalized gold nanoparticles and their biomedical applications. *Nanomaterials 1*, 31–63.
141. Usman, M. S., Zowalaty, M. E., Shameli, El. K., Zainuddin, N., Salama, M., Ibrahim, N. A. (2013). Synthesis, characterization, and antimicrobial properties of copper nanoparticles, *International Journal of Nanomedicine.*, *8*, 4467–4479.
142. Venieria, D., Fraggedakia, A., Kostadimaa, M., Chatzisymeona, E., Binasb, V., Zachopoulosb, A., Kiriakidisb, G., & Mantzavinos, D. (2014). Solar light and metal-doped TiO_2 to eliminate water-transmitted bacterial pathogens: Photocatalyst characterization and disinfection performance. *Applied Catalysis B: Environmental*. *154*. 93–101.
143. Vidic, J., Stankic, S., Haque, F., Ciric, D., Goffic, R. L., Vidy, A., Jupille, J., & Delmas, B. (2013). Selective antibacterial effects of mixed ZnMgO nanoparticles. *Journal of Nanoparticle Research.*, *15*, 1–10.
144. Wahab, R., Kim, Y. S., Mishra, A., Yun, S. I., Shin H. S. (2010). Formation of ZnO micro-flowers prepared via solution process and their antibacterial activity. *Nanoscale Res. Lett.*, *5*, 1675–1681.
145. Wang, S., Lawson, R., Ray, P. C., & Yu, H. (2011). Toxic effects of gold nanoparticles on Salmonella typhimurium bacteria. *Toxicology and Industrial Health*, *27*, 547–554.
146. Wani, I. A., & Ahmad, T. (2013). Size and shape dependent antifungal activity of gold nanoparticles: A case study of Candida. *Colloids Surface B Bio interfaces.*, *101*, 162–170.

147. Waschinski, C. J., Tiller, J. C. (2005). Poly (oxazoline) with telechelic antimicrobial functions. *Biomacromolecules*, *6*, 235–243.
148. Watnick, P., & Kolter, R. (2000). Biofilm, city of microbes. Journal of Bacteriology, *182*, 2675–2679.
149. Wu, H. Q., WX, W.ei; Shao, M. W., Gu, J. S., Qu, M. Z. (2002). Synthesis of copper oxide nanoparticles using carbon nanotubes as templates. *Chemical Physics Letters*, *364*, 152–156.
150. Yamamoto, O. (2001). Influence of particle size on the antibacterial activity of zinc oxide. *Int. J. Inorg. Mater. 3*, 643–646.
151. Yamamoto, O., Ohira, T., Alvarez, K., & Fukuda, M. (2010). Antibacterial characteristics of $CaCO_3$–MgO composites. *Materials Science and Engineering. B*, *173*, 208–212.
152. Zaleska, A. (2008). Doped-TiO_2: a review. *Recent Patents in Engineering*, *2*, 157–164.
153. Zarei, M., Jamnejad, A., & Khajehali, E. (2014). Antibacterial Effect of Silver Nanoparticles against Four Foodborne Pathogens. *Journal of Microbiology*, *7*, e78720.
154. Zhang, L., Ding, Y., Povey, M., & York, D. (2008). ZnO nanofluids—A potential antibacterial agent. *Progress in Natural Science*, *18*, 939–944.
155. Zhang, L., Jiang, Y., Ding, Y., Povey, M., & York, D. (2007). Investigation into the antibacterial behavior of suspensions of ZnO nanoparticles (ZnO nanofluids). *J. Nanopart. Res. 9*, 479–489.
156. Zhang, L., Pornpattananangkul, D., Hu, C. M. J., Huang, C. M. (2010). Development of nanoparticles for antimicrobial drug delivery. *Current Medicinal Chemistry*, *17*, 585– 594.
157. Zhang, P., Li, B., Du, J., & Wang, Y. (2017). Regulation the morphology of cationized gold nanoparticles for effective gene delivery. *Colloids Surface B Bio Interfaces*, *28*, 18–25.
158. Zheng, K., Setyawati, M. I., Leong, D. T., & Xie, J. (2017). Antimicrobial gold nanoclusters. *ACS Nano*, *11*, 6904–6910.
159. Zhiruiguo, S., Song Zhang, Y., & Ninggu, J. (2009). Biosynthesis of Gold Nanoparticles Using the Bacteria Rhodopseudomonascapsulata. *Materials Letters*, *61*, 3984–3987.
160. Zhou, C., Wang, M., Zou, K., Chen, J., Zhu, Y., Du, J. (2013). Antibacterial polypeptide-grafted chitosan-based nanocapsules as an armed carrier of anticancer and antiepileptic drugs. *ACS Macro Lett. 2*, 1021–1025.

4
CHAPTER

MULTI-UTILITY OF GOLD NANOPARTICLES IN CANCER

Ashok K. Nayak[1,2] and Alfa S. Mohanty[1,3]

[1] DNA Nanotechnology and Application Laboratory, CSIR-Institute of Minerals and Materials Technology, Bhubaneswar, 751013, India

[2] Department of Biotechnology, Ravenshaw University, Cuttack, 753003, India

[3] Academy of Scientific and Innovative Research, New Delhi, India

CONTENTS

ABSTRACT

The current strategies in cancer therapeutics include surgical removal of solid tumor tissue followed by chemotherapy. The intravenous injection of anticancer chemotherapeutic drugs elicits cytotoxicity and undesirable side effects in a systemic tumor as well as healthy tissues. In this regard, nanotechnology-based anticancer therapeutic modalities implement efficient target therapy by addressing safety and efficacy. Especially gold nanoparticles (GNPs) proved to be a promising agent because of multidimensional utilizations such as targeting ligand, bio-imaging, carrying antisense drugs and integration of other surface modifiers for successful cancer therapy. This chapter illustrates the utilization of GNPs as an anticancer agent. The mechanism of action such as hypothermal and as drug carriers has been discussed in association with clinical trials as cancer therapeutics. Moreover, integration of GNPs with natural techniques led to the development of biogenic nano-materials with medicinal importance, mostly as anticancer modalities. The future of GNPs as cancer therapeutics seems brighter as a vigorous study by interdisciplinary researchers is going on. However, many hindrances yet have to be overthrown to proof GNPs as magic wands against cancer.

Keywords: Gold nanoparticles, Chemotherapy, Drug delivery, Cancer, Green Synthesis

4.1. INTRODUCTION

Cancer belongs to the class of diseases that is fatal to human health and life in most of the developing countries. According to one survey, around 8.2 million cancer-related deaths occurred during 2012, worldwide (Torre et al., 2015). It has been estimated that new cases of all types of cancer worldwide will rise up to 15.4 million in the year 2020. Cancer develops due to aberrant behavior in cell signaling, gene expression, and apoptosis. These abnormalities promote rapid growth of cancer which leads to mortality in the affected patients. Current strategies in cancer therapy are restricted to chemotherapy, radiotherapy, and surgery. Radiotherapy and surgery are implemented as primary treatment for local and non-metastatic cancer. However, they induce metastatic growth of tumor probably due to inflammatory responses during wound healing (Coffey et al., 2003). Both radiation therapy and chemotherapy have nonspecific targets as they kill tumor tissue as well as healthy cells. On the other hand, chemotherapeutic drugs are mostly used against metastatic cancers which inhibit fast-growing cancer cells by targeting cell signaling molecules. However, in most instances, they elicit unwanted side effects by affecting normal rapid growing cells like bone marrow, gastrointestinal tract and hair follicles (Saini et al., 2012). Among other methods, delivery of small antisense oligonucleotides to target corresponding oncogenic miRNA or mRNA through RNA interference (RNAi) technology proved to be a potent anti-tumor therapeutics (Nahar et al., 2017). But these small DNA/RNA molecules have been associated with certain clinical concerns like poor intracellular delivery, instability in *in-vivo* environment and off-target effects. At present, all the aforementioned cancer therapeutics have been performed based on the pathological stages that are determined by radiological and histopathological diagnostic tools. However, the present strategies of early-stage cancer diagnosis and successful treatment face many hindrances. Despite many technological advances in conventional surgery, radiation and chemotherapy, the eradication of cancer do not meet the optimal efficiency. The major challenges that are yet to be achieved in cancer therapeutics are associated with specific accumulation of anti-cancer agents in cancer tissue without disturbing physiology of healthy cells, limited cytotoxicity, recessive to drug resistance and most importantly multi-utility drugs for diagnosis and therapeutics. So it is necessary to develop new therapeutic modalities with advanced technologies to successfully counter all cancer types and diagnostic tools to confirm their complete removal. Taking into account these visions, oncology researchers have so far showcased an array of nano-architectural drug and delivery vehicles such as liposomes, carbon dots, quantum dots and metallic nanoparticles (Biju, 2014).

Nanotechnology is an interdisciplinary research field involving chemistry, biology, pharmaceutical science for creation and functionalization of materials at the nanometer level. Nanoparticles are inherently smaller in size usually in the range of 1–100 nm comparable to intracellular biological molecules like proteins, antibodies, and surface receptor molecules (Daraee et al., 2014). Metal nanoparticles possess some unique characteristics like tunable optical behavior, due to availability of free marching conduction band electrons. When a particle is struck with a light source of wavelength larger than its factual size, the outer orbital electrons can induce coherent localized oscillation of those conducting band electrons leading to localized surface plasmon resonance (LSPRS) (Mayer and Hafner, 2011). The nanoparticles with such physical properties and biocompatibility introduce a new window of opportunity for the development of an advanced tool for early detection, precise diagnosis and target specific cancer therapeutics. The fundamental advantages of nanoparticle-mediated cancer therapeutics are the multi-functionality of nanoparticles where the primary objective is to bypass the biological barriers, attach to target marker and report therapeutic efficacy. Among the various forms of metallic nanoparticles, gold nanoparticles (GNPs) with their inherent chemical and physical properties are the most promising and widely investigated entities (Cannor et al., 2005). Till today, a number of gold based nanoarchitecture like gold nanospheres, nanorods, nanocages and nanoshell (Huang et al., 2007; Javier et al., 2008; Von et al., 2009; Dickerson et al., 2008; Chen et al., 2007; Skrabalak et al., 2008; Gobin et al., 2007) have been implemented as cancer therapeutics. The surface plasmon resonance phenomena of gold nanoparticles have been explored for imaging cancer cells using the techniques of microscopy and optical coherence tomography. Moreover, upon illumination with light, these nanoparticles absorb the light energy and undergo local heating, which makes them the suitable candidate for photothermal ablation of cancer cells. These GNPs showed higher chemical stability, non-toxicity, and ease in surface functionalization as compared to other nano-modalities (Chen et al., 2016). Additionally, the GNPs electrostatically interact with biological macromolecules like DNA, protein, antibodies and other functional units without interfering with their action. So these molecules can be easily conjugated with GNPs and successfully delivered into the affected tissues as an efficient therapeutic molecule. GNPs function as potent drug delivery vehicle when attached with target-specific ligands. The effectiveness of GNPs also varies with their source of synthesis. GNPs synthesized through conventional chemical method is also supplemented with natural products. In the latter condition, GNPs have been synthesized from the precursor molecule chloroauric acid ($HAuCl_4$) by using biological reducing and stabilizing molecules (Paul et al., 2018). On the other hand, the involvement of green chemistry in the synthesis of gold nanoparticles has gained significant attention over conventional synthesis processes due to minimization of hazardous chemicals and enhanced biocompatibility. Over the past decade, there has been an increased prominence of green synthetic processes for GNPs synthesis. These aim at the utilization of non-toxic, environmentally benign and renewable materials such as biomolecules, plants/animal products, and microorganisms as the reducing and stabilizing agents. The as-formed GNPs offer a neutral nano platform for combinatorial approach including imaging, sensing and therapeutic molecule with drug delivery facility in cancer ablation (Figure 4.1).

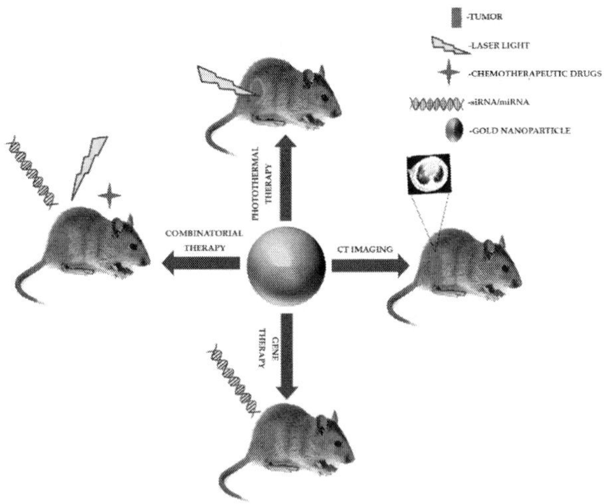

Figure 4.1. Schematic presentation of different approaches of gold nanoparticles mediated cancer ablation.

In this chapter, a brief discussion on the synthesis of gold nanoparticles of various morphologies through conventional and green synthesis method has been presented. Afterward, their characterization and application towards the cancer treatment has been elaborately discussed. This chapter highlights the multi-dimensional utility of GNPs in cancer tissue which will help researchers to get a broad knowledge about GNPs as an anticancer agent.

4.2. SYNTHESIS OF GOLD NANOPARTICLES

4.2.1. Conventional Methods

Though gold is one of the most ancient subjects of scientific investigation in material science, the renaissance occurred with the work of English scientist Michael Faraday in 1857. He reported the formation of a red color solution of colloidal gold when an aqueous solution of chloroaurate ($AuCl_4^-$) was reduced with phosphorous in CS_2, a two-phase system (Faraday, 1857). Thereafter in the 20th century, the breakthrough work of Schmid (2001) and Bethell et al. (1996) influenced a broad range of researchers to develop various methods for preparing GNPs. The initially reported GNPs were of spherical shape. With due course of time, various types of GNPs were synthesized and characterized based on their shape, size, and physiochemical properties. In addition to nanospheres, other forms of GNPs such as nanorods, nanocages, and nanoshell underpin the goal of using GNPs in biomedical application (Zhang, 2015).

There exist a variety of synthesis methods for the fabrication of divergent GNPs. The common oxidation states of gold are +1 for aurous compound and +3 for auric compounds. Among the conventional methods of GNP synthesis, the procedure introduced by Turkevich et al. (citrate reduction method) in 1951 is a milestone in this field. They have reported the formation of monodisperse gold colloids of ~20 nm in diameter by the citrate reduction of an aqueous solution of gold. Then, predetermine sized (between 20–150 nm) GNPs were synthesized by Frens (1973) via a controlled titration method by varying the ratio of metal precursor (gold chloride) and reducing/stabilizing (trisodium citrate) agents. Usually, citrate at lower concentration will yield larger nanoparticles. In 1993, Geirsig and Mulvaney reported a two-phase method inspired by faraday's work in which they were capable of synthesizing thermally and ambient environment stabilized monodisperse GNPs. In some cases, the thiols of various chain lengths were used for preparation of well-controlled size (< 10 nm in diameter) GNPs (Giersig and Mulvaney, 1993). The most impactful method for synthesis of GNPs is the Brust-Schiffrin method (1994). This method allows facile synthesis of air-stable GNPs in common organic solvents without any aggregation or decomposition. These GNPs can be easily functionalized with a variety of molecules and organic ligands to achieve multimodal applications. In this method, $AuCl_4^-$ is transferred from the aqueous medium to toluene with the help of tetraoctyl ammonium bromide as the phase transfer reagent followed by their reduction using the $NaBH_4$ in the presence of dodecanethiol. The organic phase alters its color from orange to a deep brown as soon as $NaBH_4$ is added. GNPs with cuboctahedral and icosahedral configuration are observed with a size distribution of 2–2.5 nm. Moreover, larger thiol/gold molar ratio and quick addition of reductant at lower temperature yield smaller (< 2 nm) monodisperse GNPs. Later, Burst modified this method to a single-phase system by using p-mercaptoethanol for reducing and stabilizing GNPs (Brust et al., 1998). Other sulfur-containing ligands like di and tri-thiols, disulfides and xanthates are also used to synthesize stable gold nanospheres. Typically, gold nanospheres exhibit single absorption peak between 510–550 nm wavelengths. The position and width of the absorption peak is highly related to the particle size and their distribution, i.e., on increasing the size of particles, a red shift in the absorbance is observed. Gold nanorods were synthesized using a wide variety of methods.

In the template method, electrochemical deposition of gold occurs within the opening of nanoporous polycarbonates or alumina template membrane (Muench et al., 2017). The pore diameter defines the dimension of the gold nanorod and the length is determined by the amount of gold deposited within the openings. However, this method has disadvantages of low yield as only a single layer of the nanorod is synthesized. The most conceived method for gold nanorod synthesis is the seed-mediated growth method (Nikoobakht and El-Sayed, 2003). Usually, 0.5 mM auric acid is reduced in 0.2 M cetyltrimethylammonium bromide (CTAB) by ice-cold 0.01 M sodium borohydride at room temperature. This process yields small gold nanospheres of size < 5 nm as seed nanoparticles. The growth solution has been prepared by mixing 1 mM auric acid in 0.2 M CTAB and 0.15 M benzyl dimethyl-hexadecyl ammonium chloride (BDAC). Then 4 mM silver salt is added followed by the addition of 0.0788 M ascorbic acid to the growth solution with constant stirring to reduce $HAuCl_4$ to $HAuCl_2$. Eventually, the seed solution is mixed with the growth solution and is kept undisturbed for nanorod growth.

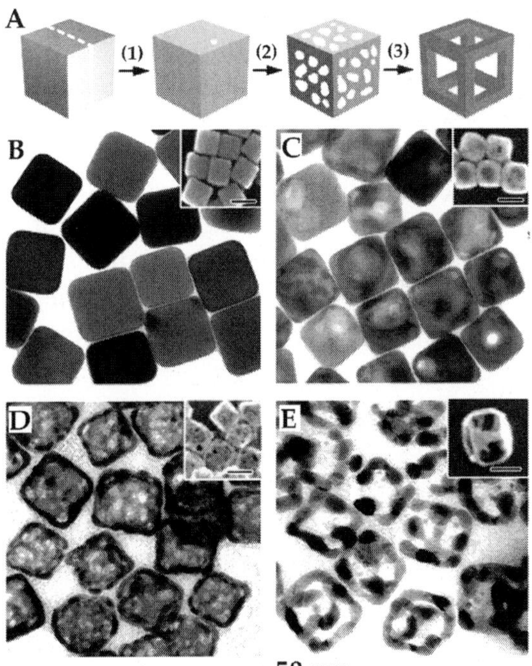

Figure 4.2: (A) Illustration summarizing cubic Au nanoframe formation. Coloration indicates the conversion of a Ag nanocube into a Au/Ag nanocage and

then a predominately Au nanoframe. Beginning with Ag nanocubes, Au/Ag nanoboxes are prepared by galvanic replacement (step 1). Then a wet etchant removes the remaining Ag to form a porous nanocage (step 2), which with more etchant evolves into a cubic nanoframe (step 3). (B-E) TEM and SEM (insets) of (B) 50 nm Ag nanocubes, (C) Au/Ag nanoboxes prepared by galvanic replacement, and (D) nanocages and (E) nanoframes prepared with $Fe(NO_3)_3$ as a Ag etchant (From Skrabalak et al., Accounts of chemical research, 41, 1587, 2008).

Gold nanorod with different aspect ratio is obtained by varying the volume of silver ion. The gold nanorod thus obtained exhibits typical absorption maximum at 800 nm wavelength. Though the mechanism of gold nanorod growth is not fully understood, it is speculated that auric acid binds to CTAB by electrostatic interaction and forms CTAB-capped seed nanoparticles. The $AuCl_2$-CTAB complex binds to tips of the seed nanoparticles leading to nanorod formation and growth. Besides the above-mentioned methods, some other methods such as bio-reduction, growth on a mica surface and photochemical synthesis have also been investigated (Cai et al., 2008). Gold nanocages are a novel class of nanostructures that contains hollow nanometric shaped porous wells (Skrabalak et al., 2008). These are synthesized via galvanic replacement reaction between truncated silver nanocubes and chloroauric acid (Figure 4.2). Here, the silver nanocubes are synthesized by polyol reduction in which ethylene glycol reduces $AgNO_3$ to yield silver atoms and subsequently nanocrystals or seeds. Finally, the addition of silver atoms to the seeds generates nanocubes in the presence of polyvinylpyrrolidone. The silver nanostructure used as the template is then transformed into gold nanocages via the galvanic replacement. The dimension of gold nanocage is controlled by varying the molar ratio of silver to $HAuCl_4$.

Gold nanoshells can be designed and synthesized with variable surface plasmon resonance peaks that range between visible to near-infrared (NIR) region. The synthesis is a four-step method which includes: (1) the fabrication of monodisperse silica cores, (2) functionalization of core surface with positively charged amino groups, (3) adsorption of 3–5 nm tetrakis(hydroxymethyl) phosphonium chloride (THPC) derived gold colloids onto the functionalized silica surface followed by the (4) reduction to form the gold nanoshells (Oldenburg et al., 1998). All the above described GNPs synthesized through conventional chemical synthesis method have scintillating features for their use in cancer therapeutics.

4.2.2. Green Methods

Like the conventional methods, in this process the GNPs are synthesized following a two-step chemical reaction; (1) reduction of Au^{3+} to Au^0 followed by the (2) agglomeration/stabilization of the tiny particles to form colloidal nanoparticles. However, in this process, no additional stabilizing/capping agents are used and the plant extract/animal products act both as the reducing and stabilizing agent. Generally, the synthesis can be achieved by using plants (whole plant or plant parts like leaf, stem, root, fruit, flower, etc.), animal products and microorganisms (bacteria, fungi, algae and actinomycetes, etc.) as the reducing and capping agent. In case of plant-mediated synthesis process, the plant material is collected, washed, dried, boiled in water, and then filtered (Kumar and Yadav, 2009). The filtrates are then used as the reducing agent directly in different concentration depending upon their reducing ability. Microorganism based synthesis can be carried out either by intracellular or extracellular method. However extracellular synthesis process is more adapted due to easier downstream processing for metal recovery in comparison to intracellular synthetic methods (Narayanan and Sakthivel, 2010). In this process, the microbial cells cultured, centrifuged for biomass removal and the supernatant is used as the reducing agent. The bio-reducing agents with auric acid in an optimized condition give rise to GNPs which can be monitored from the change in the color of the reaction solution.

GNPs synthesized using *Stenotrophomonas maltophilia* bacterium was reported by Nangia (2009). In this study, the nicotinamide adenine dinucleotide phosphate (NADPH)-dependent enzyme present in the bacterial strain is responsible for the reduction of gold ions through electron shuttling mechanism (Nangia et al., 2009). *Escherichia coli* DH5a as a bio-reducer for GNP synthesis was observed where the synthesized nanoparticles attached to the bacterial surface used for direct electrochemistry of hemoglobin (Du et al., 2007). A meticulous literature review indicates plant-mediated formulation of gold nanoparticles is preferable as animal products and microorganisms can raise some biosafety issues. Further, microorganisms need more incubation time for reduction of metal ions and expensive downstream processing. So the involvement of plants in biogenic gold nanoparticles synthesis, their optimization, and applications, has opened a new door for nanotechnology researchers and is known as phyto-nanotechnology (Singh et al., 2016).

Synthesis of gold nanoparticles of different shape and size is strongly influenced by several parameters such as the concentration of reducing agent and metal ions, reaction temperature, pH, reaction time and reaction medium, etc. It has been observed that the synthesis at low pH leads to less nucleation and particle agglomeration (Ahmed et al., 2016). In most cases, aqueous medium is preferred to synthesize GNPs but sometimes other solvents were also used where synthesis cannot be processed in water. Therefore optimization of synthesis process is a crucial step for getting GNPs of desirable morphology and application. Effect of temperature and extract concentration was observed in case of GNPs synthesized using Magnolia kobus and Diospyros kaki leaf extract (Song et al., 2009). In this work, the nanoparticles were found to be small spherical shaped at higher extract concentration and temperature while bigger hexagons, pentagons, triangular and spherical shapes obtained at lower temperature and leaf extract concentration (Song et al., 2009).

Application of GNPs is varied depending upon their diversified physicochemical properties such as size, shape, surface area, etc. Size of the particle is the most vital attribute in nanoscience research as lots of evidence embellished that the biological interactions like cellular uptake, the effect of adsorbed biomolecule and the ability of particles to translocate within the cells are size dependent. According to the literature survey, the shape of the nanoparticles play a vital role for their biomedical application, i.e., drug-loading capacity, high MTD, long circulation times, enhanced cellular uptake, large tumor accumulation, long tumor inhibition time and so on which encourage the researcher to synthesize non-spherical structures for future cancer therapeutics (Truong et al., 2015).

4.3. APPLICATION OF GOLD NANOPARTICLES FOR CANCER THERAPY

Gold nanoparticles have unique features that differ from their bulk counterparts. A high surface area to volume ratio in GNPs facilitates the presence of abundant active sites for the interaction with biomolecules at cellular level. Moreover, the intrinsic properties of GNPs can be precisely modulated without changing their chemical composition. In this context, gold nanorods and nanoshells can be fine-tuned at their optical resonance wavelength for shifting surface plasmon to the NIR region. At this wavelength, they absorb light most intensely and convert it to heat energy. From a clinical application point of view, this form of non-invasive heating of cancer tissue provides a new modality as cancer therapeutics (Dykman and Khlebtsov, 2012). At the same time, it is requisite to exploit the unique pathophysiologic and anatomic properties of cancer cells that differentiate them from normal healthy cells for the development of effective anticancer strategies. This can be achieved either by the passive accumulation of the therapeutic agent in tumors or active targeting of cancer-specific marker molecules. The passive accumulation of therapeutic agents in cancer cells is facilitated by enhanced permeation and retention (EPR) phenomena. In case of rapidly growing tumors, there is an increase in nutrition and oxygen demand promoting the localized growth of blood vessels and is called angiogenesis (Carmeliet and Jain, 2000). The newly formed blood vessels are immature channels with leaky inter-endothelial junctions and incomplete or absent basement membrane which facilitates extravasation of nanoparticles (< 400 nm) into tumor environment. This differential accumulation of nanoparticle in tumors and normal healthy cells justify their therapeutic values. Furthermore, gold nanoparticles can be functionalized with the conjugation of various tumor-specific marker molecules such as aptamers, peptides, antisense molecules (siRNA or miRNA) and antibodies enhancing the therapeutic index through target specific pathway (Tiwari et al., 2011). Thus gold nanoparticles present an efficient mean of cancer ablation.

4.3.1. Photothermal Therapy

Photothermal therapy is a non-invasive and inexpensive method of cancer treatment where tissues are exposed to a higher temperature to undergo apoptotic cell death. Since the tumor cells are highly metabolically active, they are more susceptible to this method in comparison to normal healthy cells. Among various nanoparticles, GNPs are more suitable for this application as their optical properties can be tuned by manipulating their shape, structure and lateral dimension to absorb maximum near-infrared light (650–1350 nm). Light energy of this wavelength is suitable for photothermal therapy because it can deeply and safely penetrate into distant localized tumor tissues passing through normal healthy cells (Nam et al., 2018). Moreover, GNPs can passively accumulate in tumor cells via EPR effect by exploiting the leaky and unorganized tumor vasculature. GNPs of ~10–200 nm in diameter when illuminated with light of wavelength smaller than the particle size, induce LSPR (which are localized oscillation of free conduction band electrons) (Nehl et al., 2006). Since these electrons are highly polarizable, they absorb the light and move under the influence of external field leading to a net charge difference at the surface of the nanoparticle. The LSPR depends largely on shape, size and surface properties of GNPs along with the dielectric properties of the surrounding environment (Zhang et al., 2014). The GNPs absorb and scatter the electromagnetic radiation which allows GNPs to be used in photothermal therapy. Due to surface plasmon resonance, heat is generated for which this is regarded as the plasmonic photothermal therapy. Additionally, GNPs efficiently accumulate in tumor tissues and generate sufficient heat upon illumination within a time frame of ~1 ps which is then efficiently transferred to a surrounding tumor through phonon-phonon interaction due to their high metallic conductance. The efficiency of various GNPs as a therapeutic module via photothermal therapy is because of their tunable plasmonic resonance, high photothermal conversion, surface functionalization to modulate biostability and tumor tissue encapsulation, etc. (Wang et al., 2013). Based on these requirements, uniquely shaped GNPs including gold nanospheres, nanorods, nanocages and nanoshells have been explored as suitable candidates for photothermal therapy in tumors (Kennedy et al., 2011). The GNPs that have to be used in photothermal therapy must absorb light within NIR wavelength range which is also termed as a therapeutic window in the electromagnetic spectrum because the light of this wavelength can safely penetrate into 2–3 cm depth of tissue to deeply rooted GNPs containing tumor tissues without affecting healthy cells.

For a conducive thermal therapy, the GNPs should possess high photothermal conversion efficiency which is largely determined by the morphological outline of the GNPs. It is worth mentioning that the photothermal conversion efficiency does not indicate the ability of nanoparticles to generate heat rather it describes how the nanoparticle disposes of the energy either by absorption or scattering. There are different (direct and indirect) methods to quantify the heat generated during the laser-nanoparticle interaction. The direct method includes laser heating in a water droplet, cuvette in vacuum and cuvette in standard room temperature and pressure whereas photoacoustic imaging provides an indirect method of measuring photothermal activity (Qin et al., 2016). According to a seminal report on the quantitative determination of photothermal heat generation between gold nanospheres and nanorods, gold nanospheres most efficiently radiate heat energy which satisfies the theoretical value (Qin et al., 2016). However, gold nanorods generate only 30% of the predicted theoretical value at absorption maxima. This report suggested less than 10% change for gold nanospheres and more than 70% reduction for gold nanorods in optical absorption due to change in polydispersity. So, in general, for a monodisperse solution of gold nanospheres with small diameter have higher conversion efficiency than those of larger ones whereas nanorods, nanocages, and nanostars are more efficient photothermal transducers than the spherical ones due to larger absorption surface area.

In addition to plasmon resonance tunability and photothermal conversion efficiency, the successfulness of photothermal therapy is really determined by the ability of GNPs to extravasate from the vasculature to tumor tissue (Huang et al., 2012). This is also influenced by the shape and size of GNPs that will determine the even distribution inside tumor tissue. The bio-stability and distribution of GNPs are distorted by robust uptake and clearance by the reticuloendothelial system (RES) which constitutes the combinatorial action of macrophages present in liver, spleen and lymph nodes (Chen et al., 2016). The intravenous introduction of GNPs promotes protein corona formation by the serum proteins. This alters the biological identity of GNPs that leads to uptake by macrophages and thus clearance from the blood system. Small GNPs (< 5 nm), although bypass the uptake by RES system, are usually cleared by the filtration unit of the kidney, whereas the particles with larger dimension facilitates their accumulation in spleen and liver. So it is necessary to determine the optimal size and shape for efficient penetration and retention depending on various tumor characteristics. This is worth mentioning that the EPR effect is heterogeneous in various tumor tissues. To circumvent the issues related to non-specific protein corona formation, reduced circulation time and biocompatibility, the GNPs are often functionalized with compounds like polyethene glycol (PEG) that mimic the highly crowded cellular environment (Ma et al., 2017) whereas the stoichiometry ratio of PEG and GNPs influence the reduction in protein corona formation.

Various GNPs affected photothermal therapy interferes with an array of molecular and physiological pathways leading to biological changes in the tumor tissue. These changes can cause apoptosis and enhance the efficacy of subsidiary chemotherapeutics by limiting the required dosage. On the contrary, this can maximize the cytotoxic effects of photothermal therapy itself. The mechanism behind photothermal therapy by GNPs relies mostly on increased temperature up to 40–45°C that is several degrees higher than the physiological temperature (37°C) around the GNPs localized area inducing an increase in permeability of vasculature and cell membrane present within the tumor environment (Spyratou et al., 2017). In addition, photothermal therapy can cause denaturation of intracellular proteins, impairment in DNA/RNA synthesis and drastic change in gene expression leading to cell death via apoptosis (Pérez-Hernández et al., 2014). However, at certain instances, high energy irradiation in photothermal therapy can result in undesirable side effects like cellular necrosis by triggering the release of damage-associated molecular patterns (DAMPs) which can eventually induce inflammation and secondary tumor growth (Figure 4.3) (Melamed et al., 2015). So, to induce apoptosis through photothermal therapy, it is required to have precise control over the concentration of GNPs as well as irradiation condition. Overall, there are many designing criteria that have to be manipulated to synthesize GNPs specific for photothermal ablation of cancer.

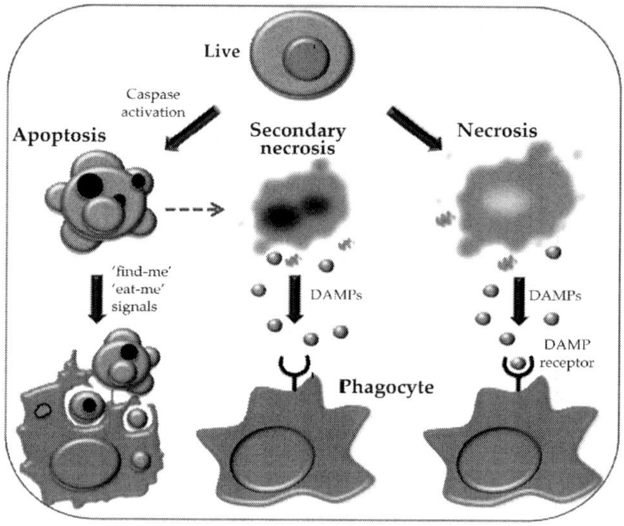

Figure 4.3: Apoptosis and necrosis are the two mechanisms of cell death induced by photothermal therapy. A cell undergoing apoptosis maintains its membrane integrity and produces "eat me" signals to mark the cell for phagocytosis without incurring inflammation. Apoptosis may lead to secondary necrosis in which the cell experiences loss of membrane integrity and release of damage-associated molecular patterns (DAMPs), but without activating phagocytosis. Conversely, primary necrosis destroys plasma membrane integrity to cause the release of DAMPs, leading to an inflammatory response (From Melamed et al., ACS Nano, 9, 6, 2015).

The effective utilization of distinct GNPs with various shape and size has been discussed below. The first report on GNPs mediated photothermal cancer therapy came through the work of Hirsch and co-workers in 2003. This group used core-shell containing 110 nm silica core surrounded by 10 nm gold shells and was functionalized with the coating of PEG. These gold nanoshells with λ_{max} at 820 nm, when incubated with SK-BR-3 human breast cancer cells for 1h and exposed to laser light (820 nm, 35 W/Cm^2 for 7 min), caused cell death. To validate the potential in *in-vivo* cancer therapy, Db 17-Prkd c SCID/J mice with 1 cm tumors (in hind legs) were treated with the gold nanoshells. Then, NIR irradiation (820 nm, 4W/Cm^2 for 6 min) caused elevated temperature (37.4 ± 6.6°C) in the tumor region as compared to laser alone (9.1 ± 4.7°C) which did thermal damage to tumors. In another work, the same group have demonstrated the gold nanoshell mediated thermal destruction of CT26.WT murine colon tumor tissue at the dorsal region of BALB/cAnNHsd mice (O'Neal et al., 2004). Here they irradiated (808 nm, 4 W/cm^2 for 3 min) the tumor tissues after 6 h of intraperitoneal administration of gold nanoshells. This resulted substantial increase in the temperature (T = 37.4 ± 6.6°C) which was sufficient enough to cause tumor cell death within 4–6 min. The tumors vanished after 10 days of initial treatment and the mice remained healthy for next 90 days in comparison to control mice which survived up to 10 days only. The interesting aspect of this report was the real-time magnetic resonance thermal imaging (MRTI) during the laser illumination. These gold nanoshells containing spherical silica cores with thin overlay of gold shells can be fine-tuned by playing with the core diameter to shell thickness ratio. For example, nanoshells with 120 nm core diameter and 15 nm thick shells showed maximum absorption at a wavelength of 800 nm and successfully used for photothermal therapy (Day et al., 2011). The clinical translation of gold nanoshells mediated photothermal cancer therapy was subsequently reported by Stern et al. (2008). They administrated gold nanoshells systemically into mice containing human prostate cancer xenograft. Gold nanoshells (at two different doses 7 µl/g and 8.5 µl/g) accumulated passively in tumor cells by the EPR effect and when irradiated for 3 min, it caused elevated temperature leading to tumor cell death. In a similar study of nanoshell mediated thermal ablation of intracranial tumors, an orthotopic canine model was used (Schwartz et al., 2009). Gold nanoshells were delivered by intravenous infusion and they passively accumulated in the intracranial tumors. This was followed by NIR irradiation at 808 nm using optical fiber, 3.5-W average dose for 3 minutes which elevated the temperature of tumors to 65.8 ± 4.1°C. The histopathological study of treated brain section confirmed the selective and localized thermal destruction caused by the gold nanoshells. Gold nanoshells have also demonstrated *in vivo* photothermal therapeutic efficacy in allografted tumors in dogs (Schwartz et al., 2009). In a recent report, photothermal chemotherapy of gold nanoshells-drug conjugates have been employed for enhanced tumor destruction (Liu et al., 2012). 160 nm gold-silica nanoshells containing chemotherapeutic Docetaxel and functionalized with transferring were injected into Balb/c mice having MCF-7 tumors. NIR irradiation (808 nm, 2 W/Cm^2 for 3 min) caused increased temperature by 15.2°C resulting tumor cell death. This treatment method using drug/nanoparticle conjugates was found to be more effective as compared to either drug or nanoparticle treatment. These large size silica based gold nanoshells, as opposed to other GNPs also provides opportunities in scattering based imaging in addition to photothermal therapy.

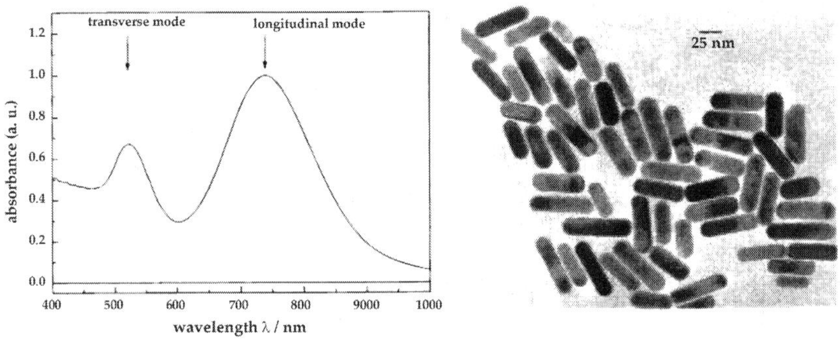

Figure 4.4: Left: Experimental UV-vis absorption spectrum of a gold nanorod sample with an average aspect ratio of 3.3. The band at 525 nm is referred to as the transverse plasmon resonance, while the one centered at 740 nm is called the longitudinal plasmon absorption. Right: TEM image of the same solution (From Link et al., Journal of Physical Chemistry B, 103, 3073, 1999).

Gold nanorods that were developed during the same period as gold nanoshells have been found to be one of the most effective photothermal anticancer therapeutic agents. They show some unique physiochemical properties like strong absorbance, NIR radiation scattering, and plasmon resonance tunability by simple manipulation in their aspect ratio (length divided by width), etc. (Huang et al., 2009). Their functionality is facilitated by small size (~ 10 nm × 50 nm), high absorption coefficient and narrow spectral bandwidth. Because of the rod shape, gold nanorods show two absorption peaks, one corresponding to the longitudinal and the other due to the transverse resonance (Figure 4.4) (Link et al., 1999). The longitudinal resonance can scan the visible and NIR wavelength whereas the transverse resonance occurs at ~520 nm. Recent studies on gold nanorods have shown greater absorption cross-section considering particle size and heating per unit gram of gold, which is around six times faster than in the case of gold nanoshells. Huang et al. (2010) theoretically and experimentally observed the temperature change by NIR radiation of PEGylated gold nanorods. They irradiated evenly distributed gold nanorods in 16 mm well plates with a continuous wave laser (800 nm, 20 W/cm^2 for 15 min) and monitored the temperature change at different positions of the well. Increase in optical density of gold nanorods lead to increased temperature (up to certain high concentration). Studies have been conducted to optimize photothermal conversion efficiency which is the ratio of absorption to extinction for greater photothermal heating efficiency. A seminal study on the photothermal conversion efficiency of gold nanorods was reported by Wang and his group (2012). An increase in temperature from 55°C to 76°C was observed when CTAB coated gold nanorods having width between 10–20 nm and length between 20–65 nm were irradiated (809 nm for 30 min) by continuous wave laser (LSPR at 810 nm). However, when the LSPR was shifted 50 nm towards lower or higher wavelength (blue shifted or red shifted respectively), temperature increase was less (<5°C) which was due to decreased NIR absorption and extinction. Gold nanorods were found to mediate cancer cell death in photothermal therapy via similar mechanism as observed in other GNPs. Tong et al. (2007) studied the mechanism of gold nanorod mediated photothermal destruction of KB tumor cells. They treated cancer cells with folate functionalized gold nanorods (aspect ratio, AR-3.7) for 6 h to facilitate cell surface binding or for 17 h to ensure intracellular encapsulation. The photothermal damage after NIR irradiation (765 nm, 6–60 W/cm^2 for 81.4 s) highlighted the effect of gold nanorod positioning on cell damage as surface-bound nanorods found to be more efficient photothermal transducers. Further ethidium bromide staining of the cells showed membrane cavity formation and actin staining revealed disintegration of actin filaments due to an influx of extracellular Ca^{2+}leading to membrane blebbing and finally apoptosis. In another study, photothermal heating of gold nanorods (AR- 4.0) was optimized by manipulating laser energy to induce apoptosis preferably over necrosis (Li et al., 2010). HeLa cancer cells were incubated with transferrin functionalized gold nanorods for 6 h. This was followed by irradiation with a femtosecond pulsed scanning laser (800 nm, 13.9–55.6 W/cm^2) and staining with Annexin V-Cy3.18 and propidium iodide to differentiate between apoptosis and necrotic cell death, respectively. It was observed that at 13.9 W/cm^2 and 9.5 W/cm^2, apoptosis occurred whereas increasing the power to 55.6 W/cm^2 resulted in necrosis. This report dictated the role of laser power in influencing physiological pathways of cell death suggesting to design GNPs for photothermal treatments to trigger apoptosis. More recent studies are focused on tumor cell-specific designed gold nanorods which are achieved by surface functionalization of gold nanorods with tumor cell targeting agents. Based on such concept, Ali et al. (2014) demonstrated increased photothermal therapy of two different sizes of GNPs (3.5 × 17 nm and 10 × 47 nm) functionalized with rifampicin. Rifampicin is an antibiotic that facilitates internalization of gold nanorods into human squamous carcinoma (HSC) cells. When the cells containing rifampicin coated gold nanorods were irradiated, smaller gold nanorods showed ~ 80% cell death in comparison to ~ 35% without rifampicin. However, larger gold nanorods seemed to be less efficient with ~ 65% cell death in presence of rifampicin and ~ 20% cell death without it. In a different approach, Wang et al., reported gold nanorods (AR-3.3) functionalized with an aptamer (Sgc8) linked to the photosensitizer chlorin e6 (2012). The aptamer could bind to protein tyrosine kinase 7, highly expressed in CCRF-CEM acute lymphoblastic leukaemia T-cells. Subsequent irradiation with white light for 2 h and then with NIR laser (812 nm for 10 min) elevated the cellular temperature from 25°C to 55°C leading to 60% cell death. In a recent study, Sun et al., used 64Cu labeled 25 nm arginine-glycine-aspartic acid peptide-modified gold nanorods against U87 human glioblastoma tumors embracing Athymic nude mice (2014). Upon irradiation, increased temperature by 27°C and tumor size decreased by 50 % was observed (Figure 4.5). Such gold nanorod based system proved to be highly beneficial for clinical applications as the therapy can be limited to the diseased region containing gold nanorods and also preventing unintended inflammatory responses and side effects in the associated body parts. Gold nanospheres are the final group of GNPs discussed for photothermal application in cancer therapy. Small size and comparatively simple method of synthesis made these GNPs as a potent agent for hyperthermia application. They have unique surface plasmon response and show absorption peak in the visible region at ~ 530 nm. The absorption peak largely depends on the size of gold nanospheres which can be precisely controlled by varying the citrate to gold ratio during synthesis (Piella et al., 2016). For instance, the absorption maxima show slightly red shift when the size of the gold nanospheres increases. This group of GNPs have high photothermal conversion efficiency owing to their surface Plasmon resonance. They absorb visible light efficiently and quickly convert into heat energy. Here the size also plays a crucial role as photothermal conversion efficiency changes from 50 % to 80 % with an increase in size (up to 20 nm) and concentration. El-Sayed and co-workers (2010) first reported the use of anti-epidermal growth factor receptor functionalized gold nanospheres for imaging and photothermal therapy in oral cancer cells. They used a continuous argon laser (514 nm, 13–76 W/cm^2 for 4 min) to irradiate gold nanospheres internalized HSC-3 and HOC-313 oral cancer cells demonstrating cancer cell death. Moreover, the localization of gold nanospheres inside the cancer cells also influences the efficacy of photothermal destruction. In a seminal report by Huang et al. (2010) gold nanospheres were functionalized with two different peptides first with an arginine-glycine-aspartate peptide (RGDRGDRGDRGDPGC) that could bind to alpha V beta 3 integrin receptors and limited its localization to cytoplasm and the second with the nuclear localization peptide sequence (GGGPKKKRKVGG). Human oral cancer cells (HSC-3) were incubated with this peptide functionalized gold nanospheres for 24 h followed by irradiation using continuous wave laser (514 nm, 25 W/cm^2 for 5 min) or nanosecond pulsed laser (532 nm, 0.3–0.8 mg for 6–7 ns). It was observed that high threshold energy (0.8 mJ) was required for cytoplasm targeted gold nanospheres to cause cancer cell death in comparison to nuclear-targeted one (0.3 mJ). However, a contrasting result was observed with laser irradiation where lower threshold energy (60 J) was required for cytoplasm targeted gold nanospheres in comparison to nuclear-targeted GNPs (210 J) for the photothermal destruction of cancer cells. Pitsillides et al. (2003) also demonstrated the use of gold nanospheres to treat cancer cells.In this report, lymphocytes were irradiated with short pulsed laser (20 ns) in presence of gold nanospheres which increased the plasma membrane permeability leading to cell death. Zharov et al. (2005) reported similar enhanced photothermal cancer therapy. In both these methods, the particle size largely influenced the photothermal effect. Gold nanospheres with size from 10–30 nm showed the strongest effect. Though the theoretical calculation indicated higher photothermal activities for larger gold nanospheres, they showed reduced effect (Zharov et al., 2005). In another study, a continuous wave argon ion laser operating at 514 nm was utilized to treat oral squamous carcinoma cells. In addition to these cell culture models, gold nanospheres also showed hyperthermal therapeutic efficacy in various *in vivo* models. Shao et al. (2013) have used PEG and TNF-α functionalized gold nanospheres (33 nm, λ_{max}-535) with pulsed laser irradiated (690 nm, 1 J/cm^2 for 1 min) and observed a tumor growth suppression in A/J mice with SCK mammary tumors. A robust method of combinatorial approach including photothermal photodynamic therapy was reported by Liu et al. (2013) where chlorine e6 loaded gold nanospheres (26 nm, λ_{max}-650–800 nm) were irradiated (671 nm, 2 W/cm^2 for 6 min) to cause tumor suppression in athymic nude mice with MDA-MB-435 breast cancer tumors (Figure 4.6). Another study by Xing et al. (2018) demonstrated the efficacy of PEG-coated gold nanospheres combined with laser light (532 nm, 10–30 J/cm^2 for 5–30 ms) to cause photothermolysis of blood vessels in tumor tissues taking a rat model. These potential GNPs provide a unique theranostic system to encounter modern days challenge in cancer ablative therapy.

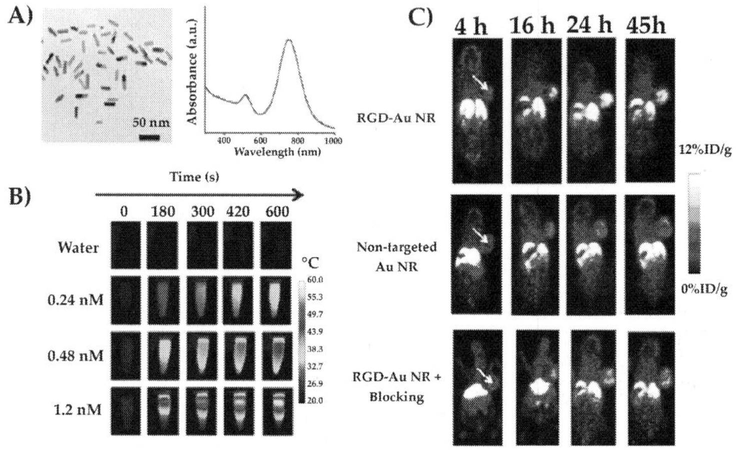

Figure 4.5: (A) TEM and UV-vis spectra of RGD-[64Cu]Au NR808. (B) Temperature mapping during the 808 nm laser irradiation (2 W/cm^2) of an RGD-[64Cu]Au NR808 suspension in distilled water by real-time thermal imaging using a thermal camera (FLIR). (C) Representative whole-body coronal PET images of U87MG tumor-bearing mice at 4, 16, 24, and 45 h after intravenous injection of gold nanorods (From Sun et al., ACS Nano, 8, 8438, 2014).

4.3.2. GNPs-siRNA/miRNA Conjugated Gene Therapy

Gene therapy is one of the modern day technologies with promising effect to deal with the treatment of various genetic and acquired diseases that are incurable by conventional drugs. RNA interference (RNAi) is found to be a remarkable tool in gene therapy through effective knocking down of target post-transcriptional mRNA or miRNA molecules that regulate specific molecular and biochemical pathways at the translational level. Small interfering RNA (siRNA) and antisense oligonucleotides (ASO) which are usually a short stretch of nucleic acid (20–25 nucleotide) represent the candidates that regulate target gene expression through RNAi pathway.

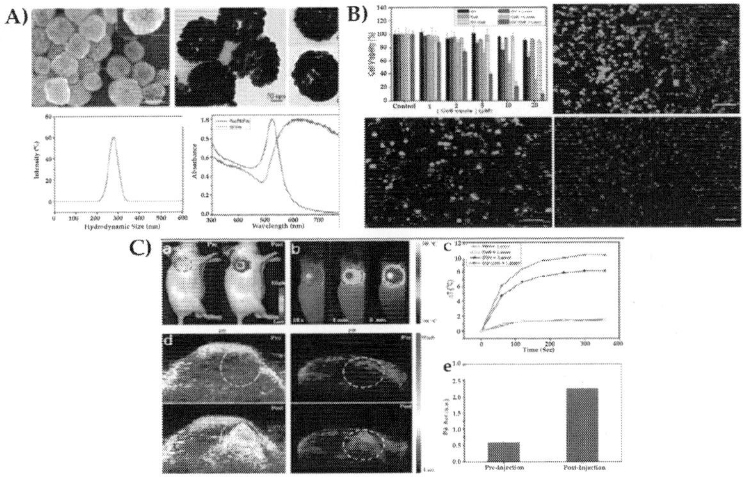

Figure 4.6: (A) SEM and TEM images of plasmonic gold vesicles (GVs) self-assembled from GNPs, Size distribution of GVs by DLS and UVvis spectra of BCP-tethered GNPs and GVs. (B) MDA-MB-435 cell viability at different concentrations of GVs, Ce6, and GV-Ce6 for 12 h at 37 C with or without. Fluorescence images of Calcein AM and Ethidium homodimer-1 costaining cancer cells incubated with GVs (36.8 μg/mL), Ce6 (10 μM), and GV-Ce6 at a Ce6 concentration of 10 μM for 12 h at 37 C after irradiation. Scale bars: 50 μm. (C) In vivo NIR fluorescence image of MDA-MB-435 tumor-bearing mice at preinjection and postinjection of GV-Ce6 (From Liu et al., ACS Nano, 7, 5320, 2013)

In fact, there are certain genes which drastically suffer alterations in their expression which leads to cellular transformation ultimately to tumor growth. ASOs function by hybridizing with target mRNA, signalling its cleavage and finally blocking target gene's expression by inhibiting translation of mRNA. Similarly, in siRNA mediated therapy, double-stranded RNA duplexes are transfected into the cells. Inside cellular environment, they are cleaved by an enzyme, Dicer and form a complex with RNA induced silencing complex (RISC) (Ariyoshi et al., 2017.). Then the sense strand is cleaved, antisense strand binds to the target complementary mRNA, consequently degrading it and finally leading to translational repression of vital oncogenic gene expression. The efficacy of these therapeutic molecules in modulating cancer biology has been extensively studied. In 2001, Elbashir et al., for the first time demonstrated the role of artificial siRNA in silencing mRNA in mammalian cells. However, their successfulness in various cancer types at the clinical trial level has yet to be established. The naked siRNA molecule is highly susceptible to RNase mediated degradation showing a lower residential period in biological fluids (Zheng et al., 2018). Moreover, the high anionic charge renders siRNA molecules to penetrate the cell membrane and also activate the non-specific immune system. When systematically administered, naked siRNA molecules underwent degradation by nuclease, and get accumulated in kidney and excreted in urine due to the small size of these nucleic acid molecules. To circumvent the aforementioned limitations, various delivery molecules like cationic peptides, liposomes and virus-based methodologies were introduced. Virus-based siRNA delivery methods suffered from drawbacks due to large-scale production, host immunogenicity and deleterious response. Among the diverse non-viral carrier molecules, lipid-based vehicles (lipofectamine and transfectant) showed high experimental improvements in delivering therapeutic genes because of the advantages associated with facile synthesis, ease in the functional modification, experimental repeatability and reduced cytotoxicity (He et al., 2018). However, harsh chemical reactions for manufacturing, low storage stability and limited in vivo tracking/ monitoring remain as issues to address. In this context, rigid inorganic nanoparticles (specifically GNPs) proved to be monumental in their unique optical, electric and magnetic properties. Apart from this, GNPs provide activated surface for covalent attachment of synthetic and biological compounds via interaction between sulphur and gold (S-Au binding) (Majoral et al., 2018). Especially for gene therapy application, oligonucleotides can be chemically modified to acquire thiolated groups which will help them to adhere on to the surface of GNPs without altering their biological activity (Hartl et al., 2018). On the other hand, unmodified oligonucleotides can bind to cationic GNPs through ionic interactions. Moreover, recent studies established the functionalization of GNPs by di-block oligonucleotides with adenines as anchoring moieties without opting for thiol modification (Zhu et al., 2016). This anchored nucleotide (adenine) mediated method is a cost effective synthesis process that provides an improved bio stability, surface density, hybridization thermodynamics and tunable kinetics (can be better regulated by varying poly A length and non-specific interaction with GNP surface is eliminated). For successful implementation of GNPs conjugated siRNA delivery, it is a necessary step to quantify accurately the amount of siRNA loaded on the nanoparticle surface for precise dose therapy. Additionally, the loading density is found to influence the cellular uptake of GNPs as high-density favours an increased cellular uptake. So, quantitative characterization is essential to establish the fundamental understanding of the interaction between siRNA coated GNPs and cell. There are several methods which have been used to measure quantitatively or qualitatively siRNA bound to GNPs. Dynamic light scattering and zeta potential measurements confirm siRNA loading since the binding of siRNA to GNPs changes the hydrodynamic radius and surface charge (Jensen et al., 2013).These methods provide the qualitative confirmation of siRNA molecules present on the surface of the gold nanoparticles. For quantitative characterization, a simple method based on spectrophotometer is used. In this method, the siRNA remained in solution after nanoparticle conjugation is determined by observing the absorbance at 260 nm and calculating the amount of siRNA using Beer-Lambert law (Raja et al., 2015). This is the simplest method assuming no loss during processing. Another method for quantitative measurement is a real-time polymerase chain reaction (RT-PCR) based approach (Kim et al., 2006). Here the quantitative detection of GNPs bound nucleic acid has been achieved by targeting the bound oligonucleotides with primers and probe binding sites, followed by amplification of the nucleic acid molecule by PCR and real-time measurements of fluorescence (emitted during the reaction). This method allows for quantitative measurements of surface oligonucleotides for a wide range of GNPs size even at their lower concentration. A method based on the use of fluorophore labeled oligonucleotides to quantify the loading has also been practised (McKenzie et al., 2009). Here quantification of fluorophore attached oligonucleotides is achieved by using the endonuclease, DNase I that catalyses the hydrolysis of phosphodiester linkage preferentially adjacent to the pyrimidine nucleotides in DNA backbone. Breakage of any phosphodiester bond in the DNA molecule would allow the release of the fluorescent molecule into the solution which can be monitored quantitatively. Melamed et al. (2017) reported a unique method to quantify the siRNA loaded on GNPs by using a fluorescence-based approach instead of modification with the fluorophores. In this method, the antisense RNA strand was first hybridized with sense RNA strand that was modified with a thiol group at 3' end. This thiol modification facilitated the loading of siRNA duplex on the GNP via gold-thiol interaction. For quantification, the antisense strand was separated from the GNP bound sense strand by adding urea. The antisense strand containing supernatant was then incubated with Oli Green® dye which was a weak fluorescent solution but produced strong fluorescence when bound to single-stranded nucleic acid. Comparing the fluorescence intensity with a standard curve of known antisense RNA concentration the amount of antisense could be deduced. Next the thiolated sense strand was released by breaking the thiol-gold bond using β-mercaptoethanol. The siRNA strands were collected and quantified similarly. This method provided a versatile tool for quantification of both sense and antisense strand attached on GNPs (Figure 4.7).

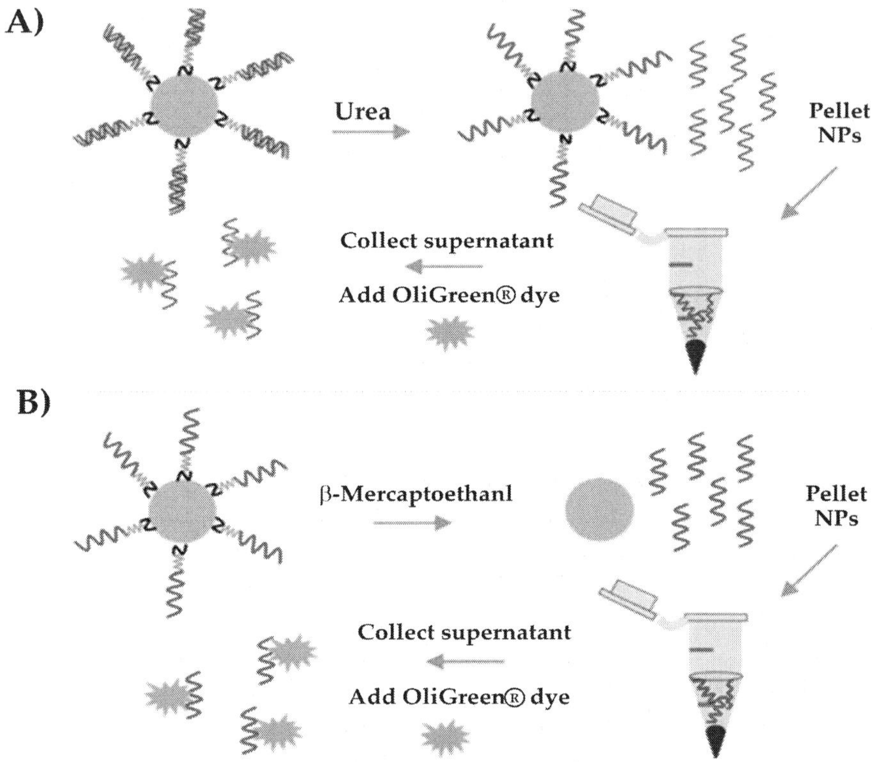

Figure 4.7: Schematic representing the procedure to quantify siRNA duplexes on nanoparticles (NPs). (A) Antisense RNA strands (blue) are dehybridized from sense RNA strands (red) on nanoparticles by incubating in 8 M urea at 45°C. The sense-loaded nanoparticles are pelleted by centrifugation, and the supernatant containing the antisense strands is collected. Antisense RNA strands within the supernatant are measured using components of the QuantiT™OliGreen® kit. (B) Sense RNA strands (red) are removed from the nanoparticles' surfaces by breaking the gold–thiol bond with β Mercaptoethanol. The nanoparticles are pelleted by centrifugation, and the supernatant containing the sense RNA strands is collected for analysis of RNA content with the Quant-iT™ OliGreen® kit. (From Melamed et al., Biomedical Nanotechnology: Methods and

Protocols, 1, 2017).

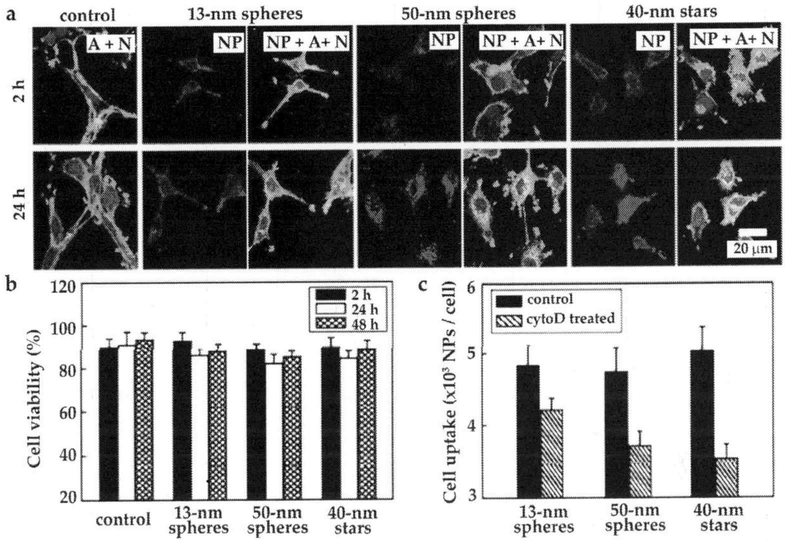

Figure 4.8: Intracellular distribution of NP-siRNA constructs and cell morphology changes induced by larger constructs. (a) Confocal fluorescence images of cells treated with PBS (control) or three different formulations of constructs with the same concentration of NPs (0.2 nM); The actin (A) cytoskeleton and nucleus (N) were stained with Alexa Fluor® 594 Phalloidin (green) and DAPI (blue), respectively; nanoconstructs were labeled with Cy5 (red). The scale bar is 20 μm for all images. (b) MTS assay of the cell viabilities of U87 cells treated with three different formulations of NP-siRNA/Cy5 (0.2 nM) or PBS as the control; (c) Cell uptake inhibition by CytoD for three formulations. Following 30-min pre-treatment with CytoD, U87 cells were incubated with NP-siRNA in presence of CytoD for another 30 min and cell uptake was measured by ICP-MS. As a control, U87 cells were also treated with constructs (From Zhu et al., Bioconjugate Chemistry, 28, 1791, 2017).

The physical properties of GNPs that influence cellular uptake are size, shape and surface chemistry. It has been reported that citrate-capped or protein functionalized 50 nm GNPs showed the highest uptake by cervical cancer cells (Chithrani et al., 2006). On the contrary, cellular uptake by breast cancer cells was highest for ultra-small (2 nm) GNPs functionalized with the drug, tiopronin (Huang et al., 2012). Taking shape of GNPs into account, it has been established that gold nanorods show lower cellular uptake efficiency as compared to gold nanospheres (Cho et al., 2011).

Recently, a study based on the influence of shape and size of GNPs on siRNA conjugated construct delivery revealed that, over a longer period of time (16–24 h), gold nanospheres (50 nm) and gold nanostars (40 nm) showed higher cellular uptake in comparison to gold nanospheres of 13 nm in human primary glioblastoma U87 cells (Figure 4.8) (Yue et al., 2017). Confocal fluorescence microscopy and immunostaining studies revealed that after 24 h, 13 nm gold nanospheres-siRNA conjugates were localized in endosomes where 50 nm spheres and 40 nm gold stars neither localized in endosomes or lysosomes. TEM images provided insight into aggregation of GNPs outside the endocytic vesicles. GNPs provide an excellent surface for modification with targeting ligands specific for tumor cells (Nicol et al., 2015). Most cancer cells overexpress certain cell surface markers like epidermal growth factor receptors (EGFR), E-cadherins, carcino embryonic antigens, epithelial cell adhesion molecule and urokinase-type plasminogen activator receptor, etc. Zhu and co-workers (2017) reported a pH-responsive siRNA delivery system based on gold nanoshells to target human hepatocellular carcinoma cells (HepG2).They functionalized the gold nanoshells with PEGylated 2-amino-2- deoxy- D- glucose [DG, a glucose analogue which can selectively bind to glucose transporter 1 (GLUT 1)]. GLUT 1 is highly expressed on the plasma membrane of tumors and act as a prognostic indicator of tumorigenesis. Moreover, DG can also activate cell autophagy and activate endoplasmic reticulum stress (ERS) response inhibiting the tumor growth. The DG-PEG complex was attached to lipoic acid (LA), lysine (Lys) and 9 poly D-arginine (9R) via a hydrazone bond (DG-PEG-LA-Lys-9R). The hydrazone bond acted as the pH-responsive linkage. This 9R/DG-gold nanoshell complex was used to deliver siRNA to target mRNA of cyclooxygenase-2 (Cox-2, a rate-limiting enzyme in the conversion of arachidonic acid to prostaglandin). Studies have revealed that over expression of cox-2 induces tumorigenesis (Majumder et al., 2018). This siRNA-gold nanoshell complex showed promising inhibition effect on tumor cell growth. The major hurdles that come across the siRNA delivery vehicle are cellular uptake, endosomal or lysosomal escape and successful dissociation of siRNA from the carriers. To construct such an efficient carrier molecule Zhao et al. (2012) assembled poly (allylamine hydrochloride) (PAH) and siRNA alternatively on the surface of GNP (11.8 ± 0.9 nm) by ionic layer-by-layer method. This structure was stabilized by denatured bovine serum albumin. Playing with the surface electric potential, it has been shown that GNP- PAH molecules with the surface potential of $\pm$ 25 mV exhibited highest cellular uptake efficiency and cytotoxicity in MCF-7 breast cancer cells. When siRNA targeting EGFR was conjugated with GNP-PAH and delivered it showed improved silencing effect by downregulating EGFR. Similar to siRNA, miRNAs are noncoding RNAs that are essential for normal cellular function and their deregulation lead to different diseases including cancer. The miRNAs are either down regulated or up regulated with respect to tumorigenesis. So, restoration of down-regulated miRNAs to the normal level (miRNA replacement therapy) has been practised to reverse disease condition partially or fully. On the contrary, exogenous antisense oligonucleotides complementary to target miRNA (antimiR) have been employed to down regulate over expressed oncogenic miRNAs (Rupaimoole and Slack, 2017). GNPs have been successfully employed for delivering miRNA and antimiR oligonucleotides. Ghosh et al. (2013) reported a unique method by utilizing the cysteamine functionalized GNPs to deliver chemically unmodified miRNAs into four different cancer cell types. They tested a different combination of GNP, miRNA and PEG to formulate the optimal polyelectrolyte complex. It was found that miR1-GNP10-S-PEG 0.5 complex showed no cytotoxicity, easily taken up by cancer cells through endocytosis and efficiently released the cargo oligonucleotides altering the target entity expression. In a different approach, oncogenic miRNAs which promote tumorigenesis has been knocked down using anti-miRs. According to a report, polyethyleneimine (PEI) modified and folate receptor-targeted PEGylated gold nanocages proved to be an efficient delivery vehicle to carry antimiR-181b into an SMMC-7721 tumor bearing nude mice (Huang et al., 2016). The miR-181b usually enhances the activity of matrix metalloproteinase 2 (MMP2) and matrix metalloproteinase 9 (MMP9) which cause growth, clonogenic survival, migration and invasion of hepatocellular carcinoma cells. The *in vitro* targets of miR-181b with this gold nanocage complex showed a drastic decrease in tumor growth. Moreover, NIR irradiation (808 nm, 1.25 W/cm^2 for 5 min) enhanced the antitumor efficacy of this GNP complex through photothermal therapy. Overall, GNPs proved to be the best delivery vehicle in gene therapy against cancer in addition to other antitumor properties which makes them suitable agents for a combinatorial therapy approach in cancer ablation.

4.3.3. CT Imaging

The primary objective in developing efficacious nanomedicine in cancer therapy is the competency to exhibit multidimensional function like imaging, targeting and monitoring in *in vivo* conditions by a single entity. Though many complex molecules and nanoparticles have been studied to achieve this goal, the unique physical and optical properties of GNPs led to utilize a single platform to perform multiple functions. GNPs have shown successful outcomes in photothermal therapy, drug delivery, as a contrasting agent for optical computed tomography (CT), X-ray and photoacoustic imaging applications, etc. CT is the most common imaging method in medical science and the inclusion of GNPs has expanded this technology beyond its present capabilities. CT provides greater visualization of bone structures because of the inherent contrast between electron dense bones and the surrounding permeable soft tissues. However, it is difficult to differentiate between the soft tissues with similar density using CT. The principle behind CT is that, different tissues contribute different degree of X-ray attenuation which is based on the following equation (Sayyed et al., 2016):

$$I = I_0 e^{-\mu x} \quad (1)$$

where 'I_0' is the intensity of incident X-ray, 'I' is the transmitted X-ray intensity, 'x' is the thickness of the absorbing medium and 'μ' is the mass attenuation coefficient. The prime factor influencing the μ value is the photoelectric effect which is proportionally dependant on the third power of the atomic number of the constituent material. So, to get a good contrast in CT images, it is essential to select a material having a high atomic number. The currently used contrasting agent in CT is iodine (atomic number-53) which has lower atomic number than gold (atomic number-79). So gold can induce greater X-ray attenuation. Furthermore, iodine provides shorter imaging period because of its rapid clearance by the kidneys. On the contrary, GNPs are inert to biomolecules, can overcome the biological barriers and confine its presence to the intravascular space for prolonged time (Curry et al., 2014). By taking advantage of passive accumulation of GNPs in cancer cells through trans-endothelial pores, GNPs with certain size range has been explored as the next generation CT contrast agents. Besides its high atomic number and inertness in a biological environment, ease in fabrication and surface functionalization of GNPs as discussed in earlier sections provides an advantage over currently used hydrophilic poly-iodinated compounds (Gaikwad et al., 2018). For example, Wang et al. (2010) reported the synthesis of size-tunable, highly stable, monodisperse and hydrophilic GNPs by regulating the nucleation and growth of gold nucleus. They synthesized homogenous Au^0 nuclei and maintained uniform growth rate for every Au nucleus resulting monodisperse solution of GNPs with 6.6 nm diameter. The GNP showed yield at a higher percentage (90%) and storage stability for months as they were stabilized by ω-carboxylic acid functionalized alkanethiolate ligands. When the X-ray attenuation of these water-soluble GNPs was examined on a CT scanner, they exhibited much greater X-ray attenuation than commercial iodinated contrast agent, Ousu at the same concentration. The mechanism behind such enhanced X-ray attenuation is based on the higher theoretical X-ray absorption coefficient of gold in comparison to iodine (5.16 and 1.94 cm^2/g respectively at 100 KeV) (Wang et al., 2010). Furthermore, increased X-ray attenuation was observed with decrease in GNP core size from 12 to 6.6 nm. This is attributed to the larger gold surface area of small GNPs. So, GNPs with smaller dimensions are preferred for highly efficient CT imaging. Some of the most important characteristics of a smart contrasting agent are lack of toxicity, a prolonged residential period in the blood and ease in elimination from the body. In this regard, GNPs were found to be the most appropriate modality as they were cleared through renal excretion thus avoiding the fatal accumulation of metal ions in different tissues. In addition to this, functionalization of GNPs surface for active targeting of cancer cells requires minimal dosages of contrasting agents. Focusing on these objectives, highly stable and functionalized GNPs have been reported to elucidate the main issues involved in *in vivo* bio-distribution. Silvestri et al. (2016) synthesized GNPs of various shape by fine-tuning the surface coatings and core morphology through 'one-pot' method. Quasi-spherical, spherical and anisotropic PEGylated GNPs were fabricated in this 'one-pot' reaction by modulating the relative amount of the contributing agents. Micro-CT imaging after intravenous administration of GNPs showed higher X-ray absorption and prolonged persistence in blood by anisotropic GNPs which pave the way to develop innovative contrasting agents for CT imaging.

4.3.4. Combinatorial Therapy

Cancer is a complex disease manifested by the action of multiple genes depending on the tissue types. So it is always a challenge to develop a single approach to control the network of multiple pathways that depicts the fate of cancer cells. However, recent studies on understanding the biochemical and molecular mechanism of tumorigenesis allows the evolution of cancer therapy from unidirectional approach to combined therapeutics. Use of multiple drug units with different molecule targets will more efficiently eliminate cancer cells which show resistance to one of the drugs present in the drug mixture (Kyriazi et al., 2018). Moreover, team of drugs with different mechanism of action in the most appropriate way for combining therapy such as mix up of cytotoxic anticancer drugs, biotherapies (antibody and gene therapy, etc.) and imaging reagents. In this regard, GNPs allow the loading of different molecules for cancer cell imaging and therapeutic showing their multi functionality. For example, simultaneous utilization of gold nanorods for CT imaging and photothermal cancer therapy in both *in vitro* and *in vivo* conditions has been established. Hybrid monosaccharide imprinted gold nanorods are prepared and studied their efficacy in cancer cell imaging and targeting photothermal therapeutics. When the solution containing sialic acid (SA) imprinted gold nanorods was irradiated (750 nm, 1 W/cm^2 for 6 min), an increase in temperature from 23°C to 60°C occurred. Moreover, the SA-imprinted fluorescein isothiocyanate doped gold nanorods could distinguish between human hepatoma carcinoma cells (HepG-2) and normal hepatic cells (L-02). The HepG-2 cells exhibited strong fluorescence in comparison to L-02 cells in the presence of gold nanorods. To evaluate the plasmonic effects, both HepG-2 and L-02 cells were incubated with 0.1 mg/mL SA-imprinted gold nanorod followed by irradiation with 750 nm laser for different duration. As a result, HepG-2 cells were found dead over 6 min of irradiation where L-02 cells showed higher viability (> 85%). These SA-imprinted gold nanorods exhibited a significant decrease in tumor tissue in HepG-2 tumor bearing mice (Yin et al., 2017). This study opened a new access to design cancer specific plasmonic GNPs for real time imaging and photothermal therapy of cancer.

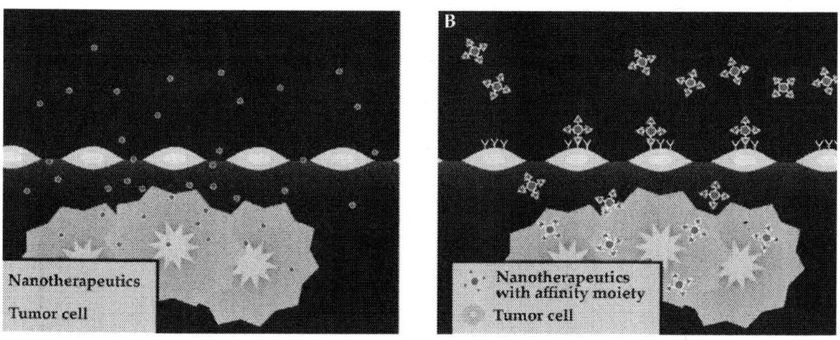

Figure 4.9: Schematic views of nanoparticle delivery. (A) Nanoparticle delivery by passive targeting. In this mode, nanoparticles pass through the leaky vasculature and enter tumor tissues. (B) Nanoparticle delivery by active targeting. In this mode, the surface of nanoparticles is coated/conjugated with an active targeting moiety, such as peptide, antibody, or aptamer. Binding of the targeting moiety with cell surface molecules facilities tumor tissue entry of the nanoparticle (From Shen et al., Breast Cancer Research, 15, 205, 2013).

Hyperthermia treatment by GNPs often encounters problems by the acquisition of thermo-tolerance against heat stress. In such cases, photothermal therapy can be coupled with other methods like gene delivery to get synergistic effect against cancer. Successful attempts have been made in light triggered gene delivery to induce bimodal cancer treatments (Vijayaraghavan et al., 2015). Here, ultralow dose of NIR irradiation is applied which results in the production of reactive oxygen species (ROS). At the same time, silencing of superoxide dismutase 1 (SOD 1) resulted complete destruction of deep tissue buried tumors. Though, only RNAi based treatments have partial antitumor effects, but is more potent and long lasting therapy when combined with chemotherapy. Shen et al. (2013) have used siRNA targeting MDR1 gene in tumors sensitizing to chemotherapeutic drugs. They have broadly studied the targeted delivery of nanoparticles by both active and passive targeting process (Figure 4.9). During the last decade, many multimodal GNPs have been designed to target cancer cells. These GNPs provide nano platforms which can be exploited for combination therapy.

4.4. GREEN SYNTHESIZED GNPS IN CANCER THERAPY

Gold nanoparticles play a crucial role in cancer diagnosis and therapeutics due to its diversified physicochemical properties and recent studies have shown that green synthesized gold nanoparticles can inhibit the growth of cancerous cells in a time and dose-dependent manner. Extensive research on the usage of biogenic gold nanoparticles in cytotoxicity studies was reported suggesting the efficacy of GNPs as an anticancer agent. GNPs synthesized using bacitracin reported to show less cytotoxicity in HepG2 and HEK293 cell lines suggesting their excellent biocompatibility from bio safety point of view (Li et al., 2018). *Acalypha indica*leaf extract mediated biosynthesized GNPs exhibit significant cytotoxic effect against human breast cancer cell lines (Krishnaraj, 2014). The in-vitro cytotoxicity assay of casein hydrolytic peptide-mediated GNPs demonstrated low viability towards the human prostate cancer cell line as compared to the normal mouse hepatocyte cell line. Chaga mushroom mediated GNPs was found to inhibit the growth of human breast cancer cell line and human stomach cancer cell line *in vitro* in a dose-dependent manner where most of the particles were spherical, some had triangle and rod shapes with sizes ranging from 11.0 to 37.7 nm (Lee et al., 2015). Aqueous leaves and stem extract of *Hibiscus sabdariffa* based spherical shaped gold nanoparticles were found to be having selective cytotoxic activity against glioblastoma multiforme cell line in normal and hyperglycemic condition (Mishra et al., 2016). The above reports suggest the usage of biosynthesized GNPs as potential anticancer agents. Zebrafish embryo model was screened for nanotoxicity of GNPs modified with folic acid and coupled with 6-mercaptopurine was found to be an efficient anticancer nano conjugate system for drug delivery in HEp-2 cell line (Ganeshkumar et al., 2012). The above reports suggest that green GNPs plays a crucial role in *in vitro* condition by serving as a better candidate for the treatment of cancer cells. Cytotoxicity of green synthesized GNPs in different cell lines*in vitro* reported extensively. Table 4.1 gives a brief idea about some more bio-mediated gold nanoparticle synthesis, optimization using various parameters, morphology and cytotoxic effects on different cell lines.

Table 4.1: Some other Studies on Biogenic Gold Nanoparticles as Anticancer Agents

Biomediated reducers	**Optimization parameters**			**Morphology**		**Cancer cell line**	**References**
Plant and part	**pH**	**Temp (°C)**	**Incubation time**	**Size (nm)**	**Shape**		
Juglans regia Green Husk		45	10 min	14.32±3.24	Spherical	3T3, HT-29	Izadiyan et al.2018
Zataria multiflora Leaves	Neutral	room	rapid	10–42	Spherical	HeLa	Baharara et al., 2016
Tabernaemontana divaricata Flowers	Neutral	room	rapid	100	Spherical	MCF-7	Preetam Raj et al., 2016
Peltophorum pterocarpum Leaves	Neutral	room	rapid	54.2	Spherical	A549,B16	Mukherjee et al., 2016
Barleria cristata Leaves	Neutral	room	rapid	40	Rod, round, triangular	HeLa	Basker et al., 2016
Carica papaya Leaves	Neutral	60	5 min	15–28	Spherical	MCF-7 HepG2	Muthukumar et al., 2016
Catharanthus roseus Leaves	Neutral	60	5 min	15–28	Spherical	MCF-7 HepG2	Muthukumar et al,2016
Argemone,exicana Leaves	Neutral	room	rapid	Polydis-persed	Spherical	MCF-7	Varun & Sudha, 2015
Argemone mexi-cana Leaves	Neutral	room	rapid	Polydis-persed	Spherical	HepG2	Sellappa et al,2015
Artabotrrys hexa-petalus Whorl plant	Neutral	room	rapid	~30	Spherical	MCF-7	Priya & Iyer, 2015

Comellia sinensis Whorl plant	Neutral	room	rapid	~30	Spherical	MCF-7	Priya & Iyer, 2015
Coriandrum sativum Whorl plant	Neutral	room	rapid	~30	Spherical	MCF-7	Priya & Iyer, 2015
Phyllanthus amarus Whorl plant	Neutral	room	rapid	~30	Spherical	MCF-7	Priya & Iyer, 2015
Mentha arvensis Whorl plant	Neutral	room	rapid	~30	Spherical	MCF-7	Priya & Iyer, 2015
Syzygium aromaticum Whorl plant	Neutral	room	rapid	~30	Spherical	MCF-7	Priya & Iyer, 2015
Mimusops elengi Whorl plant	Neutral	room	rapid	~30	Spherical	MCF-7	Priya & Iyer, 2015
Punax ginseng Root	Neutral	room	rapid	16.2	Spherical	SiHa, CaSki	Mukherjee et al.,2015
Diospyrus ferrea Whorl plant	Neutral	room	rapid	7–90	Spherical, rods	HepG2	Verrub, 2015
Gymnema sylvestre Leaves	Neutral	room	30 min	26	Spherical	Hep2, HaCaT	Nakkala et al.,2015
Hygrophilla spinosa Leaves	Neutral	room	10 min	50–80	Spherical triangular	HeLa	Koperuncholan, 2015
Gymnema sylvestre Leaves	Neutral	room	30 min	72.8	Spherical	HT29	Arunachalam et al.,2014
Syzygium aromaticum Fruits	Neutral	room	1 min	2–20	Spherical	SU-DHL-4	Parida et al.,2014b
Theobroma cacao Cocoa seeds	Neutral	37	15 min	150–200	Anisotropic spherical	A431, MDA-MB231, L929, NIH-3T3	Fazal et al.,2014
Xanthan gum Gum	Neutral	80	2 hr	15–20	Spherical	A549	Pooja et al.,2014
Olea europaea Leaves	Neutral	room	rapid	20–100	Spherical	K562	Parida et al.,2014a
Abelmosschus esculentu Pulp	Neutral	room	18 hr	14	Spherical	PBL Jurkat	Mollick et al.,2014
Achyranthes aspera Seed epicotyls	Neutral	room	1 hr	9	Spherical, hexagonal, triangular	HeLa	Alagar et al.,2014

Curcuma pseudo-montana Rhizome	Neutral	room	30 min	20	Spherical	T47D	Muniyappan & Nagarajan,2014
Couroupita guianensis Leaves	Neutral	room	5 min	Polydispersed	Spherical, triangular, tetragonal, pentagonal	HL-60	Geetha et al.,2013
Couroupita guianensis Flowers	Neutral	room	5 min	7–48	Spherical, triangular, tetragonal, pentagonal	HL-60	Geetha et al.,2013
Olax scandens Leaves	Neutral	room	25 min	5–15	Spherical, rod	A549, COLO 205, MCF-7	Mukherjee et al.,2013
Lantana montevidensis Leaves	Neutral	room	~1 hr 30 min	70.16	Spherical	A549, B16, CHO, MCF-7	Mukherjee et al.,2012
Comellia sinensis Leaves	Neutral	25	5 min	15–45	Spherical	PC-3, MCF-7	Nune et al.,2008
Bauhinia tomentosa Leaves	Neutral	room	rapid	31	Spherical	HEp-2	Mukundan et al.,2005
Punica granatum Fruit peel	Neutral	room	1 min	70	spherical	MCF-7	Gao et al.,2004

Protein Tyrosine Phosphatase 1B inhibitory activity evinced by biosynthesized GNPs using *Psidium guajava* has been reported. These GNPs were found to be stable at various physiological medium like human serum albumin, bovine serum albumin, histidine, cysteine, and buffers of different pH indicating the higher in vitro stability to take part in therapeutic applicabilities (Basha et al., 2010). Studies have revealed that biomediated GNPs were associated with production of reactive oxygen species (ROS) resulting oxidative stress in cancer cells. According to Mukherjee et al. (2015), *Lantana montevidensis* leaf extract mediated GNPs generated oxidative stress by producing ROS in lung cancer cells. In this study, researchers observed that ROS production may be due to the capping behavior of phyto-constituents on the surface of nanoparticles. Some of the anticancer drugs like oltipraz are available in market which has the mechanistic function of free radical generation leading to chemoprevention of cancer (Velayutham et al., 2005). ROS mediated cell damage can be prevented by glutathione (GSH) that on oxidation converted to GSSH (glutathione disulfide) in presence of free radicals. These oxidative stresses due to ROS production and oxidation of GSH are considered to play important role in anti-cancer activity (Liu et al., 2011; Mukherjee et al.,

2014). Gum karaya based GNPs as anticancer drug carrier for gemcitabine hydrochloride showed better anticancer activity, colony formation inhibition and ROS induction against human lung cancer cells than bare gemcitabine hydrochloride (Pooja et al., 2015). Induction of apoptosis through caspase 3 and 9 activation and protein p53 upregulation is another pathway to study anticancer mechanism of biogenic GNPs. cancer cells treated with GNPs were found to be damaging the nuclei leading to condensed and fragmented chromatin. Green synthesiszed GNPs using *Zataria multiflora* leaves can inhibit the cell proliferation in a dose dependent manner and could induce apoptosis in cells through caspase activation as reported by Baharara et al., in 2016. Biomediated GNPs are also used for photo-thermal diagnosis of cancer. Anisotropic GNPs synthesized using cocoa seeds were used as photo thermal agents for *in vitro* biocompatibility studies in both cancerous and normal human cell line (Fazal et al., 2014). GNPs synthesized using *Jatropa curcas* seed shells were loaded with curcin were found to be very efficient for selective ablation of glioma on NIR laser irradiation (Mohamed et al., 2014).

In vitro and *in vivo* tumor imaging using fluorescent gold nanocluster was reported where GNPs were synthesized using egg white, egg yolk and serums as both reducing and stabilising agents. The synthesized GNPs showed sublime biocompatibility for cancer cells tracking (Li et al., 2015). *Olax scadens* leaves mediated GNPs are found to be having self-fluorescence behavior and used as both diagnostic and therapeutic agent (Mukherjee et al., 2013). This fluorescence ability was also maintained inside cancer cells when treated withA549 and MCF-7 cell lines. Fazal et al., synthesized GNPs using cocoa seeds which exhibit X-ray contrast in computed tomography (CT). Nune et al., in 2009, reported cinnamon coated GNPs as efficient CT/optical contrast-augmentation agents for cancer diagnosis. Recently, mannan (polysaccharide) based gold nanoparticles were used as X-ray CT contrasting agent for targeting popiteal lymph node (Uthaman et al., 2018). The above advanced characteristics of biosynthesized GNPs opened a new direction for utilization of GNPs in targeted cancer diagnosis. During drug delivery, bio-mediated GNPs accumulate more in sub-G1 phase and G0/G1 phase in comparison to other phase of cell cycle. Recent study revealed that the GNPs synthesized using *Butea monosperma* and *Peltophorum pterocarpum* leaves coated with doxorubicin accumulated in sub-G1 phase of B16F10 and in G0/G1 of A549 cell line, respectively leading to inhibition of cancer cell growth (Patra et al., 2015; Mukherjee et al., 2016).

Drug resistance can be overcome by nanoparticle conjugated drug delivery system. In this regard, a recent study reported using *P.pterocarpum* leaf extract mediated GNPs loaded with doxorubicin were found to show better result in comparison to baredoxorubicin in A549 and B16F10 cell lines (Mukherjee et al., 2016). GNPs synthesized using peanut in the presence of cysteamine were attached to plasmid DNA for enhancing drug delivery applications (Raju et al., 2014). *Punica granutum* fruit peels extract mediated biocompatible GNPs showed in vitro cytotoxicity against breast cancer cells. In this report, GNPs conjugated with 5-Fu (5-Fu@PAu) showed better therapeutic efficacy compared to only 5-Fu during in vitro release of drug (Ganeshkumar et al., 2013). Green synthesis of GNPs using citrate and *Zingiber officinale* extract adsorb very low amount of protein. These nanoparticles were found to be non-reactive at physiological condition without activating platelet and complement system in human blood. Citrate capped GNPs were aggregated and showed less stability when compared with GNPs synthesized with *Z. officinale* extract (Kumar et al., 2011).

4.5. CONCLUSION

The increased appearance of cancer, limitation in the conventional therapy like high cost and unwanted spurious side effects have raised several challenges for the researchers to develop alternative biocompatible and cost effective nanomedicine for cancer ablation. In this regard, nanotechnology promises in translating the knowledge about cancer from molecular approach to clinical diagnostic and therapeutic application. Multifunctional nano platforms are the key advantage of nanoparticles over conventional approaches. Cancer cell targeting ligands, imaging probes, chemotherapeutic drugs, gene silencing tools and other functional anticancer moieties can all be integrated into the nanoparticle to deliver targeted molecular imaging and therapeutic applications in cancer. GNPs which have been established for their biocompatibility, tuneable optical properties and ease of surface functionalization show a greater promise as contrast agents in molecular CT imaging, absorptive agents in photothermal therapy and delivery agent in transferring anticancer drugs. Owing to their morphological nano dimension, GNPs can easily penetrate into cells and preferentially accumulate in cancer cells due to the EPR effect. Moreover, their inherent and conjugated affinity towards proteins, nucleic acid and drugs can be effectively utilized to actively target cancer cells overexpressing cell surface receptors. The biosynthesized multifunctional GNPs provide an alternative method for developing eco-friendly and cost effective nanomedicine and likely to revolutionize the face of nanomedicine in the coming decades in cancer therapeutics. The early promising outcomes and continued inter-disciplinary interactions between researchers from optical, manufacturing engineers, biologists, chemists and clinicians are likely to harbinger a new era of nanomedicine application in oncology.

ACKNOWLEDGMENTS

Authors like to thank the support of Council of Scientific and Industrial Research, Academy of Scientific and Innovative Research, Department of Science and Technology and Ravenshaw University, India.

REFERENCES

1. Abadeer, N. S., & Murphy, C. J. (2016). Recent progress in cancer thermal therapy using gold nanoparticles. *The Journal of Physical Chemistry C, 120*, 4691–4716.
2. Abadeer, N. S., Brennan, M. R., Wilson, W. L., & Murphy, C. J. (2014). Distance and Plasmon Wavelength Dependent Fluorescence of Molecules Bound to Silica-Coated Gold Nanorods. *ACS Nano, 8*, 8392–8406
3. Ahmed, S., & Ikram, S. (2016). Biosynthesis of gold nanoparticles: a green approach. *Journal of Photochemistry and Photobiology B: Biology, 161*, 141–153.
4. Alagar, M., Anand, M., Selvaraj, V., & Ranjitha, J. (2014). Green phyto-synthesis of gold nanoparticles using *Achyranthes asperalinn* seed-epicotyls layer extracts and its anticancer activity. *Asian J Pharm C lin Res, 7*,136–139.
5. Ali, M. R., Panikkanvalappil, S. R., & El-Sayed, M. A. (2014). Enhancing the efficiency of gold nanoparticles treatment of cancer by increasing their rate of endocytosis and cell accumulation using rifampicin. *Journal of the American Chemical Society, 136*, 4464–4467.
6. Ariyoshi, J., Matsuyama, Y., Kobori, A., Murakami, A., Sugiyama, H., & Yamayoshi, A. (2017). Effective Anti-miRNA Oligonucleotides Show High Releasing Rate of MicroRNA from RNA-Induced Silencing Complex. *Nucleic Acid Therapeutics, 27*, 303–308.
7. Arunachalam, K. D., Arun, L. B., Annamalai, S. K., & Arunachalam, A. M. (2014). Biofunctionalized gold nanoparticles synthesis from Gymnema sylvestre and its preliminary anticancer activity. *Int. J Pharm Sci, 6*, 423–430.
8. Baharara, J., Ramezani, T., Divsalar, A., Mousavi, M., Seyedarabi, A 2016. Induction of apoptosis by green synthesized gold nanoparticles through activation of caspase-3 and 9 in human cervical cancer cells. *Avicenna Journal of Medical Biotechnology, 8*, 75–83.
9. Basha, S. K., Govindaraju, K., Manikandan, R., Ahn, J. S., Bae, E. Y., & Singaravelu, G. (2010). Phytochemical mediated gold nanoparticles and their PTP 1B inhibitory activity. *Colloids and Surfaces B: Biointerfaces, 75*, 405–409.

10. Baskar, S., Selvan, G., Anbarasu, R., & Raja, V. (2016). Green synthesis of gold nanoparticles (Au-NPs) using Barleria cristataand study their pharmacological applications. *World J Pharm Res*, *5*, 1072–1085.
11. Bethell, D., Brust, M., Schiffrin, D. J., & Kiely, C. (1996). From monolayers to nanostructured materials: an organic chemist's view of self-assembly. *Journal of Electroanalytical Chemistry*, 1996, *409*, 137–143.
12. Biju, V. (2014). Chemical modifications and bioconjugate reactions of nanomaterials for sensing, imaging, drug delivery and therapy. *Chemical Society Reviews*, *43*, 744–764.
13. Brust, M., Bethell, D., Kiely, C. J., & Schiffrin, D. J. (1998). Self-assembled gold nanoparticle thin films with nonmetallic optical and electronic properties. *Langmuir*,*14*, 5425–5429.
14. Brust, M., Walker, M., Bethell, D., Schiffrin, D. J., & Whyman, R. (1994). Synthesis of thiol-derivatised gold nanoparticles in a two-phase liquid–liquid system. *Chemical Communications*. 801–802.
15. Cai, W., Gao, T., Hong, H., & Sun, J. (2008). Applications of gold nanoparticles in cancer nanotechnology. *Nanotechnology, science and applications*, *1*, 17.
16. Carmeliet, P., & Jain, R. K. (2000). Angiogenesis in cancer and other diseases. *Nature*, 2000, 407, 249.
17. Chen, G., Roy, I. Yang, C., & Prasad, P. N. (2016). Nanochemistry and nanomedicine for nanoparticle-based diagnostics and therapy. *Chemical reviews*, *116*, 2826–2885.
18. Chen, J., Wang, D., Xi, J., Au, L., Siekkinen, A., Warsen, A., & Li, X. (2007). Immuno gold nanocages with tailored optical properties for targeted photothermal destruction of cancer cells. *Nano Letters*, *27*, 1318–1322.
19. Chen, W., Zhang, S., Yu, Y., Zhang, H., & He, Q. (2016). Structural-Engineering Rationales of Gold Nanoparticles for Cancer Theranostics. *Advanced Materials*, *28*, 8567–8585.
20. Chithrani, B. D., Ghazani, A. A., & Chan, W. C. (2006). Determining the size and shape dependence of gold nanoparticle uptake into mammalian cells. *Nano Letters*, *6*, 662–668.
21. Cho, E. C., Zhang, Q., & Xia, Y. (2011). The effect of sedimentation and diffusion on cellular uptake of gold nanoparticles. *Nature*

Nanotechnology, *6*, 385.

22. Coffey, J. C., Wang, J. H., Smith, M. J. F., Bouchier-Hayes, D., Cotter, T. G., & Redmond, H. P. (2003). Excisional surgery for cancer cure: therapy at a cost. *The Lancet Oncology*, *4*, 760–768.
23. Connor, E. E., Mwamuka, J., Gole, A., Murphy, C. J., Wyatt, M. D. (2005). Gold nanoparticles are taken up by human cells but do not cause acute cytotoxicity. *Small*, *1*, 325–327.
24. Curry, T., Kopelman, R., Shilo, M., & Popovtzer, R. (2014). Multifunctional theranostic gold nanoparticles for targeted CT imaging and photothermal therapy. *Contrast Media & Molecular Imaging*, *9*, 53–61.
25. Daraee, H., Eatemadi, A., Abbasi, E., Fekri Aval, S., Kouhi, M., & Akbarzadeh, A. (2014). Application of gold nanoparticles in biomedical and drug delivery. *Artificial Cells, Nanomedicine, And Biotechnology*, 2014, 44, 410–422.
26. Day, E. S., Thompson, P. A., Zhang, L., Lewinski, N. A., Ahmed, N., Drezek, R. A., Blaney, S. M., & West, J. L. (2011). Nanoshell-mediated photothermal therapy improves survival in a murine glioma model. *Journal of Neuro-Oncology*, *104*, 55–63.
27. Dickerson, E. B., Dreaden, E. C., Huang, X., El-Sayed, I. H., Chu, H., Pushpanketh, S., & El-Sayed, M. A. (2008). Gold nanorod assisted near-infrared plasmonic photothermal therapy (PPTT) of squamous cell carcinoma in mice. *Cancer letters*, *269*, 57–66.
28. Du, L., Jiang, H., Liu, X., & Wang, E. (2007). Biosynthesis of gold nanoparticles assisted by *Escherichia coli* DH5α and its application on direct electrochemistry of hemoglobin. *Electrochemistry Communications*, *9*, 1165–1170.
29. Dykman, L., & Khlebtsov, N. (2012). Gold nanoparticles in biomedical applications: recent advances and perspectives. *Chemical Society Reviews*, *41*, 2256–2282.
30. El-Sayed, I. H., Huang, X., & El-Sayed, M. A. (2006). Selective laser photo-thermal therapy of epithelial carcinoma using anti-EGFR antibody conjugated gold nanoparticles. *Cancer Letters*, *239*, 129–135.
31. Elbashir, S. M., Harborth, J., Lendeckel, W., Yalcin, A., Weber, K., & Tuschl, T. (2001). Duplexes of 21-nucleotide RNAs mediate RNA interference in cultured mammalian cells. *Nature*, *411*, 494.
32. Faraday, M. (1857). LIX. Experimental relations of gold (and other

metals) to light—The bakerian lecture. The London, Edinburgh, and Dublin Philosophical Magazine and Journal of Science. *14*, 512–539.

33. Fazal, S., Jayasree, A., Sasidharan, S., Koyakutty, M., Nair, S. V., & Menon, D. (2014). Green synthesis of anisotropic gold nanoparticles for photothermal therapy of cancer. *ACS applied materials & interfaces*, *6*, 8080–8089.
34. Frens, G. (1973). Controlled nucleation for the regulation of the particle size in monodisperse gold suspensions. *Nature physical science*, *241*, 20.
35. Gaikwad, H. K., Tsvirkun, D., Ben-Nun, Y., Merquiol, E., Popovtzer, R., & Blum, G. (2018). Molecular Imaging of Cancer Using X-ray Computed Tomography with Protease Targeted Iodinated Activity-Based Probes. *Nano letters*, *18*, 1582–1591.
36. Ganeshkumar, M., Sastry, T. P., Kumar, M. S., Dinesh, M. G., Kannappan, S., & Suguna, L. (2012). Sun light mediated synthesis of gold nanoparticles as carrier for 6-mercaptopurine: Preparation, characterization and toxicity studies in zebrafish embryo model. *Materials Research Bulletin*, *47*, 2113–2119.
37. Ganeshkumar, M., Sathishkumar, M., Ponrasu, T., Dinesh, M. G., Suguna, L 2013. Spontaneous ultra-fast synthesis of gold nanoparticles using Punica granatum for cancer targeted drug delivery. *Colloids and Surfaces B: Biointerfaces*, *106*, 208–216.
38. Gao, X., Cui, Y., Levenson, R. M., Chung, L. W., & Nie, S. (2004). In vivo cancer targeting and imaging with semiconductor quantum dots. *Nat Bio Technol*, *22*, 969–976.
39. Geetha, R., Ashokkumar, T., Tamilselvan, S., Govindaraju, K., Sadiq, M., & Singaravelu, G. (2013). Green synthesis of gold nanoparticles and their anticancer activity. *Cancer Nanotechnol*, *4*, 91–98.
40. Ghosh, R., Singh, L. C., Shohet, J. M., & Gunaratne, P. H. (2013). A gold nanoparticle platform for the delivery of functional microRNAs into cancer cells. *Biomaterials*, *34*, 807–816.
41. Giersig, M., & Mulvaney, P. (1993). Preparation of ordered colloid monolayers by electrophoretic deposition. *Langmuir*, *9*, 3408–3413.
42. Gobin, A. M., Lee, M. H., Halas, N. J., James, W. D., Drezek, R. A., & West, J. L. (2007). Near-infrared resonant nanoshells for combined optical imaging and photothermal cancer therapy. *Nano Letters*, *7*,

1929–1934.

43. Hartl, C., Frank, K., Amenitsch, H., Fischer, S., Liedl, T., & Nickel, B. (2018). Position Accuracy of Gold Nanoparticles on DNA Origami Structures Studied with Small-Angle X-ray Scattering. *Nano Letters*, 2018.
44. He, S., Fan, W., Wu, N., Zhu, J., Miao, Y., Miao, X., Li, F., Zhang, X., & Gan, Y. (2018). Lipid-Based Liquid Crystalline Nanoparticles Facilitate Cytosolic Delivery of siRNA via Structural Transformation. *Nano Letters*, *18*, 2411-2419.
45. Hirsch, L. R., Stafford, R. J., Bankson, J. A., Sershen, S. R., Rivera, B., Price, R. E., Hazle, J. D., Halas, N. J., & West, J. L. (2003). Nanoshell-mediated near-infrared thermal therapy of tumors under magnetic resonance guidance. *Proceedings of the National Academy of Sciences*, *100*, 13549–13554.
46. Huang, H. C., Rege, K., & Heys, J. J. (2010). Spatiotemporal temperature distribution and cancer cell death in response to extracellular hyperthermia induced by gold nanorods. *ACS Nano*, *4*, 2892–2900.
47. Huang, K., Ma, H., Liu, J., Huo, S., Kumar, A., Wei, T., Zhang, X., Jin, S., Gan, Y., Wang, P. C., & He, S. (2012). Size-dependent localization and penetration of ultrasmall gold nanoparticles in cancer cells, multicellular spheroids, and tumors in vivo. *ACS Nano*, *6*, 4483–4493.
48. Huang, S., Duan, S., Wang, J., Bao, S., Qiu, X., Li, C., & Hu, Y. (2016). Folic-Acid-Mediated Functionalized Gold Nanocages for Targeted Delivery of Anti-miR-181b in Combination of Gene Therapy and Photothermal Therapy against Hepatocellular Carcinoma. *Advanced Functional Materials*, *26*, 2532–2544.
49. Huang, X., Kang, B., Qian, W., Mackey, M. A., Chen, P. C., Oyelere, A. K., & El-Sayed, M. A. (2010). Comparative study of photothermolysis of cancer cells with nuclear-targeted or cytoplasm-targeted gold nanospheres: continuous wave or pulsed lasers. *Journal of Biomedical Optics*, *15*, 058002.
50. Huang, X., Neretina, S., El-Sayed, M. A. (2009). Gold nanorods: from synthesis and properties to biological and biomedical applications. *Advanced Materials*, *21*, 4880–4910.
51. Huang, X., Qian, W., El-Sayed, I. H., El-Sayed, M. A. (2007). The potential use of the enhanced nonlinear properties of gold nanospheres

in photothermal cancer therapy. *Lasers in Surgery and Medicine*, *39*, 747–753.

52. Izadiyan, Z., Shameli, K., Hara, H., Taib, S. H. M. (2018). Cytotoxicity assay of biosynthesis gold nanoparticles mediated by walnut (Juglans regia) green husk extract. *Journal of Molecular Structure*, *1151*, 97–105.
53. Javier, D. J., Nitin, N., Levy, M., Ellington, A., & Richards-Kortum, R. (2008). Aptamer-targeted gold nanoparticles as molecular-specific contrast agents for reflectance imaging. *Bio-Conjugate Chemistry*, *19*, 1309–1312.
54. Jensen, S. A., Day, E. S., Ko, C. H., Hurley, L. A., Luciano, J. P., Kouri, F. M., Ko, C. H., Hurley, L. A., Luciano, J. P., Kouri, F. M., Merkel, T. J., Luthi, A. J., Patel, P. C., Cutler, J. I., Daniel, W. L., Scot, A. W., Rotz, M. W., Meade, T. J., Giljohann, D. A., Mirkin, C. A., & Stegh, A. H. (2013). Spherical nucleic acid nanoparticle conjugates as an RNAi-based therapy for glioblastoma. *Science Translational Medicine*, *5*, 209ra152.
55. Kennedy, L. C., Bickford, L. R., Lewinski, N. A., Coughlin, A. J., Hu, Y., Day, E. S., West, J. L., & Drezek, R. A. (2011). A new era for cancer treatment: gold-nanoparticle-mediated thermal therapies. *Small*, *7*, 169–183.
56. Kim, E. Y., Stanton, J., Vega, R. A., Kunstman, K. J., Mirkin, C. A., & Wolinsky, S. M. (2006). A real-time PCR-based method for determining the surface coverage of thiol-capped oligonucleotides bound onto gold nanoparticles. *Nucleic Acids Research*, *34*, e54-e54.
57. Koperuncholan, M. (2015). Bioreduction of chloroauric acid ($HAuCl_4$) for the synthesis of gold nanoparticles (GNPs): a special empathies of pharmacological activity. *Int J Phytopharmacy*, *5*, 72–80.
58. Krishnaraj, C., Muthukumaran, P., Ramachandran, R., Balakumaran, M. D., & Kalaichelvan, P. T. (2014). Acalypha indica Linn: biogenic synthesis of silver and gold nanoparticles and their cytotoxic effects against MDA-MB-231, human breast cancer cells. *Biotechnology Reports*, *4*, 42–49.
59. Kumar, K. P., Paul, W., Sharma, C. P 2011. Green synthesis of gold nanoparticles with *Zingiber officinale* extract: characterization and blood compatibility. *Process Biochemistry*, *46*, 2007–2013.
60. Kumar, V., & Yadav, S. K. (2009). Plant-mediated synthesis of silver

and gold nanoparticles and their applications. *Journal of chemical Technology and Biotechnology*, *84*, 151–157.

61. Kyriazi, M. E., Giust, D., El-Sagheer, A. H., Lackie, P. M., Muskens, O. L., Brown, T., & Kanaras, A. G. (2015). Multiplexed mRNA Sensing and Combinatorial-Targeted Drug Delivery Using DNA-Gold Nanoparticle Dimers. *ACS Nano, 12,* 3333-3340.
62. Lee, K. D., Nagajyothi, P. C., Sreekanth, T. V. M., & Park, S. (2015). Eco-friendly synthesis of gold nanoparticles (Au NPs) using Inonotus obliquus and their antibacterial, antioxidant and cytotoxic activities. *Journal of Industrial and Engineering Chemistry*, *26*, 67–72.
63. Li, J. L., & Gu, M. (2010). Surface plasmonic gold nanorods for enhanced two-photon microscopic imaging and apoptosis induction of cancer cells. *Biomaterials*. *31*, 9492–9498.
64. Li, L., Liu, X., Fu, C., Tan, L., & Liu, H. (2015). Biosynthesis of fluorescent gold nanoclusters for in vitro and in vivo tumor imaging. *Optics Communications*, *355*, 567–574.
65. Li, X., Wang, Z., Li, Y., Bian, K., Yin, T., & Gao, D. (2018). Self-assembly of bacitracin-gold nanoparticles and their toxicity analysis. *Materials Science and Engineering: C*, *82*, 310–316.
66. Lin, J., Wang, S., Huang, P., Wang, Z., Chen, S., Niu, G., Li, W., He, J., Cui, D., Lu, G., & Chen, X. (2013). Photosensitizer-loaded gold vesicles with strong plasmonic coupling effect for imaging-guided photothermal/photodynamic therapy. *ACS Nano*, *7*, 5320–5329
67. Link, S., Mohamed, M. B., & El-Sayed, M. A. (1999). Simulation of the optical absorption spectra of gold nanorods as a function of their aspect ratio and the effect of the medium dielectric constant. The *Journal of Physical Chemistry B*, *103*, 3073–3077.
68. Liu, H., Liu, T., Wu, X., Li, L., Tan, L., Chen, D., & Tang, F. (2012). Targeting gold nanoshells on silica nano rattles: a drug cocktail to fight breast tumors via a single irradiation with near-infrared laser light. *Advanced Materials*, *24*, 755–761.
69. Liu, S., Zeng, T. H., Hofmann, M., Burcombe, E., Wei, J., Jiang, R., Kong, J., & Chen, Y. (2011). Antibacterial activity of graphite, graphite oxide, grapheme oxide, and reduced graphene oxide: membrane and oxidative stress. *ACS Nano*, *5*, 6971–6980.
70. Ma, N., Wu, F. G., Zhang, X., Jiang, Y. W., Jia, H. R., Wang, H. Y., Li, Y. H., Liu, P., Gu, N., & Chen, Z. (2017). Shape-dependent radio

sensitization effect of gold nanostructures in cancer radiotherapy: comparison of gold nanoparticles, nanospikes, and nanorods. *ACS Applied Materials & Interfaces*, *9*, 13037–13048.

71. Majoral, J. P., Zablocka, M., Caminade, A. M., Balczewski, P., Shi, X., & Mignani, S. (2018). Interactions gold/phosphorus dendrimers. Versatile ways to hybrid organic–metallic macromolecules. *Coordination Chemistry Reviews*, *358*, 80–91.
72. Majumder, M., Dunn, L., Liu, L., Hasan, A., Vincent, K., Brackstone, M., Hess, D., & Lala, P. K. (2018). COX-2 induces oncogenic micro RNA miR655 in human breast cancer. *Scientific Report*, *8*, 327.
73. Mayer, K. M., Hafner, J. H 2011. Localized surface plasmon resonance sensors. *Chemical Reviews*, *111*, 3828–3857.
74. McKenzie, F., Steven, V., Ingram, A., & Graham, D. (2009). Quantitation of biomolecules conjugated to nanoparticles by enzyme hydrolysis. *Chemical Communications*. 2872–2874.
75. Melamed, J. R., Edelstein, R. S., & Day, E. S. (2015). Elucidating the fundamental mechanisms of cell death triggered by photothermal therapy. *ACS Nano*, *9*, 6–11.
76. Melamed, J. R., Riley, R. S., Valcourt, D. M., Billingsley, M. M., Kreuzberger, N. L., & Day, E. S. (2017). Quantification of siRNA duplexes bound to gold Nanoparticle surfaces. *Biomedical Nanotechnology: Methods and Protocols*, 2017, 1–15.
77. Mishra, P., Ray, S., Sinha, S., Das, B., Khan, M. I., Behera, S. K., & Mishra, A. (2016). Facile bio-synthesis of gold nanoparticles by using extract of Hibiscus sabdariffa and evaluation of its cytotoxicity against U87 glioblastoma cells under hyperglycemic condition. *Biochemical Engineering Journal*, *105*, 264–272.
78. Mohamed, M. S., Veeranarayanan, S., Poulose, A. C., Nagaoka, Y., Minegishi, H., Yoshida, Y., & Kumar, D. S. (2014). Type 1 ribotoxin-curcin conjugated biogenic gold nanoparticles for a multimodal therapeutic approach towards brain cancer. *Biochimica et Biophysica Acta (BBA)-General Subjects*, *1840*, 1657–1669.
79. Mollick, M. M. R., Bhowmick, B., Mondal, D., Maity, D., Rana, D. S. K., Chattopadhyay, S., Roy, S., Sarkar, J., Acharya, K., Chakrabortye, M., & Chattopadhyay, D. (2014). Anticancer (in vitro) and antimicrobial effect of gold nanoparticles synthesized usingAbelmoschus esculentus (L.) pulp extract via a green route. *RSC Adv. 4*, 37838– 37848.

80. Muench, F., Schaefer, S., Hagelüken, L., Molina-Luna, L., Duerrschnabel, M., Kleebe, H. J., Broïz, J., Vaskevich, A., Rubinstein, I., & Ensinger, W. (2017). Template-Free Electroless Plating of Gold Nanowires: Direct Surface Functionalization with Shape-Selective Nanostructures for Electrochemical Applications. *ACS Applied Materials & Interfaces*, *9*, 31142–31152.

81. Mukherjee, S., Dasari, M., Priyamvada, S., Kotcherlakota, R., Bollu, V. S., & Patra, C. R. (2015). A green chemistry approach for the synthesis of gold nanoconjugates that induce the inhibition of cancer cell proliferation through induction of oxidative stress and their in vivo toxicity study. *J Mater Chem B*, *3*, 3820–3830.

82. Mukherjee, S., Sau, S., Madhuri, D., Bollu, V. S., Madhusudana, K., Sreedhar, B., Banerjee, R., & Patra, C. R. (2016). Green synthesis and characterization of monodispersed gold nanoparticles: toxicity study, delivery of doxorubicin and its bio-distribution in mouse model. *J Biomed Nanotechnol*, *12*, 165–181.

83. Mukherjee, S., Sushma, V., Patra, S., Barui, A. K., Bhadra, M. P., Sreedhar, B., & Patra, C. R. (2012). Green chemistry approach for the synthesis and tabilization of biocompatible gold nanoparticles and their potential applications in cancer therapy. *Nanotechnology*, *23*, 455103.

84. Mukundan, D., Mohankumar, R., & Vasanthakumari, R. (2005). Green synthesis of gold nanoparticles using leaves extract of Bauhinia tomentosa Linn and in-vitro anticancer activity. *Int J Innov Res Sci & Eng*. 2347–3207.

85. Muniyappan, N., & Nagarajan, N. (2014). Green synthesis of gold nanoparticles using Curcuma pseudo montana essential oil, its biological activity and cytotoxicity against human ductal breast carcinoma cells T47D. *J Environ Chem Eng.*, *2*, 2037–2044.

86. Muthukumar, T., Sambandam, B., Aravinthan, A., Sastry, T. P., & Kim, J. H. (2016). Green and synthesis of gold nanoparticles their enhanced synergistic antitumor activity using HepG2 and MCF7 cells and its antibacterial effects. *Process Biochem.*, *51*, 384–391.

87. Nahar, S., Nayak, A. K., Ghosh, A., Subudhi, U., & Maiti, S. (2018). Enhanced and synergistic downregulation of oncogenic miRNAs by self-assembled branched DNA. *Nanoscale*, *10*, 195–202.

88. Nakkala, J. R., Mata, R., Bhagat, E., & Sadras, S. R. (2015). Green synthesis of silver and gold nanoparticles fromGymnema sylvestreleaf extract: study of antioxidant and anticancer activities. *J Nanopart Res.*,

17, 1– 15.

89. Nam, J., Son, S., Ochyl, L. J., Kuai, R., Schwendeman, A., & Moon, J. J. (2018). Chemo-photothermal therapy combination elicits anti-tumor immunity against advanced metastatic cancer. *Nature Communications*, *9*, 1074.
90. Nangia, Y., Wangoo, N., Goyal, N., Shekhawat, G., & Suri, C. R. (2009). A novel bacterial isolate Stenotrophomonas maltophilia as living factory for synthesis of gold nanoparticles. *Microbial Cell Factories*, *8*, 39.
91. Narayanan, K. B., & Sakthivel, N. (2010). Biological synthesis of metal nanoparticles by microbes. *Advances in Colloid and Interface Science*, *156*, 1–13.
92. Nehl, C. L., Liao, H., & Hafner, J. H. (2006). Optical properties of star-shaped gold nanoparticles. *Nano Letters*, *6*, 683–688.
93. Nicol, J. R., Dixon, D., & Coulter, J. A. (2015). Gold nanoparticle surface functionalization: a necessary requirement in the development of novel nano therapeutics. *Nanomedicine*, *10*, 1315–1326.
94. Nikoo bakht, B., & El-Sayed, M. A. (2003). Preparation and growth mechanism of gold nanorods (NRs) using seed-mediated growth method. *Chemistry of Materials*, *15*, 1957–1962.
95. Nune, S. K., Chanda, N., Shukla, R., Katti, K., Kulkarni, R. R., Thilakavathy, S., & Katti, K. V. (2009). Green nanotechnology from tea: phytochemicals in tea as building blocks for production of biocompatible gold nanoparticles. *Journal of Materials Chemistry*, *19*, 2912–2920.
96. O'Neal, D. P., Hirsch, L. R., Halas, N. J., Payne, J. D., & West, J. L. (2004). Photo-thermal tumor ablation in mice using near infrared-absorbing nanoparticles. *Cancer Letters*, *209*, 171–176.
97. Oldenburg, S. J., Averitt, R. D., Westcott, S. L., & Halas, N. J. (1998). Nanoengineering of optical resonances. *Chemical Physics Letters*, *288*, 243–247.
98. Parida, U. K., Biswal, S. K., & Bindhani, B. K. (2014). Green synthesis and characterization of gold nanoparticles: study of its biological mechanism in human SUDHL-4 cell line. *Adv. Biol. Chem.*, *4*, 360–375.
99. Patra, S., Mukherjee, S., Barui, A. K., Ganguly, A., Sreedhar, B., & Patra, C. R. (2015). Green synthesis, characterization of gold and silver

nanoparticles and their potential application for cancer therapeutics. *Mater Sci Eng C*, *53*, 298–309.

100. Paul, P., Chatterjee, S., Pramanik, A., Karmakar, P., Bhattacharyya, S. C., & Kumar, G. S. (2018). Thionine Conjugated Gold Nanoparticles Trigger Apoptotic Activity Toward HepG2 Cancer Cell Line. *ACS Biomaterials Science & Engineering*, *4*, 635–646.
101. Pérez-Hernández, M., del Pino, P., Mitchell, S. G., Moros, M., Stepien, G., Pelaz, B., Parak, W. J., Gálvez, E. M., Pardo, J., & de la Fuente, J. M. (2014). Dissecting the molecular mechanism of apoptosis during photothermal therapy using gold nanoprisms. *ACS Nano*, *9*, 52–61.
102. Piella, J., Bastús, N. G., & Puntes, V. (2016). Size-Controlled Synthesis of Sub-10-nanometer Citrate-Stabilized Gold Nanoparticles and Related Optical Properties. *Chemistry of Materials*, *28*, 1066–1075.
103. Pitsillides, C. M., Joe, E. K., Wei, X., Anderson, R. R., & Lin, C. P. (2003). Selective cell targeting with light-absorbing microparticles and nanoparticles. *Biophysical Journal*, *84*, 4023–4032.
104. Pooja, D., Panyaram, S., Kulhari, H., Rachamalla, S. S., & Sistla, R. (2014). Xanthan gum stabilized gold nanoparticles: characterization, biocompatibility, stability and cytotoxicity. *Carbohydr Polym. 110*, 1–9.
105. Pooja, D., Panyaram, S., Kulhari, H., Reddy, B., Rachamalla, S. S., & Sistla, R. (2015). Natural polysaccharide functionalized gold nanoparticles as biocompatible drug delivery carrier. *International Journal of Biological Macromolecules*, *80*, 48–56.
106. Preetam, J. P., Khusro, A., & Panicker, S. G. (2016). In-vitro anticancer and antioxidant activity of gold nanoparticles conjugate with Tabernaemontana divaricataflower SMs against MCF-7 breast cancer cells. *Korean Chem Eng Res.*, *54*, 75–80.
107. Priya, M. K., & Iyer, P. R. (2015). Anticancer studies of the synthesized gold nanoparticles against MCF 7 breast cancer cell lines. *Appl Nano. 5*, 443–448.
108. Qin, Z., Wang, Y., Randrianalisoa, J., Raeesi, V., Chan, W. C., Lipiński, W., & Bischof, J. C. (2016). Quantitative comparison of photothermal heat generation between gold nanospheres and nanorods. *Scientific Reports*. *6*, 29836.
109. Raja, M. A. G., Katas, H., & Wen, T. J. (2015). Stability, intracellular delivery, and release of siRNA from chitosan nanoparticles using

different cross-linkers. *PloS one*, 10, e0128963.

110. Raju, D., Vishwakarma, R. K., Khan, B. M., Mehta, U. J., & Ahmad, A. (2014). Biological synthesis of cationic gold nanoparticles and binding of plasmid DNA. *Materials Letters*, *129*, 159–161.
111. Rupaimoole, R., & Slack, F. J. (2017). MicroRNA therapeutics: towards a new era for the management of cancer and other diseases. *Nature reviews Drug discovery*, *16*, 203–222.
112. Saini, R. K., Chouhan, R., Bagri, L. P., & Bajpai, A. K. (2012). Strategies of targeting tumors and cancers. *Journal of Cancer Research Updates*, *1*, 129–152.
113. Sayyed, M. I., Elmahroug, Y., Elbashir, B. O., & Issa, S. A. (2016). Gamma-ray shielding properties of zinc oxide soda lime silica glasses. *Journal of Materials Science: Materials in Electronics*, *28*, 4064–4074.
114. Schmid, G. (2001). Metals. *Nanoscale Materials in Chemistry*, 15–59.
115. Schwartz, J. A., Shetty, A. M., Price, R. E., Stafford, R. J., Wang, J. C., Uthamanthil, R. K., Pham, K., McNichols, R. J., Coleman, C. L., & Payne, J. D. (2009). Feasibility study of particle-assisted laser ablation of brain tumors in orthotopic canine model. *Cancer Research*, *69*, 1659–1667.
116. Sellappa, S., RafiqKhan, M., Selvaraj, V., & Vijayakumar, S. (2015). Cytotoxic effect of green synthesized gold nanoparticles using Argemone Mexicana leaf against HEPG2 cells. *Indo Amer J Phar Res.*, *5*, 3394–3398.
117. Shao, J., Griffin, R. J., Galanzha, E. I., Kim, J. W., Koonce, N., Webber, J., & Day, E. S. (2013). Photothermal nanodrugs: potential of TNF-gold nanospheres for cancer theranostics. *Scientific Reports*, *3*, 1293.
118. Shen, H., Mittal, V., Ferrari, M., & Chang, J. (2013). Delivery of gene silencing agents for breast cancer therapy. *Breast Cancer Research*, *15*, 205.
119. Silvestri, A., Zambelli, V., Ferretti, A. M., Salerno, D., Bellani, G., & Polito, L. (2016). Design of functionalized gold nanoparticle probes for computed tomography imaging. *Contrast Media & Molecular Imaging*, *11*, 405–414.
120. Singh, P., Kim, Y. J., Zhang, D., & Yang, D. C. (2016). Biological synthesis of nanoparticles from plants and microorganisms. *Trends in Biotechnology*, *34*, 588–599.
121. Skrabalak, S. E., Chen, J., Sun, Y., Lu, X., Au, L., Cobley, C. M., &

Xia, Y. (2008). Gold nanocages: synthesis, properties, and applications. *Accounts of Chemical Research*, *41*, 1587–1595.

122. Song, J. Y., Jang, H. K., & Kim, B. S. (2009). Biological synthesis of gold nanoparticles using Magnolia kobus and Diopyros kaki leaf extracts. *Process Biochemistry*, *44*, 1133–1138.
123. Spyratou, E., Makropoulou, M., Efstathopoulos, E. P., Georgakilas, A. G., & Sihver, L. (2017). Recent Advances in Cancer Therapy Based on Dual Mode Gold Nanoparticles. *Cancers*, *9*, 173.
124. Stern, J. M., Stanfield, J., Kabbani, W., Hsieh, J. T., & Cadeddu, J. A. (2008). Selective prostate cancer thermal ablation with laser activated gold nanoshells. *The Journal of Urology*, *179*, 748–753.
125. Sun, X., Huang, X., Yan, X., Wang, Y., Guo, J., Jacobson, O., & Kiesewetter, D. O. (2014). Chelator-free 64 Cu-integrated gold nanomaterials for positron emission tomography imaging guided photothermal cancer therapy. *ACS Nano*, *8*, 8438–8446.
126. Tiwari, P. M., Vig, K., Dennis, V. A., & Singh, S. R. (2011). Functionalized gold nanoparticles and their biomedical applications. *Nanomaterials*, *1*, 31–63.
127. Tong, L., Zhao, Y., Huff, T. B., Hansen, M. N., Wei, A., & Cheng, J. X. (2007). Gold nanorods mediate tumor cell death by compromising membrane integrity. *Advanced Materials*, *19*, 3136–3141.
128. Torre, L. A., Bray, F., Siegel, R. L., Ferlay, J., Lortet-Tieulent, J., & Jemal, A. (2015). Global cancer statistics, 2012. *CA: A Cancer Journal For Clinicians*, *65*, 87–108.
129. Truong, N. P., Whittaker, M. R., Mak, C. W., & Davis, T. P. (2015). The importance of nanoparticle shape in cancer drug delivery. *Expert Opinion on Drug Delivery*, *12*, 129–142.
130. Turkevich, J., Stevenson, P. C., & Hillier, J. (1951). A study of the nucleation and growth processes in the synthesis of colloidal gold. *Discussions of the Faraday Society*, *11*, 55–75.
131. Uthaman, S., Kim, H. S., Revuri, V., Min, J. J., Lee, Y. K., Huh, K. M., & Park, I. K. (2018). Green synthesis of bioactive polysaccharide-capped gold nanoparticles for lymph node CT imaging. *Carbohydrate polymers*, *181*, 27–33.
132. Velayutham, M., Villamena, F. A., Fishbein, J. C., & Zweier, J. L. (2005). Cancer chemopreventive oltipraz generates superoxide anion radical. *Arch Biochem Biophys*, *435*, 83–88.

133. Verrub, R. (2015). Green synthesis of gold nanoparticles against pathogens and cancer cells. *Int J Pharm Res.*, *5*, 250–256.
134. Vijayaraghavan, P., Vankayala, R., Chiang, C. S., Sung, H. W., & Hwang, K. C. (2015). Complete destruction of deep-tissue buried tumors via combination of gene silencing and gold nanoechinus-mediated photodynamic therapy. *Biomaterials*, *62*, 13–23.
135. Von, M. G., Park, J. H., Agrawal, A., Bandaru, N. K., Das, S. K., Sailor, M. J., & Bhatia, S. N. (2009). Computationally guided photothermal tumor therapy using long-circulating gold nanorod antennas. *Cancer research*, *69*, 3892–3900.
136. Wang, J., Zhu, G., You, M., Song, E., Shukoor, M. I., Zhang, K., & Tan, W. (2012). Assembly of aptamer switch probes and photosensitizer on gold nanorods for targeted photothermal and photodynamic cancer therapy. *ACS Nano*, *6*, 5070–5077.
137. Wang, Y., Black, K. C., Luehmann, H., Li, W., Zhang, Y., Cai, X., Wan, D., Liu, S. Y., Li, M., Kim, P., Li, Z. Y. 2013. Comparison study of gold nanohexapods, nanorods, and nanocages for photothermal cancer treatment. *ACS Nano.*, *7*, 2068–2077.
138. Wang, Z., Wu, L., & Cai, W. (2010). Size-Tunable Synthesis of Monodisperse Water-Soluble Gold Nanoparticles with High X-ray Attenuation. *Chemistry-A European Journal*, *16*, 1459–1463.
139. Xing, L., Chen, B., Li, D., Wu, W., & Ying, Z. (2018). Gold nanospheres enhanced photothermal therapy in a rat model. *Lasers in Surgery and Medicine, 50*, 669-679.
140. Yin, D., Li, X., Ma, Y., & Liu, Z. (2017). Targeted cancer imaging and photothermal therapy via monosaccharide-imprinted gold nanorods. *Chemical Communications*, *53*, 6716–6719.
141. Yue, J., Feliciano, T. J., Li, W., Lee, A., & Odom, T. W. (2017). Gold nanoparticle size and shape effects on cellular uptake and intracellular distribution of siRNA nanoconstructs. *Bioconjugate Chemistry*, *28*, 1791–1800.
142. Zhang, X. (2015). Gold nanoparticles: recent advances in the biomedical applications. *Cell Biochemistry and Biophysics*, *72*, 771–775.
143. Zhang, Y., Chu, W., Foroushani, A. D., Wang, H., Li, D., Liu, J., Barrow, C. J., Wang, X., & Yang, W. (2014). New gold nanostructures for sensor applications: a review. *Materials*, *7*, 5169–5201.
144. Zhao, E., Zhao, Z., Wang, J., Yang, C., Chen, C., Gao, L., & Zhang, Q.

(2012). Surface engineering of gold nanoparticles for in vitro siRNA delivery. *Nanoscale*, *4*, 5102–5109.

145. Zharov, V. P., Galitovskaya, E. N., Johnson, C., & Kelly, T. (2005). Synergistic enhancement of selective nano photothermolysis with gold nanoclusters: potential for cancer therapy. *Lasers in Surgery and Medicine*, *37*, 219–226.

146. Zheng, M., Tao, W., Zou, Y., Farokhzad, O. C., & Shi, B. (2018). Nanotechnology-Based Strategies for siRNA Brain Delivery for Disease Therapy. *Trends in Biotechnology, 36,* 562-575.

147. Zhu, D., Song, P., Shen, J., Su, S., Chao, J., Aldalbahi, A., & Tian, Y. (2016). PolyA-mediated DNA assembly on gold nanoparticles for thermodynamically favorable and rapid hybridization analysis. *Analytical Chemistry*, 2016, 88, 4949–4954.

148. Zhu, H., Liu, W., Cheng, Z., Yao, K., Yang, Y., Xu, B., & Su, G. (2017). Targeted Delivery of siRNA with pH-Responsive Hybrid Gold Nanostars for Cancer Treatment. *International Journal of Molecular Sciences*, *18*, 2029.

5
CHAPTER

APPLICATION OF SILVER NANOPARTICLES ON VARIOUS TYPES OF CANCEROUS CELL

Arun Kumar Pradhan[1, 2] and Tanushree Mahapatra[2, 3]

[1] Centre for Biotechnology, Siksha O Anusandhan University, Bhubaneswar, India

[2] M/S Odisha NET Academy & Research Centre, Bhubaneswar, Odisha, India-751021

[3] Utkal University, Bhubaneswar-751004, India

CONTENTS

ABSTRACT

Silver nanoparticles are broadly used in the treatment/diagnosis of various types of cancers and other chronic diseases in the human body. Silver nanoparticles are composed of large number of silver oxide. Due to surface plasmon resonance, the color of the silver nanoparticles varies and mostly is bright blue with an absorbance maximum of 450 nm. In addition, they have high chemical stability, catalytic activity, electrical and thermal conductivity. These nanoparticles (NPs) can be synthesized either by physical, chemical or biological methods. The chemical synthesis includes the reduction of silver ions using different reducing agents like $NaBH_4$, $LiBH_4$, and sodium citrate, etc. in aqueous/non-aqueous medium. Also, leaves of various plants, like Ginko, Mangolia, Pine, and Persimmon, etc. are used for the commercial production of silver nanoparticles. In addition to the ease of synthesis, better optical/electrical and anticancerous/antibacterial properties make these NPs suitable for in bio-imaging, microtomical experiments and treatment of various diseases along with cancer. Moreover, to increase their cytotoxic and cellular uptake activity, silver nanoparticles are often modified with glucose, sucrose, lactose and some oligoncleotides. In this chapter, the synthesis of silver NPs, their properties and mechanism of anticancer property has been broadly presented.

Keywords: Ag NPs, Cancer, Colon cancer, Leukemia, Cervical cancer

5.1. INTRODUCTION

5.1.1. Silver Nanoparticles and Their Properties

Now-a-days the characterization and synthesis of silver nanoparticles (Ag NPs) is one of the important branch of nanotechnology. They have potential cytotoxicity against various cell lines of cancer like breast, lung, cervical, blood, colon, and gastrointestinal cancer (Yan et al., 2016).

They immediately trigger cellular damage by rupturing the cell's plasma membrane and promote apoptosis (programmed cell death) controlling both the cancerous and tumor growth. In order to improve the cytotoxicity and anti-cancerous activities, the Ag NPs are modified with glucose, lactose, etc. They possess profound biological, physical and chemical properties. It has been reported that, silver nanoparticles can be used for effective diagnosis and treatment of various diseases starting from flu to more chronic HIV (Human Immune deficiency Virus).

It has also unique optical and electronic properties. It is bright blue in color and has high thermal and electrical conductivity. Therefore, they have been used for absorption of solar energy, bio-labeling, intercalating agent in electrochemical energy storage and conversion applications, disease diagnosis and detection of cancers (Figure 5.1) (Tsibakhashvili N et al., 2010).

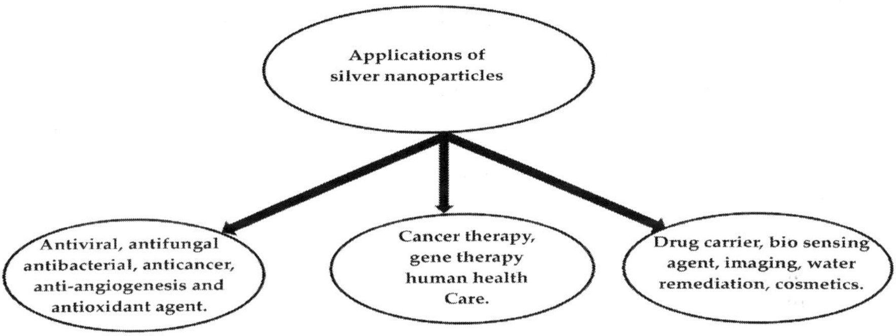

Figure 5.1: Various applications of Silver nanoparticles.

Silver nanoparticles play a very crucial role in the growing era of nanotechnology, evident from its increasing commercial demand. Size of these nanoparticles vary between the ranges of 10–100 nm having more or less a spherical shape. Silver nanoparticles are also commonly used as dyes, ligand for various biomolecules, surfactants and polymers. In medical science, they are used as carriers to facilitate targeted drug delivery for specific cells. Accumulation of silver nanoparticles at those target sites provide high payload concentration and suppresses the adversity of various diseases (Zhi, J. C. et al., 2008).

5.1.2. Synthesis of Silver Nanoparticle

Varieties of methods have been developed for the synthesis of Ag NPs, which can be categorized into three types, i.e., physical, chemical and biological methods. Each of these processes has advantages of their own with certain specific limitations as well. These are described briefly in the following sections.

5.1.2.1. Physical Method

Silver nanoparticle can be synthesized with the help of a tube furnace at atmospheric pressure, however, it requires large space, consumption of great amount of energy and significant amount of time (Kruis et al., 2000; Magnusson et al., 1999) which act as major drawbacks. Following this method, one can achieve high concentration of silver nanoparticles, which is essential for long-term research like toxicity studies (Jung et al., 2006).

Use of laser beam is also one of the physical method for the synthesis of Ag NPs (Kabashin, A. V., et al., 2003; Sylvestre, J. P., et al., 2004). The characteristics of Ag NPs depend upon many factors of laser beams like wave length, expose time period and with or without surfactants (Tarasenko, N., et al., 2006; Kawasaki, M., et al., 2006).

5.1.2.2. Chemical Method

Reduction of aqueous solution and non-aqueous solution, ultrasonic-assisted reduction, electrochemical reduction and micro-emulsion method, etc. are some of the chemical synthetic routes used for the synthesis of silver nanoparticles (Sathyavathi et al., 2010). Different reducing agents such as sodium citrate ($NaC_6H_7O_7$), polyol agent, Tollens' agent ($[Ag(NH_3)_2]^+$), ethylene glycol ($[CH_2OH]_2$), ascorbate ($C_6H_8O_6$), sodium borohydride ($NaBH_4$), etc. are used for the synthesis of Ag NPs from the silver ions in aqueous solution. Silver ions convert to metallic silver on reduction and then agglomerate to form oligomeric clusters followed by metallic colloidal silver particle formation (Wiley et al., 2005, Evanoff et al., 2004, Merga et al., 2007). Metallic colloidal silver particle act as a protective agent during nanoparticle preparation and they bind to the surface of the nanoparticle preventing agglomeration (Oliveira et al., 2005).

During chemical conversion, few key factors such as particle size, aggregation state, concentration of stabilizers, etc. are responsible for the effective stabilization of the silver nanoparticles. Iravani S et al. (2014) reported that silver nanoparticles can be chemically synthesized by several methods such as microemulsion technique, UV-initiated photoreduction, electrochemical synthesis, and microwave irradiation synthesis.

5.1.2.3 Biological or Green Synthesis Method

The chemical and physical methods for the synthesis of silver nanoparticles have various limitations, e.g., generation of various hazardous byproducts, pollutants leading to increased risk to human health and high level of energy consumption. Therefore the biological method can be an alternative for efficient and non-hazardous synthesis of silver nanoparticle. In this method, bacteria, fungi and many other medicinal plants are commonly used for the synthesis of silver nanoparticles. It is an ecofriendly, cost-effective and economical method. Some examples of biological method are biological and irradiation method, mixed-valence polyoxonctalates polysaccharides method, etc. (Iravani et al., 2014). In these methods, the size and shape of the silver nanoparticles is controlled by altering some critical conditions like temperature, pH, substrate concentration, buffer strength, electron donor like fructose or glucose, speed of centrifugation and exposure time (Iravani et al., 2014). Highly stable silver nanoparticles can be synthesized by using non-pathogenic bacteria, *Bacillus licheniformis* (Kalishwaralal et al., 2008). Silver nanocrystals of different composition have been reportedly synthesized by using *Pseudomonas stutzeri* (Klaus et al., 1999). *Escherichia coli, klebsiella pnemonia* and *Enterobacter cloacae* were used for rapid biosynthesis of silver nanoparticle by the reduction of aqueous silver ion (Shahverdi et al., 2017). Silver nanoparticles are also synthesized by using *Fusarum oxysporum* (Ahmad et al., 2003). The extracellular filterace *Cladosporium cladosporioides* biomas also used for the biosynthesis of silver nanoparticles (Balaji et al., 2009). In another work, Sadoski et al., have used *Penicillium* genus for the synthesis of silver nanoparticles (Sadowski et al., 2018).

Synthesis of silver nanoparticle by using microbes is an eco-friendly and economically feasible way. The first bacteria assisted synthesis of silver nanoparticle has been reported for *Pseudomonas stutzeri* which was obtained through its isolation from silver mine (Haefelic et al., 1984). Nitrate reductase enzyme convert nitrate into nitrite. This enzyme is useful for the production of silver nanoparticles (Pradhan N et al., 2011). Metal ions are reduced by the nitrate reductase of microbes and this reaction is coupled with oxidation- reduction of Quinone (Figure 5.2).

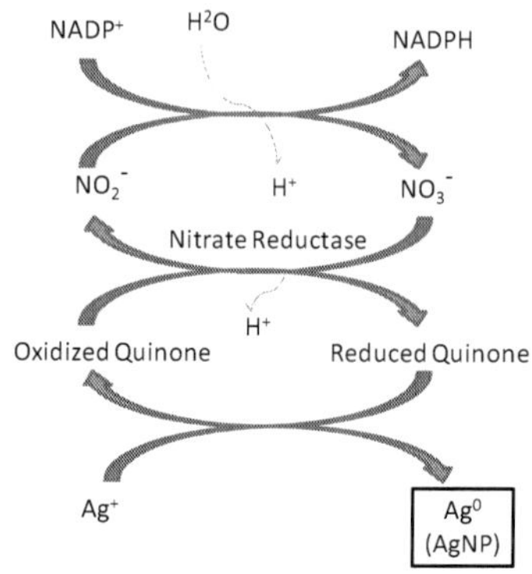

Figure 5.2: Ag NP synthesis by Nitrate reductase enzyme.

Bacillus licheniformis has also been used for the production of silver nanoparticles (Nalenthiran et al., 2009). It has been found that the dried cell of bacteria, *Lactobacillus* has the capacity to reduce the silver ions (Fujk et al., 2000). Some cyanobacteria like Spirullina and *Cholrella* have also been used for the production of silver nanoparticles. *Actinobacteria streptomyles* also helps in production of silver nanoparticles (Tsibakhashvili, N. et al., 2010). These can also be produced by using various microbes like bacteria, yeast, fungi, etc. (Mandal et al., 2006). Table 5.1 lists some of the potential microbes, i.e., bacteria, fungi, and plants from which silver nanoparticles can be prepared through non-invasive and cost-effective methods.

Table 5.1: List of Bacteria/ Fungi/ Plants Use for Synthesis of Silver Nanoparticles

Sl No.	Bacteria	Fungi	Plants
1	*Pseudomonas stutzeri*	*Verticillum sp.*	*Medicogo sativa*
2	*Bacillus megaterium*	*Aspergilus fumigatus*	*Azodirachta indica*
3	*Plectonema boryanum*	*Aspergilus flavus*	*Aloe vera*
4	*Enterobacter cloacae*	*Aspergilus clavatus*	*Piper betle*
5	*Lactobacillus fermentum*	*Fusarium solani*	*Coriandrum sativum*
6	*Proteus mirabilis*	*Fusarium semitectum*	*Desmodium triflorum*
7	*Brevibacterium casei*	*Cladosporium cladosporioides*	*Jatropha curcas*

5.1.3. Mechanism of Action of Silver Nanoparticle on Tumor Cell

P^{53} is the tumor suppressor gene whose dysfunctionality is one of the cause of cancerous growth. Expression of P^{53} gene elevates the concentration of P^{53} protein which activates proapoptotic factors like PUMA/NOXA. These factors cause the death of cells through intrinsic pathway of apoptosis (Figure 5.3). It is reported that silver nanoparticles induce the expression of P^{53} gene (Satapathy, S. R. et al., 2013; Kim, H. R. et al., 2014).

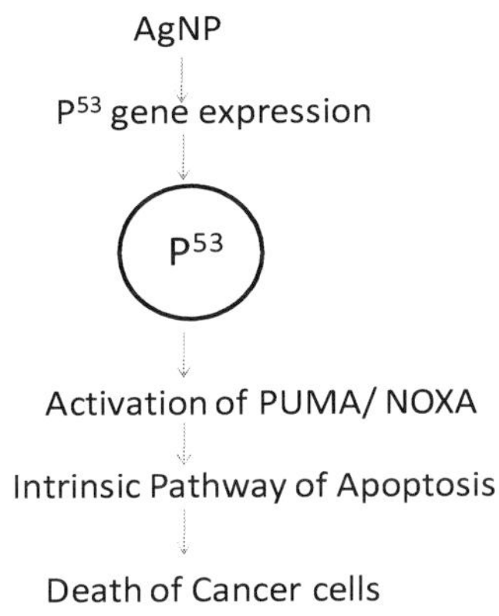

Figure 5.3: Probable mechanism of action of silver nanoparticle on tumor cell.

5.1.4. Mechanism of Silver Nanoparticle in Drug Delivery to Target Cancer Cell

The major problem with chemotherapy in cancer treatment is the gradual development in the immunity of the targeted cancerous cells towards multiple drugs (Fodale et al., 2011). Silver nanoparticles have large number of active sites which is favorable for the effective binding of various biomolecules. Silver nanoparticles can readily bind with varieties of drugs and transport them to the target cells (Ameen et al., 2007), thus preventing multidrug resistance capability of cancer cells by binding with highly selective molecules which is uniquely expressed on the target cell surface. For cancer treatment, most of the silver nanoparticles were modified with trans-activating transcriptional activator derived from specific viruses which act as cell penetrating molecules (Liu et al., 2012).

Generally, silver nanoparticles cannot be used directly for efficient cellular uptake, though cell penetrating peptide-modification is an efficient method for improving intracellular delivery of the nanoparticles. Once the nanoparticles enter the cancer cells, they cannot escape due to their large size, thus preventing multidrug resistance. Therefore, drugs are easily accumulated in target cancer cells at high concentrations. After delivering drugs with silver nanoparticles into cancer cells, the link (generally photo responsive) between drug and nanoparticle is broken by light energy (electromagnetic) (Agasti et al., 2009). The main advantage of using silver nanoparticles in drug delivery for cancer treatment is that they effectively prevent the multidrug resistance in the cancer cells.

5.2. APPLICATIONS OF SILVER NANOPARTICLES ON DIFFERENT TYPES OF CANCER

5.2.1. Silver Nanoparticle and Colon Cancer

Colon cancer is the third major type of cancer in the world. It is also known as colorectal cancer. It is a malignant cancer and mainly arises from colon polyps. Removal of colon polyps through surgical method is one of the preventive ways to suppress colon cancer. Several environmental factors such as life style, diet, heredity can be associated with colon cancer. At early stage this cancer is curable by surgery. Chemotherapy is also useful for the treatment of colorectal cancer. Nowadays, colon cancer is treated by using nanotechnology, i.e., using silver nanoparticles (Shakti Ranjan et al., 2014). Some common symptoms of colon cancer is diarrhea, constipation, rectal bleeding, narrow stools, dark color stool, change in bowel habit, etc. treatment of colon cancer cell with silver nanoparticles reduces further growth, progression and spreading of cancerous cells (Giribabu N. et al., 2016). In a particular work, Nehal Meldeeb et al., have used silver nanoparticles, extracted from the product of honeybees for the treatment of colorectal cancer (Nehal MEL-Deeb et al., 2015). Silver nanoparticles are synthesized by using various honeybees product like honey, royal jelly, honey dew and propalis. It has been found that Ag NPs of size 1–1000 nm are used for the diagnosis and treatment of various cancers and proves to be an effective, simple and cheap technique. A particular dose of Ag NPs prevent cell proliferation and are able to cross the nuclear membrane then bind with protein and DNA, inducing the cell death by chopping or disintegrating the DNA (Molina, M. A. F. et al., 2010). They increase the production of P^{53}

in colorectal cancer cells promoting the apoptosis. These Ag NPs inhibit survivin and BCl2 gene expression (Ambrosini et al., 1997). It has been reported that silver nanoparticles of dimensions within the range of 12–18 nm can be produced by using various products of honeybees and they show anticancer activity at both cellular and molecular level (Nehal MEL-Deeb et al., 2015). There have been few reports on plant derived silver nanoparticles inducing P53 dependent mitochondrial apoptotic mechanism in the cells of colorectal cancer. Shakti Ranjan et al., have used the leaf of medicinal plant, (periwinkle) for the synthesis of Ag NPs (Shakti Rnajan et al., 2014). *Catharanthus roseus* (periwinkle) is a medicinal plant and it possesses antibacterial, antiplasmodial, anti-diabetic and anticancer properties (Jacob et al., 2012). Periwinkle derived Ag NPs show prominent anticancer properties even at lower dosages as compared to other synthetic Ag NPs. These Ag NPs increase the release of cytochrome-c from mitochondria and reduce mitochondria membrane potential. They also trigger JNK cascade activation and ROS (reactive oxygen species) production that leads to rapid cell death (Shakti Ranjan et al., 2014).

Vitex negundo L. leaf mediated Ag NPs act as antitumor agent for human colon cell HCT15. They prevent proliferation of colon cancer cell line HCT15 with an IC50 of 20 mg\ml at 48 hours incubation. They promote apoptosis of cancer cell and arrest the cell at G0/G1 and G2/M phases decreasing the time of S-phase. It promotes a reduced DNA synthesis of cancer cell leading the cell death (Prabhu et al., 2012). It has been reported that silver nanoparticles decrease the growth of HCT110 colon cancer cell by increasing BAX/BCL-XL ratio and decreasing NF-KB. It increases the level of P53 which act as transcription factor for P21 that arrests the cell at G1 and S phase of silver nanoparticle causing the cell damage reducing interaction between P53 and NF-KB. NF-KB responsible for cell growth (Shakti Ranjan et al., 2013).

5.2.2. Silver Nanoparticle and Gastrointestinal Cancer

Another widely occurring form of cancer is gastrointestinal cancer which occurs mostly in the European countries. It is the fourth most occurring cancer and has high mortality rate. It is a multifactorial cancer. Various infection and environmental factors are responsible for the occurrence of this cancer. Detection and treatment method of gastrointestinal cancer is very limited. Most accepted way to cure this cancer is surgery (Syed Abdul Kuddus, 2017).

Nowadays, silver nanoparticles are frequently used for the treatment and diagnosis of gastrointestinal cancer in human because of their anticancer, apoptosis and anti-proliferating properties. In a particular work, Sirnmani et al., have reported the potential of Ag NPs extracted from neem phytochemicals for the treatment of gastrointestinal cancer (Sironmani, 2016). Neem phytochemical mediated silver nanoparticles possess bactericidal effect and few other unique chemical and physical properties (Sironmani et al., 2016). It induced apoptosis cell death mediated by Fas death receptor followed by cascade pathway of extrinsic apoptosis (Choi et al., 2012). It has been reported that oxidative stress and reactive oxidative species induce varieties of cancerous growths and diseases (Valko et al., 2007). Catalase is an enzyme reduced toxic oxygen and hydroxyl free radicals in human cells by converting hydrogen peroxide into water and oxygen (Sun Y. et al., 1990). These may be induced by using silver nanoparticle because of its antioxidant properties. Reactive oxygen species generated in the gastric cells stimulate cell proliferation and cause various oxidative stress injuries in stomach. Silver nanoparticles inhibit NF-KB pathway which promote cell growth (Schreck et al., 1992). Reactive oxygen species promote NF-KB path way. The Ag NPs modified with glucose used for the treatment of gastrointestinal cancer because it has almost three fold higher rate of success than untreated gastric cancer (Di chairo G. et al., 1984). By controlling the level of glucose utilization in the tumor cells in early stages of chemotherapy, the severity of cancerous growth in many patients can be suppresses effectively. Therefore, the monitoring of glucose utilization level could be a useful diagnostic tool for the detection of gastrointestinal cancer (Weber et al., 2006).

5.2.3. Silver Nanoparticle and Leukemia

Leukemia is the cancer of blood cells, especially the white blood cells (WBCs). It is a malignant form of cancer and has a significant 37% of the total occurrence of cancer in the world. In early stage, it does not produce any specific symptoms but later it grows rapidly to form large tumors and affects the blood cells. Leukemia is primarily of two types, i.e., myeloid and lymphoid depending on the types of WBCs being affected. Symptoms of leukemia are weight loss, discomfort, swelling of abdomen, weakness, fever, swollen lymph nodes and pain in bones and joints. Chronic leukemia is rarely hereditary and primarily occurs due to few environmental factors, e.g., smoking, chemotherapy (while treating other form(s) of cancer(s)), etc. and the treatment involves bone marrow transplantation, chemotherapy, radiation therapy, etc. Recent studies show that a macro algae *Gelidiella* mediated silver nanoparticles may be used for the diagnosis of various cancers like leukemia and colon cancer (Singh, J. et al., 2017). It is also useful for chemotherapy and causes immunotoxicity that affects the tumor development and help in patient recovery (Saraniya Devi, 2013). Nowadays, Ag NPs are being used for the preparation of anti-leukemic drug and also for the reversal of multidrug resistance of cancer cells (Govind Soni et al., 2014). One of the common causes of chronic myeloid leukemia is the Bcr-Abl fusion protein in chromosome 22 by translocation. Nanoparticle based drug delivery system can be helpful in the delivery of two types of drugs at the target sites (Bcr-Abl co-proteins) leading to the suppression of bcr-abl activity. It also induces apoptosis activity once the cells affected by leukemia are exposed to the specific drug(s) resulting in the death of the cancer cells (Sarbari Acharya et al., 2011). The reactive oxygen species could be generated upon exposing the affected cells to silver nanoparticles causing apoptosis and cytotoxicity. This Ag NPs treatment inhibited the any further growth of the acute myeloid leukemia (AML) cells (Guo D et al., 2014). The cell cycle is altered by the Ag NPs which promote the activity of vitamin C which is an antioxidant. It has been reported that PVP (Polyvinyl pyrrolidone) coated Ag NPs have cytotoxic effect on Acute Myeloid leukemia (AML) (Guo D et al., 2013). In these cases, the Ag NPs produce reactive oxygen species that affect mitochondrial membrane potential, causing DNA damage and apoptosis. Production of these reactive oxygen species with the aid of Ag NPs is helpful for cytotoxic effect against AML cells.

5.2.4. Silver Nanoparticle and Breast Cancer

Breast cancer is another malignant form of cancer which affects the cells of breast in human female (Alison, 2001). Over expression (25% to 30%) of HER2 gene causes the breast cancer. HER2 over expression is seen in both primary tumor and metastatic tumor sites (Valabrega G et al., 2006). A humanized mAb, Trastuzumab, can induce the down regulation of HER2 over expression and causes the death of breast cancer cells (Austin, C.D. et al., 2004). Nowadays, Ag NPs are predominantly used for the treatment of breast cancer. Krishnaraj et al., have synthesized Ag NPs from *Acalypha indica* that shows 40% of cell inhibition against breast cancer (Krishnaraj et al., 2014). In another work, Suman et al., have prepared Ag NPs using the roots of *Morinda citrifolia* that causes 100% death of HeLa cell lines (Suman et al., 2013). Silver nanoparticles extracted from *Eucalyptus chapmaniana* causes 85% death of cancer cells (Sulaiman et al., 2013). *Piper longum* mediated silver nanoparticles show 94.02% of cell death in the case of HEP-2 cell lines (Jacob et al., 2012). It has been reported that complete cell inhibition of human breast cancer cell could be achieved by using plant mediated silver nanoparticles as drug (cisplatin). MTT assay against MCF-7 cell line of breast cancer cells show significant cytotoxic activity of silver nanoparticles (Jannathul et al., 2015). It has been found that, albumin coated Ag NPs can be used for the treatment of MDA-MB 231 breast cancer (Marzieh et al., 2017).

It has been reported that the Ag NPs at a concentration of 1–100 mg/ml (synthesized from the leaves of *Acalypha indica*) can effectively be used against MDA-MB-231 human breast cancer cells (Krishnaraj et al., 2014). They exhibit significant cytotoxic effect and apoptosis by activating caspase-3 and DNA fragmentation in breast cancer cells. The advantage of the use of Ag NPs on the breast cancer cell is its negligible toxicity compared to other nanoparticles *Podophyllum hexandrum* leaves are also used for the synthesis of silver nanoparticles of 420 nm (Jeyaraj et al., 2013). Recently, it has been reported that MCF-7, which is another form of human breast cancer cell, shows 100% cell death when treated with 50 mg/ml concentration of silver nanoparticles (Jeyaraj et al., 2013). There are reports claiming to have derived silver nanoparticles from mushrooms that show significant cytotoxic activities against MDA-MB-231 cell line at comparatively lower concentrations than other plant derived silver nanoparticles (Gurunathan et al., 2013). Thus, it can be suggested that the cytotoxic and apoptosis nature of silver nanoparticles reduce the viability of MDA-MB-231 cell in a dose dependent manner. When silver nanoparticles act on breast cancer cells, they promote membrane blebbing, DNA fragmentation of the cancer cells which signifies apoptosis (Janicke et al., 1998). MDA-MB-231 cells treated with Ag NPs showed prominent cell death due to apoptosis. First, it causes damage to the DNA by fragmentation inducing the activation of P53 that acts as transcription factor for P21 leading to the activation of intrinsic pathway of apoptosis. Finally, it activates caspase-3 causing cell death (HU et al., 2009). The Ag NPs treated MDA-MB-231 cell showed increased level of caspase-3. So, they undergo rapid apoptosis (Gurunathan et al., 2013). They also promote apoptosis through granenzyme-B or death receptor or apoptosome (Moaddad et al., 2011). In these cases, silver nanoparticle causes cell toxicity which induced apoptosis by activating caspase-3. Recent studies show that reactive oxygen species generated by Ag NPs derived from *Ganderma neojaponicum* play a significant role in apoptosis and cytotoxicity of human breast cancer cells (Sangiliyandi et al., 2013). Whereas, the *Agaricus bisporus* mediated spherical shaped and 8–20 nm sized silver nanoparticles produced cytotoxic effect on MCF-7 human breast cancer cells (Figure 5.4) (Sonbaty et al,. 2013). The Ag NPs cause lactate dehydrogenase leakage, ROS generation, caspase activation and DNA fragmentation on cancer cells.

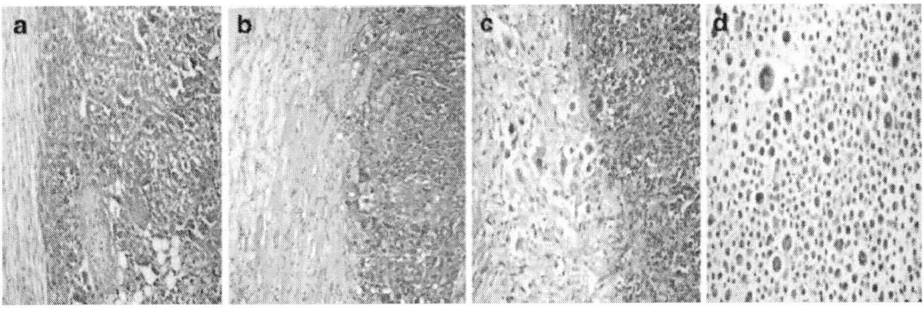

Figure 5.4: Photomicrograph of Ehrlich ascites tumor sections from mice. a Control section showing sheets of malignant cells with lot of dilated congested blood vessels (H&E ×400). b and c Section in the Ehrlich ascites tumor that received AgNPs showing malignant cells with necrotic and apoptotic cells and reduced blood vessels (H&E ×200 and 400, respectively). d Section from mice that received AgNPs and gamma radiation showing most cells in necrosis and apoptosis stage also fragmented cell constituents (H&E ×400) (From Sonbaty et al,. Cancer Mano, 293, 73, 2013).

5.2.5. Silver Nanoparticle and Lung Cancer

Lung cancer is also known as lung carcinoma and is a type of malignant cancer characterized by an uncontrolled growth of cells of lung. The main causes of lung cancer include smoking, exposure to radon gas, air pollution, pollution of asbestos, etc. some common symptoms of lung cancer are weakness, fever, regular cough, blood in cough, etc. There are several methods of treatment of lung cancer have been discovered and one of the successful technique of treatment of lung cancer is by using silver nanoparticles.

Silver nanoparticles obtained through green synthetic routes have shown effectiveness against lung cancer. Anticancer properties of these Ag NPs are useful for chemotherapy and chemoprevention of lung cancer at early stage of cancerous growth (Yan et al., 2016). These nanoparticles greatly inhibit the H1299 tumor growth. Recent study indicate that silver nanoparticles act as strong inhibitory agent which can suppress the growth of WSCLC cell (H1299) and also affects on the activities of NF-KB, bcl-2, case pace

3 and survivin. Terpenoid and peptide contents of the latex used for the synthesis of latex capped silver nanoparticles have been used as a carrier for nanoparticles. These nanoparticles produce cytotoxic effects against human lung cancer in a dose dependent manner (Uday Kumar et al., 2013).

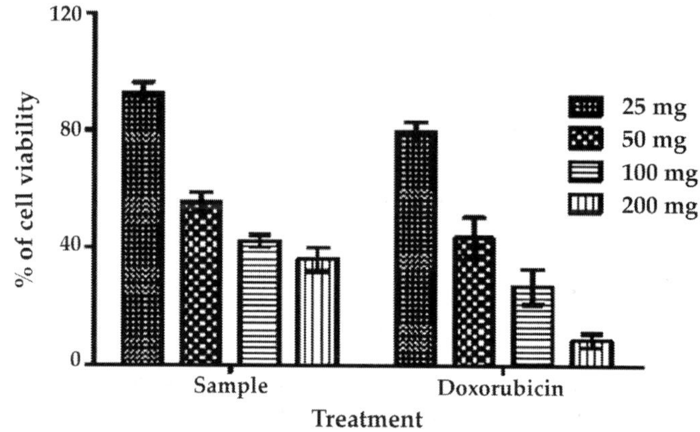

Figure 5.5: The cytotoxicity assay of biosynthesized silver nanoparticles against lung cancer cell line (A549) by MTT Assay; Doxorubicin was used as positive control. (Taken with permission from Subbaiya et al., Science engineering and management research (ICSEMR). 2014).

Silver nanoparticles increase the level of ROS in the cells that induce genotoxicity and cytotoxicity resulting in mitochondrial damage and early apoptosis occurred on target lung cancer cell line, A549 (Rasmus et al., 2010). It has been reported that, hypoxia in cancer cells of lung triggers apoptosis. Silver nanoparticles increase the expression of hypoxia inducible factor (HIF) on the lung cancer cells which induce hypoxia. Reactive oxygen species are generated by silver nanoparticle in lung cell causing high susceptibility to oxidative state (Jae-kyo et al., 2016). These Ag NPs are used as potential agent for chemoprevention and chemotherapy for the treatment of lung cancer. It initiates dose dependent apoptosis and cytotoxicity in the H1299 cell line in human lung cancer. It also inhibits NF-KB and Bcl-2 activity and increases the caspase-3 and survivin activation (Yan et al., 2016). Rhodococcus mediated silver nanoparticle used for detection and treatment of lung cancer cell line A549. The effect of silver nanoparticles against lung cancer cell line (A549) is studied by 3-(4, 5-dimethylthiazol-2-yl)-2,5-diphenyltetrazolium bromide dye assay called as MTT assay (Figure 5.5) (Subbaiya et al., 2014). This study indicates that LC_{50} value of silver nanoparticles against A549 cell line is approximately 100 μg.

5.2.6. Silver Nanoparticle and Cervical Cancer

Cervical cancer is the cancerous cell growth occurring at the cervix, particularly in the lowermost part of uterus. It is a rare type of cancer caused by Human Papilloma Virus (HPV). It is diagnosed by Pap test during the early stages that identify the changes in the cervical cell before it transforms into cancer. Some common symptoms of cervical cancer are abnormal menstruation, abdominal pain, frothy vaginal discharge, pain during sexual intercourse, etc. Common treatment for cervical cancer are chemotherapy, radiation therapy and surgery (removal of pelvic lymph node). Prevention includes avoiding sexually transmitted infections as human papilloma virus can spread through sexual contact with an infected person. Nowadays, nanotechnology has been used for the treatment of cervical cancer.

The Ag NPs that are synthesized by using leaves of *Saccharina japonica* have cytotoxicity effect on the cervical carcinoma cells. When these particles are exposed to these cancer cells, they cause apoptosis and lose their adherence property (Sreekanth et al., 2017). Appadurai et al., have synthesized the Ag NPs using the extracts of the root of *plumbago indica* (Plumbagin/PLB a nontoxic substance) and observed higher cytotoxic activity against human cervical cancer cell line. PLB prevent proliferation of cervical cancer cell. When treated with silver nanoparticles, it promotes anti-proliferating, apoptosis and anti-mitosis activities in cancer cells (Appadurai et al., 2015). Silver nanoparticles extracted from various plant derivatives may overcome the limitations of currently available cancer treatment methods (Jagpreet et al., 2017). Biosynthesis of spherical Ag NPs (having diameter of ~14 nm) from the leaves of *Podophyllum hexadrum* can be used for the treatment of human cervical cancer. The Ag NPs inhibit cellular mechanism of HeLa cells by damaging DNA and triggering apoptosis (Murugarajatal et al., 2012). Silver nanoparticles derived from the stern and bark of *Moringa olivera* has been found to be highly effective for the treatment for human cervical carcinoma cells. It induces apoptosis by producing ROS in HeLa (Vasanth, K et al., 2014).

Iresine herbstii leaf extracted silver nanoparticle exhibited strong antioxidant activity as well as cytotoxic activity against HeLa cervical cell line. This synthesis is cost efficient and ecofriendly (Dipankar et al., 2012). Green synthesis of silver nanoparticles from *Nocardiopsis sp.* can be used to treat HeLa cell (200mg/ml). Cytotoxicity of silver nanoparticles against the HeLa cell showed a dose dependent activity. IC50 value is 200 mg/ml silver nanoparticles against the HeLa cancer cell (Panchananathan et al., 2013).

It has been reported that both free and encapsulated Ag NPs are toxic for cervical cancer cell line that decreases the cell viability of cervical cancer cell (Pimental, 2013). *Argemone mexicana* leaf extracted Ag NPs are used against cervical cancer cell line SiHa due to their anticancer and cytotoxic properties (Jha AK et al., 2014).

5.3. CONCLUSION AND FUTURE ASPECTS OF SILVER NANOPARTICLES

The use of nanoparticles is gradually increasing in our day-to-day lives. From the above studies, we can assume the use of silver nanoparticles in various cancer fields like lung cancer, breast cancer, leukemia, gastrointestinal cancer, colon cancer and cervical cancer, etc. Besides Ag NPs, a number of different noble metal nanoparticles can also be used to eradicate cancer from our society. So, extensive research is essential to study the effect of nanoparticles on cancer cells as well as normal cells.

REFERENCES

1. Acharya, S., & Sahoo, S. K. (2011). Sustained targeting of Bcr-Abl + leukemia cells by synergestic action of dual drug loaded nanoparticle and its implication for leukemia therapy. *Biomaterials. 32*, 5643–5662.
2. Agasti, S. S., Chompoaser, Y. A., et al. (2009). Photoregulated release of caged anticancer drugs from gold nanoparticles. *J. Am. Chem. Soc. 131*, 5728–5729.
3. Ahmad, A., Mukherjee, P., Senapati, S., Mandal, D., Khan, M. I., & Kumar, R. (2003). Extracellular biosynthesis of silver nanoparticles using the fungus *Fusariumoxysporum colloids and Surfaces B: Biointerfaces. 28*, 313–318.
4. Alison, M. R. (2001). Cancer: encyclopedia of life sciences. Nature publishing group, London, 2001.
5. Ambrosini, G., Adida, C., & Altieri, D. C. (1997). A novel anti-apoptosis gene, survivin, *Expressed in Cancer and Lymphoma. Nat med.* 9170–921.
6. Ameen, K., Balkis, R., & Rajasekhar, T. (2007). Silver nanoparticle in mesoporous aerosol exhibiting selective catalytic oxidation of benzene in CO_2 free air. *Calalysis Letters. 119*, 289–295.
7. Appadurai, P., & Rathinasamy, K. (2015). Plumbagin- silver nanoparticle formulation enhance the cellular uptake of plumbagin and its antiproliferation activity. *IET Nano Biotechnology*, *9*, 264–272.
8. Austin, C. D., De Maziere, A. M., & Pisacane, P. I. (2004). Endocytosis and sorting of $ErbB_2$ and the site of action of cancer therapeutics trastuzumab and geldanamycin. *Mol Biol Cell*, *15*, 5268–5282.
9. Azizi, M., Ghourchian, H., Yazdian, F., Bagherifam, S., Bekhradnia, S., & Nystrom, B. (2017). Anti-cancer effect of albumin coated silver nanoparticle on MDA-MB 231 human breast cancer cell line. *Science Report. 7*, 5178.
10. Balaji, D. S., Basavaraja, S., Deshpande, R., Bedre, M. D.,Prabhakar, B. K., & Venkataraman, A. (2009). Extracellular biosynthesis of functionalised silver nanoparticle by strain of *Cladosporioids* fungus. *Colloids and surfaces B: Biointerfaces*, *68*, 88–92.
11. Choi, H. S., Seo, H. S., Kim, J. H., U. M., J. Y., Shin, Y. C., & Ko, S. G. (2012). Ethanol extract of paeuniasuffruticosaandrous (PSE) induced AGS human gastric cancer cell apoptosis and MAMZ-p^{53} pathway. *Journal of Biomedical Science*, *19*, 82.

12. Devi, J. S., & Bimba, B. V. (2013). Silver nanoparticles: anti-bacterial and invitro cytotoxic activity. *International Journal of Biological, Ecological and Environmental Sciences*, *2*, 2277–4394.

13. Di chiro, G., Brookes, R. A., Patronas, N. J., Bairamian, D., Kornblilk, P. L., Smith, B. H., Mansi, L., & Barner, J. (1984). Issue in vitro measurement of glucose metabolism of human central nervous system tumors. *Annuals of Neurology*, *15*, 1538–5146.

14. Dipankar, C., & Murugan, S. (2011). The green synthesis, characterization and evaluation of the biological activities of silver nanoparticles synthesized from *Iresineherbstii* leaf aqueous extracts. *Colloids and Surfaces B: Biointerfaces*, *98*, 112–119.

15. Evanoff & Chumanov, G. (2004). Size controlled synthesis of nanoparticle measurement of extinction cross sections. *J Phys Chem B*. *108*, 13957–13962.

16. Fodale, V., Pierobon, M., Liotta, L., & Petricoin, E. (2011). Mechanism of cell adaptation: when and how do cancer cells develop chemo resistance? *Cancer*, *17*, 89–95.

17. Foldjerg, R., Dang, D. A., & Autrup, H. (2010). Cytotoxicity and genotoxicity of silver nanoparticle in the human lung cancer cell line, A549. *Archives of Toxicology*, *85*, 743–750.

18. Fu, J. K., Liu, Y., Gu, P., Tang, D. B., Lin, Z. Y., Yan, B. Y., & Weng, S. (2000). Spectroscopic characterization on the biosorption and bioreduction of Ag (I) by Lactobacillus sp. A09. Acta, *Physico-Chemica, Sinica*. *16*, 770–782.

19. Giribabu, N., Rao, P. V., & Salleh, N. (2016). Nanoprticle and targeted systems for colon therapy. *Nano Architecture for Smart Delivery and Drug Targeting*. 695–713.

20. Gua, D., Zhao, Y., Zhang, Y., Wang, Q., Huanq, Z., Ding, Q., Guo, Z., Zhou, X., & Zhu, L. (2014). G.U.N. The cellular uptake and cytotoxic effect of silver nanoparticles on chronic myeloid leukemia cells. *J Biomed Nano Technol*, *10*, 669–678.

21. Guo, D., Zhu, L., Huanz, Z., Zhou, H., Ge, Y., Ma, W., Wu, J., Zhang, X., Zhao, Y., & Gu, N. (2013). Anti- leukemia activity of PVP-coated silver nanoparticle via generation of reactive oxygen species and release of silver oxygen species and release of silver ions. *Biomaterial*, *34*, 7884–94.

22. Gurunathan, S., Jegadeesh, R; Nurestri, A. M., Priscilla, A. J., & Vikineswary, S. (2013). Green synthesis of silver nanoparticle using *Ganodermaneo japaponicum* imazeki: a potential cytotoxic agent against breast cancer cells. *Int. J.Nanomedicine*. *8*, 4399–4413.

23. Gurunathan, S., Raman, J., Malek, S. N. B., John, P. A., & Vikineswary, S. (2013). Green synthesis of silver nanoparticle using Ganodermaneo-japonicum Imaneki: a potential cytotoxic agent against breast cancer cells. *Int J Nanomedicine*, *8*, 4399–4413.

24. Haeveli, C., Franklin, C., & Hardy, K. (1984). Plasmid determined silver resistance in *Pseudomonas* studies isolated from a silver mine. *J. Bacteriol*. *158*, 389–392.

25. He, Y., Du, Z., Ma, S., Liu, Y., Li, D., Huang, H., Cheng, S., Wu, W., Zhang, K., & Zheng, X. I. (2016). Effect of green synthesized silver nanoparticle on lung cancer cells in vitro and grown as xenograft tumor in vivo. *International journal of Nanoparticle*, *11*, 1879–1887.

26. He, Y., Du, Z., Ma, S., Liu, Y., Li, D., Huarong, H., Jiang, S., Cheng, S., Wu, W., Zhang, K., & Zheng, X. (2016). Effect of green-synthesized silver nanoparticle on lung cancer cells in vitro and grown as xenograft tumors in vivo. *Int J nanomedicine*, *11*, 1879–1887.

27. Hu, R., Yong, K. T., Roy, I., Ding, H., He, S., & Prasad, P. N. (2009). Metallic nanostructures as localized plasmon resonance enhanced scattering probes for multiplex dark field targeted imaging of cancer cell. *J. Phys. Chem. C, 113*, 2676–2684.

28. Iravani, S., Korbekandi, H., Mirmohammadi, S. V., & Zolfaghar, B. (2014). Synthesis of silver nanoparticlers: chemical, *physical and biologica; methods. Res pharm Sci.*, *9*, 385–406.

29. Jacob, S. J., Finub, J. S., & Narayanan, A. (2012). Synthesis of silver nanoparticle using piper longum leaf extract and its cytotoxic activity against Hep-2 cell line. *Colloidas surf B. Biointerfaces*, *91*: 212–214.

30. Janicke, R. U., Sprengart, M. N., Wati, M. R., & Porter, A. G. (1998). Caspase-3 is required for DNA fragmentation and morphological changes associated with apoptosis, *J. Biol. Chem.*, *273*: 935–7.

31. Jannathul, F. M., & Lalitha, P. (2015). Apoptotic efficiency of biogenic silver nanoparticles on human breast cancer MCF-7 cell lines. *Springer Prog Biomater*, *4*, 113–121.

32. Jeong, J., & Gurunathan, S. (2016). Hypoxia-mediated autophagic flux inhibits silver nanoparticle triggered apoptosis in human lung cancer cells. *Scientific Report, 6.*

33. Jeyaraj, M., Rajesh, M., Arun, R., Mubarak Ali, D., Sathishkumar, G., Sivanandhan, G., Kapildev, G., Manickavasagam, M., Premkumar, K., Thajuddin, N., & Ganapathi, A. (2013). An investigation on the cytotoxicity and caspase –mediated apoptotic effect of biologically synthesized silver nanoparticles using *Podophyllum hexandrum* on human cervical carcinoma cells. *Colloids and Surfaces B: Biointerfaces, 102*, 708–717.

34. Jeyaraj, M., Sathishkumar, G., Sivanandhan, Mubarak & Ali, D. (2013). Biogenic silver nanoparticle for cancer treatment: an experiment report. *Colloids Surf. B., 106*, 86–92.

35. Jha, A. K., & Prasad, K. (2014). Green synthesis of silver nanoparticle and its activity on siHA cervical cell line. *Adv. Mat. Lett. 5*, 501–505.

36. Jung, J., Oh, H., Noh, H., Ji, J., & Kim, S. (2006). Metal nanoparticle generation using a small ceramic heater with a local heating area. *J Aerosol Sci. 37*, 1662–1670.

37. Kabashin, A. V., & Meunier, M. (2003). Synthesis of colloidal nanoparticles during femtosecond laser ablation of gold in water. *J Appl. Phys. 94*, 7941–7943.

38. Kalishwaralal, K., Deepak, V., Ramkumarpandian, S., Nellaiah, H., & Sangiliyandi, G. (2008). Extracellular biosynthesis of silver nanoparticle using culture supernatants of Entero bacteria: *A Novel Biological Approach Process Biochemistry, 42*, 919–923.

39. Kalishwaralal, K., Deepak, V., Ramkumarpandian, S., Nellaiah, H., & Sangiliyandi, G. (2008). Extracellula biosynthesis of silver nanoparticle by the culture supernatant of Bacillus licheniformis. *Mater lett., 62*, 4411–4413.

40. Kawasaki, M., & Nishimura, N. (2006). 1064-nm laser fragmentation of thin Au and Ag flakes in acetone for highly productive pathway to stable metal nanoparticles. *Appl. Surf Sci. 253*, 2208–2216.

41. Kim, H. R., et al. (2014). Silver nanoparticles induce p53-mediated apoptosis in human bronchial epithelial (BEAS-2B) cells. *J Toxicol Sci., 39*, 401–412.

42. Klaus, T., Joerger, R., Olsson, E., & Granqvist, C. (1999). Silver-based crystalline nanoparticle, microbially fabricated. *Pro. Natl. Acad. sci. USA. 96*, 13611–13614.

43. Krishnaraj, C., Kalaichelvan, P. T., Ramachandran, R., Balakumaran, M. D., & Kalaichelvan, P. T. (2014). *Biotechnology reports. 4*, 42–49.

44. Krishnaraj, C., Muthukumaran, P., Ramachandran, R., Balakumaran, M. D., & Kalaichelvan, P. T. (2004). *Acalyphaindica*linn: biogenic synthesis of silver and gold nanoparticle and their cytotoxic effects against MDA-MB-231, human breast cancer cells. *Biotechnol Rep., 4*, 42–49.

45. Kruis, F., Fissan, H., & Rellinghaus, B. (2000). Sintering and evaporation characteristic of gas-phase synthesis of size-selected PbS nanoparticles. *Mater Science B., 69*, 329–334.

46. Kuddus, S. A. (2017). Nanoparticles to deal with Gastric cancer. *Journal of Gastrointestnal Cancer and Stromal Tumors, 2*, 2.

47. Liu, J., Zhao, Y., Guo, Q., Wang, Z., Wang, H., Yang, Y., & Huang, Y. (2012). TAT- modified nano silver for combating multidrug-resistant cancer. *Biomaterials. 33*, 6155–6161.

48. Magnusson, M., Deppert, K., Malm, J., Bovin, J., & Samuelson, L. (1999). Gold nanoparticles: production, reshaping and thermal charging. *J Nanoparticle res. 1*, 243–251.

49. Mandal, D., Bolanderi, M., Mokhopadhayayan, D., Sarkan, G., & Mukherjee, P. (2006). *Applied Micro Biao Biotechnol. 69*, 485.

50. Manivasagan, P., & Venkatesan, J. (2013). Biosynthesis, antimicrobial and cytotoxic effect of silver nanoparticles using a novel Nocardiopsis sp. MBRC-1. *Bio Med Research International*. 9.

51. Merga, G., Wilson, R., Lynn, G., Milosavljevic, B., & Meisel, D. (2007). Redox catalysis on naked silver nanoparticles. *J phys. Chem C. 111*, 12220–12206.

52. Moaddad, S., Ahari, H., Shahbazzodeh, D., Motallebi, A. A., Rahaman-Nya, J., & Shokrgozar, M. R. (2011). Toxicity study of nanosilver (Nanocid) on oseoblast cancer cell line. Iran. *Nano Lett. 16*, 11.

53. Molina, M. A. F., Gambea, E. M., Rivera, C. A. S., Flores, R. A. G. (2010). Antitumous activity of colloidal silver on MCF-7 human breast cancer cells. *Padil J Expolin cancer Res. 29*, 148.

54. Nalenthiran, P., Sambandam, A., Govindarajn, G., Nyayiru, K. U. P., Simon, C., & Muthupandian, A. K. (2009). Microbial synthesis of silver nanoparticle by *Bacillus sp. Journal of Nanoparticle Research. 11*, 18.

55. Oliveira, M., Ugarte, D., Zanchet, D., Zarbin, A 2005. Influence of synthetic parameters on the size, structure and stability of dodecanethiol-stabilized silver nanoparticle. *J Colloid Interface Sci. 292*, 429–435.

56. Pimental, R. C. (2013). Silver nanoparticles nanocarrier, synthesis and toxic effect on cervical cancer cell lines. *Bio Nanoscience, 3*, 198–207.

57. Prabhu, D., Arulvasu, C., Babu, G., Manikandan, R., & Srinivasan, P. (2012). Biologically synthesized green silver nanoparticle from leaf extract of *Vitexne gundo* L induced growth inhibitory effect on human colon cancer cell line HCT 15. *Process Biochemistry. 48*, 315–324.

58. Pradhan, N., Nayak, R. R., Pradhan, A. K., Sukla, L. B., & Mishra, B. K. (2011). In situ synthesis of entrapped silver nanoparticles by a fungus—*Penicillium purpurogenum. Nanoscience and Nanotechnology Letters. 3*, 659–665.

59. Sadowski, Z., Maliszewska, I. H., Grochowalska, B., Polowczyk, I., & Kozlecki, T. (2008). Synthesis of silver nanoparticle using microorganisns. *Materials Science-Poland., 26*, 419–424.

60. Satapathy, S. R., et al. (2013). Silver-based nanoparticles induce apoptosis in human colon cancer cells mediated through p53. *Nanomedicine (Lond). 8*, 1307–1322.

61. Satapathy, S. R., Mohapatra, P., Das, D., Siddhart, S., & Kundu, C. N. (2016). The apoptotic effect of plant based nanosilver in colon cancer cells is a P53 dependent process involving ROS and JNK cascade. *Pathology and Oncology Research., 21*, 405–411.

62. Satapathy, S. R., Mohapatra, P., Preet, R., Das, D., Sarkar, B., Choudhury, T., Wyatt, M. D., & Kundu, C. N. (2013). Silver-based nanoparticles induced apoptosis in human colon cancer cell mediated through P53. *Nanomedicine, 8*.

63. Sathyavathi, R., Krishna, M. B., Rao, S. V., Saritha, R., & Rao, D. N. (2010). Biosynthesis of silver nanoparticle using *Coriandrum sativum* leaf extract and their application in nonlinear optics. *Adv. Sci. Lett., 3*, 138–143.

64. Schreck, R., Meiar, B., Mannel, D. N., Drege, W., & Baeuerle, P. A. (1992). Dithiocarbamates as potent inhibitor of nuclear factor kappa

B Activation in intact cells. *Journal of Experimental Medicine.*, *175*, 1181–1194.

65. Shahverdi, A. R., Minaeian, S., Shahverdi, H. R., Jamalifar, H., & Nohi, A. (2007). Rapid synthesis of silver nanoparticles using culture supernanants of Enterobacteria: *A Novel Biological Approach Process Biochemistry.*, *42*, 919–923.
66. Singh, J., Singh, T., & Rawat, M. (2017). Green synthesis of silver nanoparticle via various plant extract for anti-cancer application. *Glob J Nanomed.*, *2*, 1–4.
67. Sironmani, T. A. (2016). Therapeutic potential of neem synthesis silver nanoparticle on human gastric cancer cells in vitro. *World Journal of Nano Science and Engineering*, *6*, 90–110.
68. Sonbaty, S. M. (2013). Fungus-mediated synthesis of silver nanoparticles and evolution of antitumor activity. *Cancer Mano*, *293*, 73–79.
69. Soni, G.,Yadav, K. S. (2014). Application of nanoparticle in treatment and diagnosis of leukemia. *Material Science and Engineering*, *47*, 156–164.
70. Sreekanth, T. V. M., & Pandurangan, M. (2017). Green synthesis: In vitro anticancer activity of silver nanoparticles on human cervical cancer cells. *Journal of Cluster Science*, *27*, 671–681.
71. Subbaiya, R., Preetha, L., Gayathri, S., Swarnalatha, A., & Selvam, M. M. (2014). Synthesis and charaterisation of silver nanoparticle from Rhodococcus-2891 and its anti-tumor activity against lung cancer cell line (A549). *Science Engineering and Management Research* (ICSEMR). 2014, 1-6.
72. Sulaiman, G. M., Mohammed, W. H., Marzoog, T. R., Amiery, A. A., Kadhum, A. A. H., & Mohamad, A. B. (2013). Green synthesis, antimicrobial and cytotoxic effects of silver nanoparticles using Eucalyptus chapmaniana leaves extract. *Asian Pac J Trop Biomed.*, *3*, 58–63
73. Suman, T. Y., RadhikaRajasree, S. R., Kanchana, A., Elizabeth & Been, S. (2013). Biosynthesis, characterization and cytotoxic effect of plant mediated silver nanoparticle using Mornindacitrifolia root extract colloids surf B *Biointerfaces.*, *106*, 74–78.

74. Sun, Y. (1990). Free radicles, antioxidant enzyme and carcinogenesis. *Free Radical Biology and Medicine. 8*, 585–599.
75. Sylvestre, J. P., Kabashin, A. V., Sacher, E., Meunier, M., & Luong, J. H. T. (2004). Stabilization and size control of gold nanoparticles during laser ablation in aqueous cyclodextrins. *J Am Chem Soc., 126*, 7176–7177.
76. Tarasenko, N., Butsen, A., Nevar, E., & Savastenko, N. (2006). Synthesis of nanosized particles during laser ablation of gold in water. *Appl Surf Sci., 252*, 4439–4444.
77. Tsibakhashvili, N., Kalabegishvili, T., Gabunia, V., Gintury, E., Kuchava, N., Bagdavadze, N., Pataraya, D., Gurielidzse, M., Garjiladze, D., & Lomidze, L. (2010). *Synthesis of silver nanoparticle using Bacteria Nano Studies*, *2*, 179–182.
78. Valabrega, G., Montemurro, F., & & Aglietta, M. (2007). Trastuzumab: mechanism of action, resistance and future perspectives in HER2-overexpressing breast cancer. *Annals of Oncology*, *18*, 977–984.
79. Valku, M., Leibfritz, D., Moncol, J., Crown, M. T. D., Mazur, M., & Telser, J. (2007). Free radical and Antioxidants in normal physiological function and human disease. *The International Journal of Biochemistry and Cell Biology*, *39*, 44–84.
80. Vasanth, K., Ilango, K., Ramasamy, M. K., Agarwal, A., & Dubey, P. (2014). Anticancer activity of Moringaoleifera mediated silver nanoparticle on human cervical carcinoma cells by apoptosis induction. *Colloids and Surfaces B: Biointerfaces*, *117*, 354–359.
81. Welber, W. A. (2006). Position emission tomography as an imaging biomarker. *Journal Clinical Oncology*, 29, 3282–3292.
82. Wiley, B., Sun, Y., Mayers, B., & Xi, Y. (2005). Shape controlled synthesis of metal nanostructures: the case of silver. *Chem Eur J.*, 11, 434–463.
83. Zhi, J. C., Khan, Z. L., Saxi, F., & Birch, T. D. J. (2008). Birch nanomedicine and its potential in Diabetes research and practice. *Diabetes Metab Res Rev.*, 24, 604–610.

6
CHAPTER

NANOMATERIALS: FUTURE PROSPECTIVE FOR WASTEWATER TREATMENT AND SUSTAINABLE

Minati Behera, Sanghamitra Mohapatra, and Upasana Mohapatra

Department of Environmental and Sustainability, CSIR-Institute of Minerals and Materials Technology, Bhubaneswar-751013, India

CONTENTS

ABSTRACT

Nowadays, we are facing severe scarcity of potable/fresh water for day-to-day life. The fresh water resource is continuously contaminated due to rapid population growth, industrialization, agriculture and urbanization. That means a huge amount of contaminated water (wastewater) is generated which aggravates the water scarcity combined with the deteriorating fresh water reserve. According to the World Water Assessment Programme (WWAP, 2016), approximately 70% of industrial wastewater are discharged to surface water without proper treatment or primary treatment in most of the developing countries. Due to the lack of proper treatment and sanitation, most of the wastewater can't be utilized even for agriculture in order to avoid outbreak of diseases among the farmers and consumers. The available traditional methods are not upto the mark according to the strict guidelines of environmental protection board of every country. Moreover, the existing technologies have their own disadvantages, e.g., high coast, high energy inputs, incompletely removal of pollutants and generation of secondary pollutants which are more persistence in nature. Biological methods are promising but slow in nature and the presence of non-biodegradable materials hinder growth of microorganisms which make their application infeasible at the contaminated site. Therefore, development of innovative and suitable technology for proper treatment and disinfection of wastewater is a major challenge nowadays. Recently, nanotechnology has gained importance in the field of water and wastewater treatment to secure water supply. Due to their nano dimension (1–100 nm) they exhibit a great potential in contrast to their bulk counterparts. This chapter presents a brief discussion on the preparation of nanomaterials and their application toward the wastewater treatment.

Keywords: Wastewater treatment, Carbon based nanomaterials, Metal and metal oxides, Nanocomposite, Nanomembrane

6.1. INTRODUCTION

Clean water is one of the basic requirements for survival of all living organisms. The fresh water reserve, which is ~ 1% of the net global water source, is continuously contaminated by human activities, e.g., population growth, industrialization, urbanization and agriculture. AQUASTAT of FAO's (Food and Agriculture Organization of United state) stated that, out of the 3928 km^3/year global freshwater withdrawals, 1716 km^3/year is consumed mainly by agriculture and remaining 2,212 km^3/year is discharged into the environment as wastewater in the form of municipal, industrial and agricultural effluents. According to the United Nations, every year, the industrial activities release about 300–500 million tons of heavy metal, solvent and other waste and harmful by-products to the fresh water supplies (UN, 2011). In addition to these, water contamination also occur due to some natural processes like weathering of rocks which is responsible for arsenic contamination (which is a serious issue in countries like India, Bangladesh and Nepal) (Tesh and Scott, 2014). It is predicted that in the coming era, the demand for freshwater will increase significantly, worldwide (UN, 2017). In Figure 6.1, worldwide consumption of freshwater and the percentage of wastewater discharged by the major sectors has been represented. There is an acute pressure to get clean water for day-to-day life. According to World Health Organization (WHO, 2014), 50% of the world's population will live in water stressed area in near future. A study of United Nations (UN) stated that 4,400 children under the age group of five die every day due to unclean water and sanitation. Still, the major cause of death in developing countries is waterborne diseases on a global level. So, scarcity of clean water is the major issue and challenges for the researchers. However, in contrast to this a huge amount of contaminated water (wastewater) is generated and discharged to the natural water bodies which affect the flora and fauna of the ecosystem. It has been estimated that, approximately 80% of wastewater get discharged to the environment without proper treatment (UN, 2012; UN, 2015).

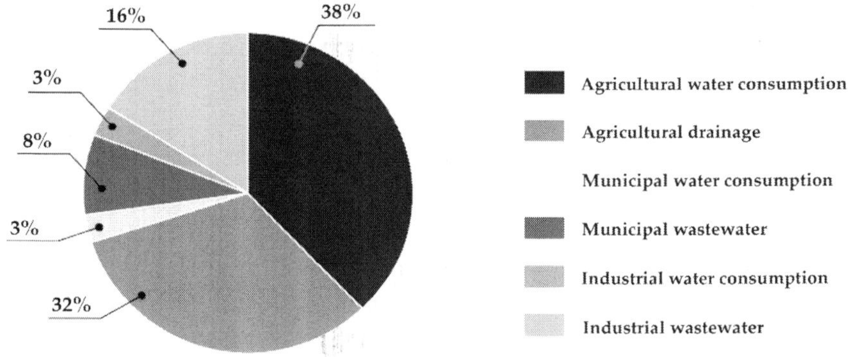

Figure 6.1: Worldwide freshwater withdrawals, consumption and wastewater generation by major water use sector (From Circa, 2010).

Sato et al., reported that about 70% of wastewater was treated by the developed countries while the percentage drops to 38% in the case of upper-middle-income countries, 20% in the case of middle-income countries and only 8% in the case of low-income countries (2013). This information suggests that the people of rural area due to lack of proper treatment and sanitation facilities, are directly exposed to untreated water. Figure 6.2 represents the overall difference in current levels of wastewater treatment which will place a high financial burden on low-income and middle-income countries.

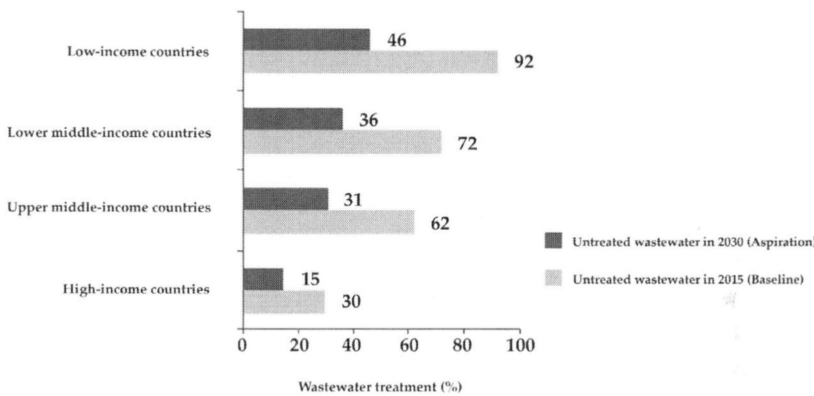

Figure 6.2: Untreated wastewater percentage of countries with different income levels in 2015 and aspiration for 2030, (From UN-Water 2017: Sato et al., Agri. Water. Mang. 130, 1, 2013).

In developing countries, most of the deaths occur due to infectious waterborne diseases which are still a global issue. The introduction of disinfection processes, e.g., chlorination, ozonation and ultraviolet irradiation have significantly reduced the waterborne diseases but have triggered the generation of few by-products which are highly carcinogenic and mutagenic in nature as well as are expensive (Sean et al., 2015; Tian et al., 2013; Chowdhury et al., 2011; Bougeard et al., 2010). Due to the lack of proper treatment and sanitation, most of the wastewater can't be utilized even for agriculture so as to avoid outbreak of diseases among the farmers and consumers. The available traditional methods are not up-to the mark according to the strict guideline of environmental protection board of every country. Moreover, the existing technologies have their own disadvantages, e.g., high coast, high energy inputs, incomplete removal of pollutants and generation of secondary pollutants which are more persistence in nature (Ferroudj et al., 2013; Anjum et al., 2016). However, even though the biological methods are promising, they are slow in nature, and also the presence of non-biodegradable materials hinders the growth of microorganisms. Also their application is not feasible at the contaminated site in field scale (Zelmanov and Semiat, 2008). Therefore, development of innovative and suitable technology for proper treatment and disinfection of wastewater is a major challenge nowadays.

6.2. NANOTECHNOLOGY IN WASTEWATER TREATMENT

Recently, nanotechnology has gained importance in the field of water and wastewater treatment to secure the quality of water supply. Compared to macro and micro-scale counterparts, nanomaterials are extensively used for wastewater treatment due to their minuscule size (1–100 nm) (Tesh and Scott, 2014), versatile physical, chemical and biological characteristics, large surface area to volume ratio facilitating higher density of surface reaction site per unit mass, greater surface free energy, etc. Therefore, they have a phenomenal surface reactivity and display high reactivity for surface mediated processes Moreover, on further reduction of particle size (when the particle size of nanomaterial approaches the electron mean-free path and wavelength scales, <30 nm), the materials show different optical/physical properties (Khin et al., 2012; Qu et al., 2013).

Nowadays, with the development of nanotechnology, its application in water and wastewater treatment is becoming imminent. Significant research have been carried out on designing of nanomaterials of different shape and size like nanowires, nanotubes, films, particles, quantum dots, and colloids, etc. (Edelstein and Cammaratra, 1998; Lubick and Betts, 2008). Till date, various studies on nanomaterials have been suggested that nanomaterials have huge capability and potential in water and wastewater treatment, particularly in the areas of adsorption, membrane process, catalytic oxidation, disinfection and sensing (Ali et al., 2012; Pendergast and Hoek, 2011, Ayati et al., 2014, Das et al., 2015). However, with advances in nanoscale science and engineering, it has been found that the water quality could be resolved by using nano absorbent, nano catalysts, bioactive nanoparticles, nanostructured catalytic membranes, nano powder, nanotubes, magnetic nanoparticles, granules, flake, high surface area metallic supramolecular assemblies with characteristic length scales of 9–10 nm including clusters, micro-molecules, nanoparticles and colloids in natural environment (Mamadou and Savage,

2005). Recently, various types of efficient, eco-friendly and cost-effective nanomaterial have been synthesized with unique functionalities for potential disinfection of industrial effluent, surface water, ground water and drinking water (Brumfiel, 2003; Theron et al., 2008; Gupta et al., 2015). The reported nanomaterials having significant role in wastewater treatment are carbon nanotubes (CNTs), graphene, ferric oxide (Fe_3O_4), manganese oxide (MnO_2), titanium oxide (TiO_2), magnesium oxide (MgO) and zinc oxide (ZnO) (Wang et al., 2013; Luo et al., 2010; Singh et al., 2013 Feng et al., 2010; Singh et al., 2011; Shen et al., 2009; Wang et al., 2012). The antimicrobial effect of various nanomaterials for killing of microorganism was represented in Figure 6.3. One of the successful and commercially available nanomaterials is the zero valent iron nanoparticles which is used in the United States of America for the treatment of ground water. Figure 6.4 represents the schematic representation of zero valent iron nanoparticles (INPs) injected into the ground as a dry powder or slurry to directly treat contaminated groundwater. Tesh and Scott (2014) reported that NPs can be significantly used as a deep underground permeable reactive barrier (PRB) in immobilized form, or migrated with the contaminated plume of water for a deep underground water treatment. Current and potential applications of nanotechnology in water and wastewater treatment was summarized in Table 6.1.

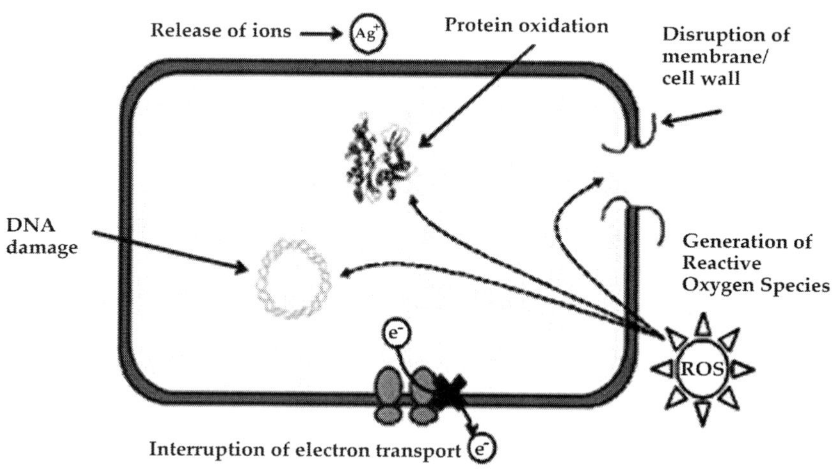

Figure 6.3: Nanoparticles exert various antimicrobial activities for microbial killing, a promising approach for wastewater disinfection, (From Li et al., 2008).

Table 6.1: Current and Potential Applications of Nanotechnology in Water and Wastewater Treatment (From Qu et al., Wat. Res. 47, 3931, 2013)

Application	Representative nanomaterials	Desirable nanomaterial properties	Enabled technologies
Adsorption	Carbon nanotubes	High specific surface area, highly assessable adsorption sites, diverse contaminant-CNT interactions, tunable surface chemistry, easy reuse	Contaminant preconcentration/detection, adsorption of recalcitrant contaminants
	Nanoscale metal oxide	High specific surface area, short intraparticle diffusion distance, more adsorption sites, compressible without significant surface area reduction, easy reuse, some are superparamagnetic	Adsorptive media filters, slurry reactors
	Nanofibers with core shell structure	Tailored shell surface chemistry for selective adsorption, reactive core for degradation, short internal diffusion distance	Reactive nano-adsorbents
Membranes and membrane processes	Nano-zeolites	Molecular sieve, hydrophilicity	High permeability thin film nanocomposite membranes
	Nano-Ag	Strong and wide-spectrum antimicrobial activity, low toxicity to human	Anti-biofouling membranes
	Carbon nanotubes	Antimicrobial activity (unaligned carbon nanotubes) Small diameter, atomic smoothness of inner surface, tunable opening chemistry, high mechanical and chemical stability	Aligned carbon nanotube membranes
	Aquaporin membranes Nano-TiO_2	High permeability and selectivity Photocatalytic activity, hydrophilicity, high chemical stability	Aquaporin membranes Reactive membranes, high performance thin film nanocomposite membranes, Forward osmosis
	Nano-magnetite	Tunable surface chemistry, superparamagnetic	
Photocatalysis	Nano-TiO_2	Photocatalytic activity in UV and possibly visible light range, low human toxicity, high stability, low cost	Photocatalytic reactors, solar disinfection systems
	Fullerene derivatives	Photocatalytic activity in solar spectrum, high selectivity	Photocatalytic reactors, solar disinfection systems

Disinfection and microbial control	Nano-Ag Carbon nanotubes Nano-TiO_2	Strong and wide-spectrum antimicrobial activity, low toxicity to humans, ease of use Antimicrobial activity, fiber shape, conductivity Photocatalytic ROS generation, high chemical stability, low human toxicity and cost	POU water disinfection, antibiofouling surface POU water disinfection, antibiofouling surface POU to full scale disinfection and decontamination
Sensing and monitoring	Quantum dots Noble metal nanoparticles Dye-doped silica nanoparticles Carbon nanotubes Magnetic nanoparticles	Broad absorption spectrum, narrow, bright and stable emission which scales with the particle size and chemical component Enhanced localized surface plasmonresonances, high conductivityHigh sensitivity and stability, rich silica chemistry for easy conjugation Large surface area, high mechanical strength and chemical stability, excellent electronic properties Tunable surface chemistry, superparamagnetism	Optical detection Optical and electrochemical detection Electrochemical detection, sample preconcentration Sample preconcentration and purification

Although, nanotechnology is promising for water and wastewater treatment, there are a number of inherent disadvantages which significantly hinder the direct use of free nanoparticles in the treatment process. Lofrano et al. (2016) reported both in fluidized system or in fixed bed, NPs tend to aggregate, resulting in severe activity loss and pressure drop. So it is a matter of concern which possess a tedious task to separate most of the exhausted NPs (except for magnetic nanoparticles) from the treated water for reuse. It seems that in terms of economical consideration it would be quite undesirable (Al-Hamadani et al., 2015; Qu et al., 2013). Dale et al. (2015) and Varma (2012) suggested that the behavior and fate of the nanomaterials in water and wastewater treatment process are not fully understood, and a major issue that could hinder their application is the impact of nanomaterials on the aquatic system and human health.

6.3. TYPES OF NANOMATERIALS USED IN WASTEWATER TREATMENT

- Carbon based nanomaterials
- Metal and metal oxides
- Nanocomposites
- Nano membranes

6.4. SYNTHESIS OF NANOMATERIALS USED FOR WASTEWATER TREATMENT

6.4.1. Carbon Based Nanomaterials

Carbon nanotube (CNT) is a tube-shaped material consisting of carbon atoms, having diameter of a few nanometers. CNTs may be single walled, i.e., consisting of a single rolled graphene sheet, called single walled nanotubes (SWNTs) or multi-walled, consisting of multiple rolled tubes of graphene, called multi walled nanotubes (MWNTs). Most commonly used method to synthesize CNTs is chemical vapor deposition (CVD) (Kumar and Ando, 2010). However, other methods such as arc discharge and laser ablation are also used (Gore and Sane, 2011). In CVD synthesis, hydrocarbon gas is decomposed to carbon in a furnace at high temperature in presence of an inert gas and carbon nanotubes grow on the substrates in presence of catalysts. The temperature of the furnace, type of hydrocarbon used and size of the catalysts play a decisive role to determine the type of CNTs to be synthesized (Zhang et al., 2011). However, the scalable production of homogeneous CNTs is associated with high cost.

Graphene is another allotrope of carbon consisting of a single layer of carbon atoms arranged in a two-dimensional hexagonal lattice. Graphene sheets were first obtained from graphite by using adhesive tape (Novoselov et al., 2004). Afterwards, different methods such as liquid phase, thermal exfoliation of graphite, laser ablation, CVD and synthesis on silicon carbide, etc. have been adopted for their scalable production (Novoselov et al., 2012).

6.4.2. Metal and Metal Oxide-Based Nanoparticles

Following procedures are broadly adopted to produce various Ni and Fe based nanoparticles.

6.4.2.1. Solvothermal Method

Solvothermal method is a heterogeneous method which is performed in closed system in the presence of solvent where the temperature is more than the solvent. Rafique et al. (2012) followed this method to prepare Ni-bipyridine and Fe-bipyridine separately by direct reaction with nickel chloride and iron chloride with 4,4-bipyridine. 100 ml of 2-propanol was used to dissolved Nickel chloride (0.1 mol) and 4,4-bipyridine (0.1 mol) separately, then dropwise 4,4-bipyridine solution was added to nickel chloride solution. The resulting mixture was vigorously stirred and refluxed at 80°C for 10 hours, then cooled down to room temperature. Green precipitation was collected at the end of the process which was washed with dry ether multiple times to remove any impurities. Finally the product was vacuum dried at 60°C for overnight. Fe-pyridine was also prepared in the same manner where white precipitation was collected at the end of the process. In argon fitted carbolite furnace both of the grinded samples were sintered and calcinated under inert atmosphere to produce the corresponding nanoparticles.

Particles of nickel oxide (NiO) can be obtained by adding the solution of 1.5g $NaHCO_3$ (in 10 ml of water) to the solution of 2.3 g $NiCl_2$ (in 10 ml water) under constant stirring till the formation of precipitate. Then the precipitates collected by centrifugation, washed repeatedly in distilled water and dried in hot air oven to remove the adsorbed moisture (Chakrabarty and Chatterjee, 2009). Composite particle of Iron and Nickel oxides (Fe_3O_4 and NiO) can be prepared by adding the oxides in equal ratio, mixing and fine grinding.

6.4.2.2. Co-precipitation Method

Co-precipitation method is preferred due to its ability to synthesize smaller size and more stable metal oxide nanoparticles as compared to solvothermal or hydrothermal method. Also control on nucleation, growth and agglomeration is required during synthesis of the nanoparticles which can only be done by co-precipitation method. (Bangash and Alam, 2004).

Patricia et al. (1999) have prepared 1:2 molar ratio of $FeCl_3$ and $FeCl_2$ in acidic solution having volumes 4 ml and 1 ml respectively can be stirred using a magnetic stirrer and then added to 1 M NaOH solution till a black precipitate is formed which is then subjected to strong magnetic field to collect the precipitate and discard the supernatant. After that, the precipitates were filtered, dried at 150°C for 2 hours and put in desiccators to remove moisture. The formation of Fe_3O_4 involves the following chemical reaction as

$$Fe^{2+} + 2Fe^{3+} + 8OH^{-} \rightarrow Fe_3O_4 + 4H_2O$$

Nickel oxide nanoparticles were also synthesized by co-precipitation method where NaHCO3 (1.5 g in 10 ml water) solution was drop-wise added to the continuously stirring aqueous solution of NiCl2 (2.3 g in 10 ml distilled water). A bluish color precipitate was formed which was centrifuged and washed repeatedly with distilled water and dried in oven at 100°C to produce the final product (Chakrabarty et al., 2009).

$$NiCl_2 + NaHCO_3 + H_2O \rightarrow NiO + NaCl + CO_2 + H_2O$$

Taman et al. (2015) have prepared nano copper oxide particles, by pouring 25 ml poly ethylene glycol in to solution of 5 gram $Cu(NO_3)_2.3H_2O$ in 250 ml deionized water. Potassium Hydroxide is then added after dispersal of poly ethylene glycol in the solution and the solution was stirred for 3 hours. The blue suspension obtained was heated at 800°C for 15 minutes. The precipitate obtained was then filtered, washed and dried for 2 hours at 200°C. Controlled hydrolysis of $TiCl_4$ is used to produce titanium oxide nanoparticles (Lee and Yang, 2005; Sun et al., 2009). For the anatase form, 0.025 M ammonium sulphate added slowly to 3.0 M $TiCl_4$ solution (prepared by adding water to $TiCl_4$ in an ice water bath) and stirred till a concentration of 0.5 M was reached. Then 2.5 M NH_4OH added to the solution and stirred till pH level of 7 was reached. On the other hand, for the rutile form, 0.5 M hydrochloric acid added slowly to 3M $TiCl_4$ (prepared by adding water to $TiCl_4$ in an ice water bath) and stirring till a concentration of 0.5 M was reached. Whereas, both the phases were obtained by mixing a small amount of ammonium sulphate to 3M $TiCl_4$,

The solutions obtained from either of the two aforementioned procedures may be gently stirred and heated at 70°C for 10 h. In order to avoid product heterogeneity, the aforementioned processes should be run slowly (Iwaszuk et al., 2013). After completion of hydrolysis, the resultant powder is then centrifuged and washed with distilled water. The as obtained product was then dried by advanced super electrical method (Ding et al., 1999) to prevent gel pore destruction and hard agglomeration, after which the samples were calcined at 400°C for 2.5 hours to obtain either anatase or rutile depending on the procedure followed.

Zinc oxide can be synthesized using various methods such as precipitation method, hydrothermal/solvothermal and sol-gel method, etc. (Kolodziejczak and Jesionowski, 2014). Following different methods with controlled variables may result in different nanoparticle size, shapes and dispersity, etc. (Espitia et al., 2012; Vaseemet al., 2010). Zinc oxide nanoparticles, having size 20–25 nm, can be synthesized using solution of $Zn(C_2H_3O_2)_2H_2O$ and xylene in methanol under a pH of 6.7—6.8 and refluxing the solution at 65°C for 6 hours till a white precipitate was formed. Then the precipitate was dried in a hot air oven at 100°C for 24 hours, crushed with mortar/pestle to get zinc oxide nano particles (Navale et al., 2015).

Chemical reduction of Silver salts (e.g., Silver nitrate ($AgNO_3$ 99.5%)) by using sodium citrate ($C_6H_5O_7Na_3$) or sodium borohydride as reducing agent and PVA(poly vinylpyrrolidone) as stabilizer is the most popular method for production of silver nanoparticles. Here, the particle size is greatly influenced by different factors such as, reaction temperature, concentrations of the metal salt, reducing agent, reaction time, etc. (Šileikaitėetal. 2006). Different concentration of $AgNO_3$ solution can be prepared by dissolving different amount of $AgNO_3$ (0.001 mg, 0.002 mg 0.003 mg) separately in 50 mL of distilled water (Kheybari et al., 2010). The solutions were then heated up to the boiling point and then 5 ml of 1% trisodium citrate was added to it in a drop-wise manner under vigorous stirring. After the solution turns light yellow, PVA was added to prevent aggregation of Ag colloids. Here, the color of the solution depends upon the initial concentration of $AgNO_3$ in the reaction solution. Lower value of $AgNO_3$ in the initial solution caused yellow color in the final solution and increase in concentration of $AgNO_3$ resulted in a brown colored final solution. The reaction involved in this process is as follows:

$$4Ag+ + C_6H_5O_7Na_3 + 2H_2O \rightarrow 4Ag^0 + C_6H_5O_7H_3 + 3Na++ H+ + O_2\uparrow$$

The Ag NPs can also be synthesized using other procedures such as thermal decomposition, electrochemical, laser ablation and chemical reduction (Zhu et al., 2001; Baeetal. 2002; Šileikaitė et al., 2006; Kheybari et al., 2010).

6.4.3. Nanocomposite

Every nanomaterials have certain drawbacks, which limit their use. For example, metal oxide NPs have limited light adsorption in the ultraviolet region due to high band gap energy. Limitation for use of nano filtration membranes is membrane fouling. Carbon nanotubes have high cost. Studies show that, for ensuring better removal efficiency and to overcome the problems with Nanomaterials, nanocomposites can be synthesized. For example, studies show that, nanocomposite of nZVI on CNTs, through chemical deposition of nZVI on CNTs, acts as an adsorbent and has good potential for quick and effective removal of nitrate in water. Also, due to its magnetic property, it can be easily separated from water (Azari et al., 2014). A 2% starch solution (prepared in acetic acid) and 2% chitosan solution (prepared in aceticacid) were mixed in 3:7 ratio under continuous stirring followed by the addition of 1% glutaraldehyde (a cross linking agent) and a few drops of glycerol (as a plasticizer). Then 2 ml of starch chitosan solution was added to the above mixture solution and put into moulds of 2.5 cm diameter for solidification. The moulds were kept in an oven at 45°C overnight (Baranet al., 2004). The nanomaterial based composites were removed from the mould the next morning and stored in sealed bags.

6.4.4. Nano Membranes

Nitrocellulose membrane filters coated with single-walled carbon nanotubes (SWNT) and SWNT based nanocomposites are used to remove and inactivate the bacterial cells during filtration and hence can be used for wastewater treatment. Studies show that, membranes coated with polyvinyl-N-carbazole (PVK):(SWNT) nanocomposite containing PVK and SWNT (in 97:3 Weight % Ratio) achieve similar or improved removal of bacteria as compared to 100% SWNT coated membranes, and since SWNTs are costly, using the nanocomposites decreases the cost of the membranes (Kang et al., 2007). Also, as SWNT is toxic, reduced SWCNT content in the nanocomposite means reduced toxicity. This has been demonstrated by conducting cytotoxicity tests on fibroblast cells which shows that, the PVK: SWNT nanocomposite is considerably less toxic to fibroblast cells as opposed to 100% SWNT.

6.5. APPLICATIONS OF NANOMATERIALS TO THE CONTAMINATED WATER

The increasing demand for clean water and rising applications of nanotechnology have drawn significant efforts to discuss various aspects of water treatment by nanomaterials. Nanotechnology has been utilized in almost all branches of science and technology and hold immense potential which can be exploited in various water treatment techniques. Functionalized nanomaterials have been developed and employed for water treatment. It has been reported that TiO_2, ZnO, ceramic membranes, nanowire membranes, polymer membranes, carbon nanotubes, submicron nano powder, metal (oxides), magnetic nanoparticles, nanostructured boron doped diamond nanoparticles were utilized in various methods such as photocatalysis, nano filtration, adsorption and electrochemical oxidation to resolve problems involving water quality in natural environment (Ali, 2012; Pendergast and Hoek, 2011; Ayati et al., 2014; Das et al., 2015). A brief description regarding the application and mechanism of wastewater treatment by nanoparticles is given in this section. Most commonly studied NPs for wastewater treatment are carbon based nano materials (prepared in different forms such as graphene, nanotubes, etc.), and metal oxides.

In recent decades, carbon based NPs have been extensively used in the field of removal of heavy metals and dye from contaminated water, because of their nontoxicity, abundance, ease synthesis, high surface area, porosity, stable structure and high sorption capacities (Glassmeyer et al., 2005; Molinari et al., 2004; Al-Degs et al., 2006; Sadegh et al., 2015a; Babel et al., 20013; Li et al., 2003; Stafiej et al., 2007). Few selected species are described along with their size distribution in the following.

At the early stage activated carbon (AC) was used as sorbents; however, due to the obstacles associated with the heavy metals and dye removal at ppb levels (parts per billion), other carbon-based nanomaterials are being used. Nowadays, AC is being used due to the weak acidic ion exchange property, high porosity, high surface area, and ease of synthesis AC gained enough attention for metal contaminants removal and adsorption of pollutants from wastewater (Lissemore et al., 2006). They can be easily prepared from any carbonaceous precursors such as coal, wood, agricultural wastes, etc. (Kotov, 2005; Li et al., 2012; Li et al., 2011a; Li et al., 2011b). The high sorption ability and cost effectiveness make AC a promising material for the removal of pollutants. Many workers have reported the synthesis of AC from diverse base materials and their application in wastewater treatment (Apul et al., 2013; Ghaedi et al., 2011; Machado et al., 2011; Bansal and Goyal, 2005; Zare et al., 2015; Kurniawan and Lo, 2009; Graydon et al., 2009; Pyrzyn´ska and Bystrzejewski, 2010; Ghaedi et al., 2013; Ying-Ying and Zhen-Hu, 2016; Dias et al., 2007). The first full-scale advanced (tertiary) wastewater treatment plant incorporating GAC (granular activated carbon) was put into operation in 1965 in South Lake Tahoe, California. Afterwards the use of GAC beds as a unit process became common in the tertiary treatment train (Hendricks, 2006). The purpose in GAC use was to reuse the municipal wastewater effluents from treatment plants for various purposes. Along with the effluent treatment, physicochemical treatment options involving PAC (poly aromatic compounds) adsorption were also tested in lieu of biological treatment. The idea was to initiate primary settling followed by coagulation and PAC adsorption, settling and filtration. However, secondary treatment could not be replaced with a merely physicochemical process because of high cost (Hendricks, 2006). Activated carbon mediated adsorption can be utilized as a separate unit process in industrial wastewater treatment. It can be placed after several physicochemical treatment steps viz., coagulation/clarification, filtration, and dissolved air flotation. The activated carbons can be beneficial before any biological treatment as they remove the compounds which might be toxic to a biological system. However, the most widely adopted procedure is to place activated carbon adsorption as a tertiary or advanced treatment step subsequent to biological treatment for removal of refractory organics. To some extent, this procedure may also be effective in the removal of inorganics. The adsorption capacity of AC toward organic pollutants depends on the accessibility of the organic molecules to the available pores (micro, meso or macro) according to their size. Under favourable experimental conditions, small molecules such as phenol can approach micro pores, natural organic matter occupy mesoporous, and bacteria access macro pores. Metallic species are smaller in size, being regularly charged in solution; therefore, electrostatic nature (Lo´pez-Ramo´n et al., 2002). The factors that primarily control the extent of adsorption are: (i) chemistry of the metal ion (speciation) or metal ion complex; (ii) pH of the solution and the point of zero charge on the surface; (iii) surface area and porosity (narrow and wider micro porosity); (iv) surface composition (oxygen functionality); and (v) the size of adsorbing species (hydrated ions in the range of 1.0–1.8 nm), mainly for carbons with significant volumes of narrow micro porosity (Figure 6.4) (Dias et al., 2007).

Carbon nanotubes (CNTs), have unique electronic, optoelectronic, as well as mechanical chemical and physical properties (Iijima 1991 Yan et al., 2008; Zare et al., 2015; Sadegh et al., 2015b). CNTs consisting of a single graphene layers are known as Single-walled carbon nanotubes (SWCNTs) whereas, multiple layers of graphene sheet assemble to form multi-walled carbon nanotubes (MWCNTs). CNTs have been applied widely to remove heavy metals and dyes in wastewater treatment (Li et al., 2003; Chen and Wang 2006; Joss et al., 2006; Lissemore et al., 2006; Miao et al., 2004; Halling-Sørensen et al., 1998; Glassmeyer et al., 2005; Molinari et al., 2004; Al-Degs et al., 2006; Sadegh et al., 2015, Babel et al., 2003; Li et al., 2003; Stafiej et al., 2007; Pyrzyn´ska and Bystrzejewski, 2010; Xu et al., 2008; Chen et al., 2008). They are regarded as one of the best adsorbents because of their large adsorption capacity for synthetic dyes in wastewater treatment (Al-Degs et al., 2006). Large quantities of individual tubes are stacked due to van der Waals force and form clustered pores in CNTs. These clustered pores are equivalent to mesoporous in size and hence provide huge outside surface areas that can be useful to remove large biological contaminants including microorganisms. The adsorption usually takes place on hollow interiors of nanotubes as they are open ended, there are interstitial pore spaces between the tube bundles, and the groves present at the boundary of nanotube bundles and the external surface of the outermost CNTs (Yan et al., 2008; Kang et al., 2008; Liao et al., 2008; Hemraj-Benny et al., 2008). However, the interior space of CNTs is not used for adsorption due to the closed caps at the end and the diameter of tubes does not accommodate a typical macromolecular sized contaminant. Whereas, the interstitial spacing present between the bundles of nanotubes show a good opportunity of adsorption only for small sized adsorbates (Yan et al., 2008). But, other possibilities are the grove edges of nanotube bundles and the external surface areas of outermost nanotubes. They have the potential adsorption sites and large pore sizes which are fully available for the micro pollutants. Therefore, the functionalized CNTs are known to be better candidate for their incorporation in the membrane. The results showed that CNT has been widely used as excellent adsorbent media for organic, inorganic, and biological water pollutants (Upadhyayula et al., 2009; Rao et al., 2007). CNTs have shown interesting adsorption capacity for removal ofa wide variety of water contaminants (Upadhyayula et al., 2009; Rao et al., 2007; Hossain et al., 2014; Yu et al., 2014; Mubarak et al., 2013; Ren et al., 2011). In a particular work, Wang et al. (2007) conducted the adsorption of Pb (II), Ag (I), Cu (II), and Co (II) ions on functionalized MWCNT and the experimental data followed Langmuir adsorption model. CNTs are aggregated due to van der Waals forces which make them different from other adsorbents such as activated carbon (AC) and carbon fibers. The AC is among one of the best adsorbents to remove many harmful contaminants but distinguished physicochemical properties and uniform atomic structure made the CNT a better adsorbent than AC (Girifalco et al., 2000; Dabrowski et al., 2005). While Singh et al. (2010) reported the adsorption properties of MWCNTs inserted into carbonized nano fibrous membranes (CNMs) for eliminating disinfection by-product i.e., monochloroacetic acid (MCAA), results showed that the MCAA elimination percentages increased from 24% to 35% with increasing MWCNT concentrations. Like the other adsorbents, factors that determine the adsorption rate are the pore volume, external surface area, functionalities, purity of CNTs, etc. Additionally, the channels and outermost surfaces of the bundles present on the surface of CNT have a great impact on the adsorption process. These pores offer available sites to both organic and inorganic contaminants. Pyrzyn´ska and Bystrzejewski (2010) have reported the advantages and drawbacks of Co (II) and Cu (II) removal using AC, CNTs, and carbon-encapsulated magnetic nanoparticles. Meanwhile, Stafiej and Pyrzynskafound (2007) out that pH and metal ion concentrations in solution play important role in adsorption. Shahryari et al. (2010) performed the same batch of experiments on MWCNTs with higher surface area of 280 m^2/g as compared to that of CNTs (160 m^2/g) used by Yao et al. (2010) and reported a higher Methylene blue adsorption of 132.6 mg/g at 310 K suggesting the effect of surface area on adsorption. In addition, cellulose grafted with soy protein isolate/hydroxyapatite rod-like nanocrystals showed a high Methylene blue adsorption capacity of 454 mg/g (Salama, 2017). Grafting of functional molecules/groups on to the CNTs surface is another way to improve their surface characteristics as it increases the surface area. Grafting can be carried out via different methods such as plasma technique, chemical modification, and microwave treatment (Shao et al., 2010a; Shao et al., 2010b; Zhang et al., 2012; Chen et al., 2012). However plasma technique is preferred due to less energy demand and environmental friendly process. Chen et al. (2012) reported the removal of heavy metals from wastewater by using modified CNTs grafted with various functional groups. Moreover CNTs modified with metal/metal oxides were also reported by many workers; such as MnO_2, Al_2O_3 and iron oxide for the removal of heavy metals from wastewater (Liang et al., 2015; Gupta et al., 2011; Tang et al., 2012; Tawabini, 2015). Oxidized CNTs have also reported to have exceptionally high sorption capacity and efficiency for the effective removal of Pb (II), Cd (II) and Cr (VI) from water (Li et al., 2002; Li et al., 2003; Robati, 2013). CNTs have also reported to have antimicrobial characteristics against a wide range of microorganisms including bacteria and viruses (Zhu et al., 2006; Deng et al., 2008; Nepal et al., 2008; Upadhyayula et al., 2008a; Akasaka and Watari, 2009; Brady-Estévez et al., 2008; Mostafavi et al., 2009; Amin et al., 2014).

Like CNTs, Graphene is another allotrope of carbon with special features that makes it highly compatible for several environmental applications. Graphene oxide (GO) has a two-dimensional structure produced by the oxidation of graphite layer via different chemical/physical methods (Kim et al., 2012; Song et al., 2014). Graphene oxide as adsorbent for the removal of heavy metals is getting more attention due to its high surface area, mechanical strength, light weight, flexibility, chemical stability, and presence of hydroxyl and carboxyl groups at its lateral and basal planes (Li et al., 2009; Lingamdinne et al., 2016b; Zare-Dorabei et al., 2016). Two main features make GO a better adsorbent than other nano materials are a single layer GO has two dimensional basal planes available for the maximum adsorption of heavy metals and it can be synthesized by a simple chemical exfoliation of graphite without any metallic catalyst or complicated instruments (Santhosh et al., 2016). Furthermore, GO does not require any post production treatment for enhancing its adsorption capacity as it already contains hydrophilic functional group (Zhao et al., 2011). Graphene based nanomaterials have been used by many researchers for the adsorption of heavy metals in wastewater (Azamat et al., 2015; Dong et al., 2015; Vu et al., 2016; Zare-Dorabei et al., 2016). GO enabled sand filter in column reactor for removal of heavy metals from wastewater was reported by Ding et al. (2014). Among different carbon based materials, rGO NSs (reduced graphene oxide nano sheets) exhibited better adsorption capacities and is cheaper than single wall carbon nanotubes (Apul et al., 2013; Kotov, 2006). The rGO NSs have already been used as adsorbents for the removal of various dyes and other organic materials from aqueous solutions (Li et al., 2012; Li et al., 2011a; Li et al., 2011b; Du et al., 2014). Modified GO with TiO_2 was used for removal of Pb^{2+}, Cd^{2+} and Zn^{2+} ions from the water by adsorption process (Lee and Yang, 2012). It was reported that adsorption capacity of hybrid composite (TiO_2 modified GO) reached up to 65.6 mg/g, 72.8 mg/g and 88.9 mg/g for Pb^{2+}, Cd^{2+} and Zn^{2+}, respectively. Graphene and its composites show a very high efficiency for the removal of heavy metals from wastewater. The few-layered rGO NSs have been synthesized by Zhao et al. (2011). They have successfully used these rGO NSs for the removal of Cd (II) and Co (II) ions from aqueous solution and observed a pH and ionic strength dependent adsorption. The oxygen containing functional groups on the surfaces of graphene oxide nano sheets play an important role during adsorption process. As (III) and As (V) adsorption by magnetite–graphene composite adsorbent was reported by Chandra et al. (2010). Removal of heavy metals from aqueous solution by self-assembled flower-like TiO_2 on exfoliated graphite oxide was reported by Lee and Yang (2012). The flower-like TiO_2 on the graphene oxide structure significantly enhanced the heavy metal removal efficiency mostly due to the surface-attached and oxygenated functional groups. The GO-TiO_2 hybrid exhibited higher adsorption capacities against heavy metal ions, after 6 and 12 h of hydrothermal treatment at 100°C. The adsorption capacities were 44.8 ± 3.4 and 88.9 ± 3.3 mg/g for the removal of Zn^{2+}, 65.1 ± 4.4 and 72.8 ± 1.6 mg/g for the removal of Cd^{2+}, and 45.0 ± 3.8 and 65.6 ± 2.7 mg/g for the removal of Pb^{2+} at pH 5.6. Whereas, colloidal GO, under similar conditions showed removal capacities of 30.1 ± 2.5, 14.9

± 1.5, and 35.6 ± 1.3 mg/g for Zn^{2+}, Cd^{2+}, and Pb^{2+}, respectively. When the hybrid was treated for longer times, increased surface area was obtained and thus increased removal capacity of heavy metal from aqueous solutions. In another work GO aerogels were prepared by a unidirectional freeze-drying method and their adsorption performance was studied for Cu^{2+} in aqueous solution by Mi et al. (2012). The resulted graphene oxide aerogels were unidirectional porous and had good adsorption ability against Cu^{2+}. Optimum adsorbent dose required for maximum Cu^{2+} removal was 0.06 g in 100 ml of aqueous solution within 15 min of equilibrium contact time. The best fitted isotherm model was Langmuir model with a maximum capacity of 29.59 mg/g at 40°C. As the adsorption of Cu^{2+} on graphene oxide aerogels strongly depends on pH, ion exchange mechanism was sought to be the best possible mechanism for adsorption. However, graphene oxide can be aggregated by Cu^{2+} interaction in aqueous solution with a huge Cu^{2+} absorption capacity (Yang et al., 2010). The maximum adsorption capacity of Cu^{2+} by graphene oxide was 46.6 mg/g, estimated by Langmuir isotherm model. The adsorption capacity shown by GO was around 10 times that of active carbon. Magnetite addition to the existing graphene increased high binding capacity as well as adsorption sites in the graphene composite. Magnetically reduced graphene oxide–bismuth oxy-bromide (MRGO–BiOBr) composite was synthesized and photocatalytic degradation of methylene blue was tested under visible light irradiation (Janani et al., 2016). They observed around 96.84% of dye degradation within a span of 2 hours which revealed that magnetically modified GO showed the highest photocatalytic dye degradation activity at76.24% as compared to 47.54 and 52.14% for pure MRGO and BiOBr, respectively. It was observed that the percentage of removal decreased from 69.89 to 49.12% with increasing initial concentration of the dye. The adsorption percentage increased from 59.44 to 96.84% while increasing the catalyst loading from 0.1 to 0.8 g and any further increase in the catalyst loading did not affect the dye concentration. As the catalyst was magnetic in nature, it was recovered by the external magnet and reused three times in the degradation reaction. Multifunctional crumpled graphene based ternary nanocomposite photocatalyst were reported for the degradation of dye in wastewater. In another report the crumpled graphene oxide was encapsulated with commercially available TiO_2 and Fe_3O_4 nanoparticles (Jiang et al., 2014). The synthesized NPs possess superior photocatalytic activity against methyl orange, and it showed more than 20-fold degradation as compared to TiO_2 alone. The synthesized graphene–TiO_2–Fe_3O_4 nanomaterial was recycled and reused for 5 consecutive cycles. By modifying GO with organics or metal oxides, more functional groups could be added to the composites which is beneficial for anionic metal removal. It was reported that in aqueous solution, Cu (II) causes GO sheets to fold due to Cu (II) – GO interaction and form large agglomerates (Yang et al., 2010). The reason behind such folding is the coordination between Cu (II) and oxygen atoms on GO. The authors had reported higher adsorption capacity of 46.6 mg/g towards Cu (II) ions, as compared to CNTs (28.5 mg/g) and AC (4–5 mg/g). When GO was modified with thiol (SH) groups by diazonium chemistry, it adsorbed six times higher concentration of Hg (II) ions than GO and AC (Gao et al., 2011). N-(trimethoxysilylpropyl) ethylene diaminetriacetic acid (EDTA-silane) assisted chelating GO was used for Pb (II) removal, and adsorption completed within 20 min, with an adsorption capacity of 479 mg/g at pH 6.8 (Madadrang et al., 2012). Graphene oxides are not only used for metal ion removal but also to remove pharmaceutical pollutants. Graphene oxide and reduced graphene oxide/magnetite (RGO–M) composites have been used in the adsorption of ciprofloxacin (CIP), norfloxacin (NOR) and tetracycline (Tang et al., 2013; Ghadim et al., 2013). In a recent study, removal of aspirin, acetaminophen, and caffeine using graphene nano platelets was reported (Al-Khateeb et al., 2014). GO/ poly (acrylic acid) grafted chitosan nanocomposite (GO/CSA) was synthesized and used for the removal of dorzolamide (dorzo), a pharmaceutical component generally found in biomedical effluents and it showed highest adsorption capacity at334 mg/g (Kyzas et al., 2014). Tetracycline remediation was also examined by using GO (Yuan et al., 2012). A brief account of graphene based removal of pharmaceutical impurities from wastewater was compiled in a review article by Carmalin et al. (2016).

GO and GNs exhibit stronger adsorption for many water pollutants as compared to AC and CNTs. The presence of strong functional groups on surface of graphene oxide (GO) enhance the adsorption potential. GO perform adsorption through both electrostatic and coordinate approaches. It showed high adsorption against cationic metals, whereas rGO NSs can be used for adsorption of both cationic and anionic species. Nevertheless, successful reduction in GO to pristine graphene material still remains as a big challenge because this reduction may cause reduced mechanical and electronic properties (Santhosh et al., 2016).

Iron oxides are very common in nature and are available in the form of Magnetite (Fe_3O_4), maghemite (γ-Fe_2O_3), and hematite (α-Fe_2O_3) ores. Iron oxides gained much importance due to their nano-scale size, high surface area to volume ratios and super para-magnetic properties (McHenry and

Laughlin, 2000; Afkhami et al., 2010; Pan et al., 2010). Iron based NPs are easy to synthesize, and have the ability to control or manipulate matter on an atomic scale that provides unparalleled versatility (Boyer et al., 2010; Dias et al., 2011). In addition to this, iron oxide NPs are less toxic, chemically inert and biocompatible which can be combined with biotechnology to yield better results and opportunity (Huang et al., 2003; Roco, 2003; Gupta and Gupta, 2005). Functionalized magnetic NPs can be implemented in catalysis, reduction, and oxidation of pollutants and for the separation of heavy metals as well as hazardous dyes. Due to the magnetic property of the nanoparticles, extraction of heavy metals and dyes from their liquid phase becomes easier. These NPs are useful as they combine the advantages of high dispersion, high reactivity, high stability under acidic conditions, and easy recyclability of the adsorbents. Superparamagnetic Fe_3O_4 nanoparticles were synthesized by co-precipitation route for the adsorption of Ni (II) ions from aqueous solutions (Gautam et al., 2014). The synthesized NPs were spherical in shape and highly crystalline in nature. $SnFe_2O_4$/activated carbon magnetic nanocomposite were synthesized via co-precipitation and applied for the adsorptive removal of the cationic dye crystal violet from aqueous solutions (Rai et al., 2015). Various parameters such as initial dye concentration, contact time, pH, and temperature were investigated to study their effect on adsorption. It was observed from the adsorption equilibrium data that the synthesized NPs follow the Langmuir model. The maximum adsorption capacity of the $SnFe_2O_4$ activated carbon magnetic nanocomposite against crystal violet was 158.73 mg/g at 323 K. The adsorption equilibrium was established within 80 min of contact time at pH 8. In another work, a novel magnetic nano adsorbent was developed from a local agricultural waste-orange peel powder for removal of toxic cadmium metal ion (Gupta and Nayak, 2011). Basing on the adsorption characteristics, Langmuir adsorption was best fitted and maximum Cd^{2+} removal capacity was 76.92 mg/g at optimal conditions. Carboxyl, amine, and thiol modified Fe_3O_4 was synthesized and used for heavy metals (Singh et al., 2011). Almost 100% removal of Cr^{3+}, Co^{2+}, Ni^{2+}, Cu^{2+}, Cd^{2+} and Pb^{2+} ions from waste-water was achieved at higher pH. The removal efficiency of As (III) by carboxyl, amine and thiol-modified Fe_3O_4 was found to be 91, 95 and 97%, respectively at pH 8. The adsorption–desorption behavior of metal ions on amine-functionalized Fe_3O_4 showed an 85% desorption ratio in the first cycle, which indicates their excellent regeneration capacity for their further use. Super magnetic ascorbic acid coated Fe_3O_4 by hydrothermal method was reported by Feng et al. (2012). These nanocomposites are reported to have diameter less than

10 nm and higher surface area about 179 m^2/g. The maximum adsorption capacity shown by the NP for As(V) and As(III) were 16.56 mg/g, and 46.06 mg/g, respectively, and they followed the Langmuir isotherm (Feng et al., 2012). Another study of Fe_2O_3 nanoparticles in the matrix of cellulose has been reported by Yu et al. (2013). The higher surface area of the produced NP was 113 m^2/g. The resulted NP showed excellent adsorption capacity to remove As(III) and As(V). According to the Langmuir isotherm model, the maximum adsorption capacity for As(III) and As(V) were 23.16 and 32.11 mg/g, respectively. Gupta et al. (2015) reported that the modification of nano-adsorbents shows high affinity for the removal of different pollutants such as Cr^{3+}, Co^{2+}, Ni^{2+}, Cu^{2+}, Cd^{2+}, Pb^{2+} and As^{3+} simultaneously from wastewater. β-Cyclodextrin (CMβCD)-monodisperse magnetite nanoparticles were reported with average nanoparticles size 10 nm, synthesized by thermal decomposition and post grafting methodology (Chalasani and Vasudevan, 2012). These CMCDFe_3O_4 were utilized in remediation of As (III), As(V), 2-naphthol, and naphthalene. Fe_2O_3 chestnut-like NPs showed the strong adsorption of As(V) (Mou et al., 2011). The NPs have shown maximum adsorption capacity of 143.12 m^2/g against As(V). From many reports, it was observed that, in comparison to bare Fe_3O_4 nanoparticles, chemically modified Fe_3O_4 nanoparticles remove pollutants more efficiently (Gupta et al., 2014; Feng et al., 2010; Wang et al., 2015, 2013; Ozmen et al., 2010; Ge et al., 2012). It was reported that for selective detection and removal of mercury ions from different sources, such as tap water, lake water and river water rhodamine hydrazide modified Fe_3O_4 microspheres (Fe_3O_4-R_6G) has been utilized has been reported (Wang et al., 2013). Fe_3O_4-gum-arabic nanocomposite for the removal of Cu(II) was reported by Banerjee and Chen (2007), in which gum arabic is attached to the surface of Fe_3O_4 via coupling between surface hydroxyl group of nanoparticles and carboxylic acid group of GA. The average diameter of particles was in the range of 13–67 nm and 5.1wt% of arabic gum was immobilized. The adsorption rate is so fast that equilibrium was achieved within 2 min with adsorption capacity 28.12, 27.64, and 27.18 mg/g in first, second, and third adsorption-desorption cycles, respectively. It was reported that iron nanoparticles with little modification used for dye removal also. The iron particles of size 40–60 nm were synthesized and used as catalyst for the degradation of methylene blue and methyl orange dyes (Shahwan et al., 2011). Other dyes like malachite green, methyl orange, methylene blue, etc. have successfully been removed (Wang et al., 2014; Wang, 2013; Luo et al., 2013).

Another iron based nanoparticle is zero valent iron (nZVI) which has been extensively studied for wastewater treatment. Pollutant remediation by nano scale zero valent iron in soils, groundwater, wastewater, and organic wastes has been widely studied over the years (Fu et al., 2014; Mueller et al., 2012; Tosco et al., 2014; Tuomi et al., 2008; You et al., 2017, Bai et al., 2012, 2013; Kumar et al., 2015). nZVI has been found to remove serious contaminants like trichloroethylene (TCE), tetrachloroethylene, polychlorinated biphenyls, and many organochlorine pesticides such as 1,1,1-trichloro-2,2-bis(4-chlorophenyl) ethane (DDT) (San Román et al., 2013; Petersen et al., 2012). Like the other nanoparticles, surface modification with various molecules such as carboxy methylcellulose, polyelectrolyte multilayers, smectite, polymethyl methacrylate have been applied for stabilizing nZVI particles, enhancing the dispersion of nZVI and efficiently sustaining their reactivity (Johnson et al., 2013; Huang et al., 2008; Jia and Wang, 2012; Wang et al., 2010a). These stabilizations were achieved by enhancing the steric or electrostatic repulsions between the particles to inhibit nZVI aggregation. Researchers have reported to use microwave energy to enhance chlorobenzene removal by microscale ZVI and nZVI and observed an increased removal capacity of 13.6 times and 3.6 times for microscale ZVI and nZVI, respectively (Lee et al., 2011). Also, the nZVI shows greater efficiency in removal of heavy metals (Li et al., 2014; Li et al., 2017; Tang et al., 2014; Wang et al., 2014; Eglal et al., 2015; Zhang et al., 2011; Giasuddin et al., 2007; Shi et al., 2011; Tomašević et al., 2014; Liu et al., 2015; Zhu et al., 2009; Fu et al., 2014).

These iron based NPs have also been demonstrated for their antimicrobial activity (Mohamed et al., 2015; Saif and Tahir, 2016; Auffan et al., 2008). Utilization of iron oxide nanoparticles based novel magnetic materials for removal of heavy metals, organic pollutants, and pathogenic microorganisms from wastewater has also been demonstrated in an efficient and cost-effective way. Surface modification with various chemicals enhanced their stability as well as efficiency. Generally, it's a big task to separate the nanoparticles after their use, but the magnetic nanoparticles are separated easily in the presence of magnetic field. These NPs provide a better opportunity towards sustainable wastewater treatment.\

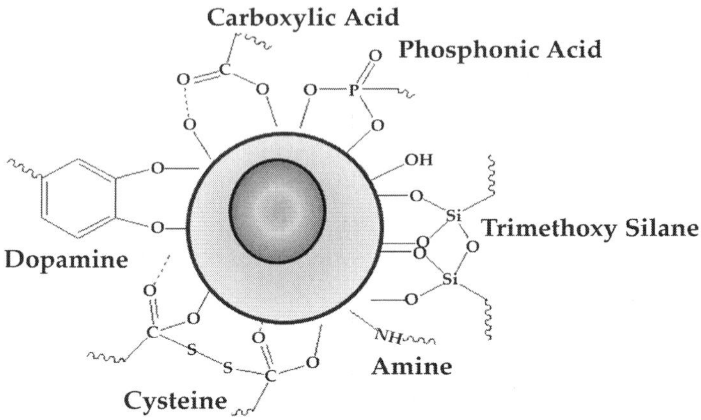

Figure 6.4: Surface of iron oxide magnetic nanoparticles having common chemical moieties for the anchoring of polymers and functional groups.

Zinc oxide (ZnO) is the most popular nanoparticle used in wastewater remediation. Due to its porous micro/nanostructure and higher specific surface area it has been considered as a good adsorbent. ZnO with various morphologies such as nano assemblies, nanoplates, microspheres with nano-sheets and hierarchical ZnO nano-rods have been extensively used as nano-adsorbent for heavy metals removal from wastewater (Ge et al., 2012; Kumar et al., 2013). The above-mentioned modified ZnO NPs have shown high heavy metal removal efficiency as compared to commercial ZnO. ZnO nano-plates and porous nano-sheet were reported for Cu (II) removal from wastewater by Wang et al. (2010). These modified ZnO nano-adsorbent exhibited better Cu (II) removal efficiency than the commercially available ZnO due to their unique micro/nanostructure. Moreover, nano-assemblies were used against a number of heavy metals such as Co^{2+}, Ni^{2+}, Cu^{2+}, Cd^{2+}, Pb^{2+}, Hg^{2+} and As^{3+} (Singh et al., 2013). They have assembled Fe_3O_4 with ZnO nanoparticles and used to remove toxic dyes such as methylene blue, methyl orange, and rhodamine B. Another study on microporous nano-assemblies have reported high affinity for the adsorption of Pb^{2+}, Hg^{2+} and As^{3+} due to their electropositive nature (Gupta et al., 2015). Removal of Pb (II) and Cd (II) from wastewater was reported by using mesoporous hierarchical ZnO nano-rods (Kumar et al., 2013). Photocatalytic degradation of methylene blue, methyl orange dye with ZnO NPs was reported by many researchers (Balcha et al., 2016; Shen et al., 2008; Welderfael et al., 2013). Balcha et al. (2016) reported that ZnO NPs were able to remove 92.5% of dye from the aqueous solution within 3 hour. In another study, Zinc oxide nanoparticles were synthesized by a simple method using leaf extract of jackfruit (Vidya et al., 2016) and shows <80% photocatalytic degradation of Rose Bengal dye within one hour. Zinc sulfide nanoparticles have also been reported to remove dyes from aqueous solutions (Zhang, 2014; Sharma et al., 2013). Zhang (2014) reported to remove methylene blue by ZnS nanocrystals and the highest photo degradation rate of methylene blue over the polyhedron-shaped ZnS reached up to almost 100%.

Magnesium oxide (MgO) has been used for the removal of a wide range of pollutants including heavy metals and dyes from contaminated water. MgO microspheres are unique in structure, which can improve the adsorption efficiency against heavy metals removal (Gupta et al., 2015). Different shapes of MgO like nanorods, nano belts fishbone fractal nanostructures, nanowires, nanotubes, nano cubes and three-dimensional entities has been developed to achieve better adsorption performances(Klug and Dravid, 2002; Zhu et al., 2001; Yin et al., 2002; Mo et al., 2005; Liang et al., 2004). Li et al. (2003) reported effective removal of Pb (II) and Cd (II) on flowerlike mesoporous MgO. According to reports, the photocatalytic degradation of the dye was maximum under UV light irradiation (97%) compared to sunlight (92%) which is due to the fact that the absorption band of MgO falls in the UV region (230 nm). The degradation efficiency of a dye indicates the remarkable ability of MgO NPs which can be attributed to the smaller crystallite size which enhances the surface to volume ratio of the catalysts and the number of surface reactive sites and hydroxyl groups.

Silver nanoparticles work as active biocides against bacteria such as *Escherichia coli*, Staphylococcus aureus, *Klebsiella pneumoniae* and *Pseudomonas aeruginosa* (Jain et al., 2005; Sons et al., 2004). Silver compounds and Silver nanoparticles are broadly used as antimicrobial agents for coliform found in wastewater (Jain et al., 2005). Due to their antibacterial activity, they are also used in healing of wounds. Silver nanoparticles typically have size of 1–40 nm with an average particle size of 2–10 micron range. The Ag NPs or nano dots or nano powder have spherical or flake shape with high surface area (specific surface area in the order of 1 m^2/g). Since the silver NPs have anti-microbial, anti-biotic and anti-fungal properties, they are used in coatings, nanofiber, first aid bandages, plastics, soap and textiles for treatment of viruses, in self-cleaning fabrics, etc.

Due to its high photocatalytic activity, reasonable price, photostability, chemical and biological stability (Guesh et al., 2016; Imamura et al., 2013; Rawal et al., 2013). The TiO_2 has been used as a photocatalyst since

1972, when electrochemical photolysis of water using TiO_2 semiconductor electrode was first observed. In wastewater, contaminant degradation can be achieved through photocatalytic degradation by gradually oxidizing the contaminants into low molecular weight intermediate products and eventually into CO_2, water and anions such as NO_3^-, PO_4^{3-} and Cl^-, in presence of light and catalyst.

When excited by UV light, charge separation occurs in TiO_2 particles generating reactive oxygen species (ROS) which completely degrades contaminants in a short period. TiO_2 nanoparticles can be used to degrade various contaminants such as, chlorinated organic compounds, polycyclic aromatic hydrocarbons, dyes, phenols, pesticides, arsenic, cyanide and heavy metals (Ohsaka et al., 2008; Ohsaka et al., 2008; Guo et al., 2015; Nguyen et al., 2016; Alalm et al., 2015; Moon et al., 2014; Kim et al., 2016; Chen et al., 2016). TiO_2 nanoparticles can kill various microorganisms such as bacteria, fungi, algae, protozoa and viruses (Foster et al., 2011) due to its photocatalytic properties. However, TiO_2 nanoparticles have certain disadvantages such as, high band gap energy (3.2 eV), It requires UV excitation to operate and the photocatalytic properties under visible light are not prominent). Studies show that, metal (e.g., Ag) doping improves visible light absorption of titanium dioxide NPs (Anpo et al., 2001) and increases photocatalytic activity under UV light illumination (Mu et al., 1989; Seery et al., 2007; Page et al., 2007; Kim et al., 2006). Also, modification of titanium dioxide NPs by non-metallic elements such as N, F, S and C have also been found to reduce the band gap significantly and improve degradation of dyes under solar light irradiation (Liu et al., 2006). Other disadvantages of titanium dioxide include difficulty in production, difficulty in recovering from treated wastewater, etc. Coupling of TiO_2 NPs with membranes such as poly (vinylidene fluoride) (Wang et al., 2013; Meng et al., 2014), polyethersulfone (Razmjou et al., 2011; Razmjou et al., 2012), polymethyl methacrylate (Hamming et al., 2009), and poly (amide-imide) (Rajesh et al., 2013), show promise for overcoming recovery problem of TiO_2 NPs. For example, the polymerization of acrylamide in an aqueous solution can be carried out to synthesize TiO_2/poly (acrylamide-co-(acrylic acid)) composite hydrogel, using N, N′-methylenebis-acrylamide as the cross-linker and ammonium persulphate as the initiator pair. Methylene blue can be removed by the photocatalysis of TiO_2 NPs. Moreover, due to the coupling with polymeric membranes, TiO_2 NPs could be easily separated from the treated system through a simple filtration (Kangwansupamonkon et al., 2010).

Most of the nanomaterials discussed before are present as nanoparticles. When concerning its application in large-scale, they have to face some inherent technical bottle necks, such as aggregation, difficult separation, leakage into the contact water, potential adverse effect imposed on ecosystem and human health. The development of nanocomposites is one of the most important approaches to forward the application of nano-particulate materials which take advantages of both the hosts and impregnated nanoparticles. In addition, nanocomposites could mitigate the release of nanoparticles into the environment, and improve the compatibility of nanotechnology with existing in frastructures. Nanocomposite is of multiphase i.e., solid material, porous media, colloids, gels and copolymers in a broad sense. Their performance mainly depend upon the selection of hosts which is more dominant the nano particles. Compared to static nanocomposite as well as with free nanomaterials, the performance and usability of nanocomposites were significantly high and improved. However, in terms of nanoparticle dispersion, stability and recyclability, nono composites performed better. Hence, nanocomposite materials could bridge the gap between nano scopic and mesoscopic scale.

6.6. CONCLUSION AND FUTURE PERSPECTIVE

Nanotechnology has gained importance worldwide and can fit to the present and future water and wastewater issues as they exhibit versatile properties as high capacity, fast kinetics, specific affinity towards targeted contaminants, enhanced photocatalytic response for a broad light spectrum, and strong anti-bacterial activity. Nanomaterials are arguably the most promising candidate for the development of next generation water treatment technology. Various nanomaterials such as nano-adsorbents (based on oxides, Fe, MnO, ZnO, MgO, CNT), photocatalysts (ZnO, TiO_2, CdS, ZnS:Cu, CdS:Eu, CdS:Mn), electrocatalysts (Pt, Pd), and nano-membranes (multi walled CNTs, electro spun PVDF, PVC, Na-TNB) have been developed and successfully implemented for wastewater treatment. Furthermore, these nano-particles can be integrated with biological processes (algal membrane, anaerobic digestion, microbial fuel cell) to improve their effectiveness in water purification. Among various nanoparticles, the development of nanocomposites takes advantage of both the hosts and the impregnated functional nanoparticles. Hosts like polymers, bio-polymers, minerals, activated carbons or membranes could facilitate the dispersion and stability of the loaded nanoparticles. They could also promote the transfer or diffusion of contaminants in the hosts, further enhancing the interfacial interaction. Moreover, nanocomposites could also be integrated with the existing infrastructure like fixed-bed column or fluidized bed to improve water quality. Still, nanotechnology has some deleterious effects, which need to be consideration as they might release into the environment during preparation and treatment processes, where they can accumulate for long time and cause serious risks. In order to decrease its toxic impact on environment there is need of future research. Intense research is required to re-evaluate the eco toxic potential for each new modification in catalyst and for existing materials. In addition, life cycle assessment of nano-materials is crucial to address their overall benefits and risks. Moreover, further research is needed on developing a cost effective, eco-friendly process of synthesizing nano-materials and testing the efficiency at large scale for successful field application.

REFERENCES

1. Afkhami, A., Saber-Tehrani, M., & Bagheri, H. (2010). Modified maghemite nanoparticles as an efficient adsorbent for removing some cationic dyes from aqueous solution. *Desalination 263*, 240–248.
2. Ahmad, M., Rajapaksha, A. U., Lim, J. E., Zhang, M., Bolan, N., Mohan, D., Vithanage, M., & Lee, S.S. Ok, Y. S. (2014). Biochar as a sorbent for contaminant management in soil and water: a review. *Chemosphere. 99*, 19–33.
3. Ai, L., & Jiang, J. (2012). Removal of methylene blue from aqueous solution with self-assembled cylindrical graphene–carbon nanotube hybrid. Chem. Eng. J. *192*, 156–163.
4. Akasaka, T., & Watari, F. (2009). Capture of bacteria by flexible carbon nanotubes. *Acta biomaterialia*, *5*, 607–612.
5. Al-Degs, Y. S., El-Barghouthi, M. I., Issa, A. A., Khraisheh, M. A., & Walker, G. M. (2006). Sorption of Zn (II), Pb(II), and Co (II) using natural sorbents: equilibrium and kinetic studies. *Water Res. 40*, 2645–2658.
6. Al-Hamadani, Y. A. J., Chu, K. H., Son, A., Heo, J., Her, N., Jang, M., Park, C. M., & Yoon, Y. (2015). Stabilization and dispersion of carbon nanomaterials in aqueous solutions: a review. *Sep. Purif. Technol. 156*, 861–874.
7. Al-Khateeb, L. A., Almotiry, S., & Salam, M. A. (2014). Adsorption of pharmaceutical pollutants onto graphene nanoplatelets. Che. Eng. J. *248*, 191–199.
8. Alalm, M. G., Tawfik, A., & Ookawara, S. (2015). Comparison of solar TiO_2 photocatalysis and solar photo-Fenton for treatment of pesticides industry wastewater: operational conditions, kinetics, and costs, *J. of Water Process Eng. 8*, 55–63.
9. Amin, M. T., Alazba, A. A., & Manzoor, U. (2014). A review of removal of pollutants from water/wastewater using different types of nanomaterials. *Adv. in Mater. Sci. and Eng*. 2014. Article ID 825910, 24 pages, http://dx.doi.org/10.1155/2014/825910
10. Anjum, M., Miandad, R., Waqas, M., Gehany, F., & Barakat, M.A. (2016). Remediation of wastewater using various nano-materials, *Arabian J. Chem*. doi: http://dx.doi.org/10.1016/j.arabjc. 2016.10.004
11. Anpo, M., Kishiguchi, S., Ichihashi. Y., Takeuchi, M., Yamashita, H., Ikeue, K., Morin, B., Davidson, A., & Che, M. (2001). The design

and development of second-generation titanium oxide photocatalysts able to operate under visible light irradiation by applying a metal ion-implantation method, *Res. Chem. Intermediates. 27*, 459–467.

12. Apul, O. G., Wang, Q., Zhou, Y., & Karanfil, T. (2013). Adsorption of aromatic organic contaminants by graphene nanosheets: comparison with carbon nanotubes and activated carbon. *Water Res. 47*, 1648–1654
13. Auffan, M., Achouak, W., Rose, J., Roncato, M. A., Chanéac, C., Waite, D. T., Masion, A., Woicik, J. C., Wiesner, M. R., & Bottero, J. Y. (2008). Relation between the redox state of iron-based nanoparticles and their cytotoxicity toward *Escherichia coli. Environ. Sci. Technol. 42*, 6730–6735.
14. Ayati, A., Ahmadpour, A., Bamoharram, F. F., Tanhaei, B., Manttari, M., & Sillanpaa, M. (2014). A review on catalytic applications of Au/TiO_2 nanoparticles in the removal of water pollutant. *Chemosphere.* 107, 163–174.
15. Azamat, J., Sattary, B. S., Khataee, A., & Joo, S. W. (2015). Removal of ahazardous heavy metal from aqueous solution using functionalized graphene and boron nitride nanosheets: insights from simulations. J. Molecul. *Graphics Model.* 61, 13–20.
16. Azari A., Babaei A. A., Kalantary R. R., Esrafili A., Moazzen M., & Kakavandi B. (2014). Nitrate removal from aqueous solution using carbon nanotubes magnetized by nano zero-valent iron, *J. of Mazandaran University of Medical Sci.*, 23, 14–27.
17. Babel, S., & Kurniawan, T. A. (2003). Low-cost adsorbents for heavy metals uptake from contaminated water: a review. *J. Hazard. Mater.* 97, 219–243.
18. Bae, C. H., Nam, S. H., & Park, S. M. (2002). Formation of silver nanoparticles by laser ablation of silver target in NaCl solution, *Appl. Surf. Sci.* 197–198,628–634.
19. Bai, H., Kang, Y., Quan, H., Han, Y., & Feng, Y. (2012). Treatment of copper wastewater by sulfate reducing bacteria in the presence of zero valent iron. *Int. J. Miner. Processing.* (2012).,112, 71–76.
20. Balcha, A., Yadav, O. P., & Dey, T. (2016). Photocatalytic degradation of methylene blue dye by zinc oxide nanoparticles obtained from precipitation and sol-gel methods. *Environ. Sci. Pollut. Res. 23*, 25485–25493.
21. Banerjee, S. S., & Chen, D. H. (2007). Fast removal of copper ions

by gum Arabic modified magnetic Nano-adsorbent. *J. Hazard. Mater. 147*, 792–799.

22. Bangash, F. K. & Alam, S. (2004). Extent of pollutants in the effluents of Hayatabad industrial estate peshawar, *J. Chem. Society of Pakistan. 26*, 271–278.
23. Bansal, R. C., & Goyal, M. (2005). *Activated Carbon Adsorption*, CRC Press, Boca Raton, 2005, doi; 10.1201/9781420028812.
24. Baran, E. T., Mano, J. F. & Reis, R. L. (2004). Starch-chitosan hydrogels prepared by reductive alkylation cross-linking. *J. Mat. Sci.: Mat. Med. 15*, 759–765.
25. Bougeard, C. M. M., Goslan, E. H., Jefferson, B., & Parsons, S. A. (2010). Comparison of the disinfection by-product formation potential of treated waters exposed to chlorine and monochloramine. *Water Res 44*, 729–740.
26. Boyer, C., Whittaker, M. R., Bulmus, V., Liu, J. Q., & Davis, T. P. (2010). The design and utility of polymer stabilized iron-oxide nanoparticles for nanomedicine applications. *NPG Asia. Mater. 2*, 23–30.
27. Brady-Estévez, A. S., Kang, S., & Elimelech, M. (2008). A single-walled-carbon-nanotube filter for removal of viral and bacterial pathogens. *Small. 4*, 481–484.
28. Brumfiel, G. (2003). Nanotechnology: a little knowledge. *Nature. 424*, 246–248.
29. Carmalin, S., A., Lima, E. C., Allaudeen, N., & Rajan, S. (2016). Application of graphene based materials for adsorption of pharmaceutical traces from water and wastewater-a review. Desalin. *Water Treat. 57*, 27573–27586.
30. Chakrabarty, S., & Chatterjee, K. (2009). Synthesis and Characterization of Nano-Dimensional Nickelous Oxide (NiO) Semiconductor, *J. Phy. Sci. 13*, 245–250.
31. Chalasani, R., & Vasudevan, S. (2012). Cyclodextrin functionalized magnetic iron oxide nanocrystals: a host-carrier for magnetic separation of non-polar molecules and arsenic from aqueous media. *J. Mater. Chem. 22*, 14925–14931.
32. Chandra, V., Park, J., Chun, Y., Lee, J. W., Hwang, I. C., & Kim, K. S. (2010). Water-dispersible magnetite-reduced graphene oxide composites for arsenic removal. *ACS Nano. 4*, 3979–3986.
33. Chen, C., & Wang, X. (2006). Adsorption of Ni (II) from aqueous

solution using oxidized multiwall carbon nanotubes. *Ind. Eng. Chem. Fundam. 45*, 9144–9149.

34. Chen, C., Hu, J., Xu, D., Tan, X., Meng, Y., & Wang, X. (2008). Surface complexation modeling of Sr (II) and Eu(III) adsorption onto oxidized multiwall carbon nanotubes. *J. Colloid Interface Sci. 323*, 33–41.
35. Chen, C., Li, X., Zhao, D., Tan, X., & Wang, X. (2007). Adsorption kinetic, thermodynamic and desorption studies of Th(IV) on oxidized multi-wall carbon nanotubes. *Colloids Surf. A. 302*, 449–454.
36. Chen, H., Li, J., Shao, D., Ren, X., & Wang, X. (2012). Poly (acrylicacid) grafted multiwall carbon nanotubes by plasma techniques for Co(II) removal from aqueous solution. *Chem. Eng. J. 210*, 475–481.
37. Chen, Z. P., Li, Y., Guo, M., Xu, F, Wang, P., Du, Y., & Na, P. (2016). One-pot synthesis of Mn-doped TiO_2 grown on graphene and the mechanism for removal of Cr (VI) and Cr (III), *J. Hazard. Mater. 310*, 188–198.
38. Chowdhury, S., Rodriguez, M. J., & Sadiq, R. (2011). Disinfection by-products in Canadian provinces: associated cancer risks and medical expenses. *J. Hazard. Mater. 187*, 574–584.
39. Crane, R. A., & Scott T. B. (2012). Nanoscale zero-valent iron: future prospects for an emerging water treatment technology. *J. Hazard. Mater*. 211–212,112–125.
40. Dabrowski, A., Podkocielny, P., Hubicki, Z., & Barczak, M. (2005). Adsorption of phenolic compounds by activated carbon-a critical review. *Chemosphere 58*, 1049–1070.
41. Dale, A. L., Casman, E. A., Lowry, G. V., Lead, J. R., Viparelli, E., & Baalousha, M. (2015). Modeling nanomaterial environmental fate in aquatic systems. *Environ. Sci. Technol. 49*, 2587–2593.
42. Das, S., Sen, B., & Debnath, N. (2015). Recent trends in nanomaterials applications in environmental monitoring and remediation. *Environ. Sci. Pollut. Res*. 22, 18333–18344.
43. Deng, S., Upadhyayula, V. K., Smith, G. B., & Mitchell, M. C. (2008). Adsorption equilibrium and kinetics of microorganisms on single-wall carbon nanotubes. *IEEE Sensors J. 8*, 954–962.
44. Dias, A. M. G. C., Hussain, A., Marcos, A. S., & Roque, A. C. A. (2011). A biotechnological perspective on the application of iron oxide magnetic colloids modified with polysaccharides. *Biotechnol. Adv. 29*, 142–155.

45. Dias, J. M., Alvim-Ferraz, M. C., Almeida, M. F., Rivera-Utrilla, J., & Sánchez-Polo, M. (2007). Waste materials for activated carbon preparation and its use in aqueous-phase treatment: a review. *J. Environ. Manage. 85*, 833–846.

46. Ding, Z., Zhu, H, Y., Lu, G. Q., & Greenfield, P. F. (1999). Photocatalytic properties of titania pillared clays by different drying methods. *J. Colloid Interface Sci. 209*, 193–199.

47. Dong, Z., Zhang, F., Wang, D., Liu, X., & Jin, J. (2015). Polydopamine mediated surface-functionalization of graphene oxide for heavy metal ions removal. *J. Solid State Chem. 224*, 88–93.

48. Du, Q., Sun, J., Li, Y., Yang, X., Wang, X., Wang, Z., & Xia, L. (2014). Highly enhanced adsorption of Congo red onto graphene oxide/chitosan fibers by wet-chemical etching off silica nanoparticles. *Chem. Eng. J. 245*, 99–106.

49. Edelstein, A. S., & Cammaratra, R. C. (1998). Nanomaterials: synthesis, properties and applications. *CRC Press,* ISBN: 9780750305785.

50. Eglal, M. M., & Ramamurthy, A. S. (2015). Removal of Pb (II), Cd (II), Cu (II) and trichloroethylene from water by Nanofer ZVI. *J. Environ. Sci. Health*, Part A, 50, 901–912.

51. Espitia, P. J. P., Soares, N. D. F. F., Coimbra, J. S. D. R., Andrade, N. J., de, Cruz, N. J., de., Medeiros, E. A. A. (2012). Zinc oxide nanoparticles: synthesis, antimicrobial activity and food packaging applications. *Food Bioprocess Technol*, *5*, 1447–1464.

52. Feng, J., Zhang, D. D., Liu, Y. F., Bai, Y., Chen, Q. D., Liu, S. Y., & Sun, H. B. (2010). Magnetic nanofilm of Fe_3O_4 for highly efficient organic light-emitting devices. *J. Phys. Chem. C. 114*, 6718–6721.

53. Feng, Z., Zhu, S., Martins de Godoi, D. R., Samia, A. C. S., & Scherson, D. (2012). Adsorption of Cd^{2+} on carboxyl-terminated superparamagnetic iron oxide nanoparticles. *Anal. Chem. 84*, 3764–3770.

54. Ferroudj, N., Nzimoto, J., Davidson, A., Talbot, D., Briot, E., Dupuis, V., Bee, A., Medjram, M. S., & Abramson, S. (2013). Maghemite nanoparticles and maghemite/silica nanocomposite microspheres as magnetic Fenton catalysts for the removal of water pollutants. *Appl. Catal. B Environ. 136*, 9–18.

55. Foster, H. A., Ditta, I. B., Varghese, S., & Steele, A. (2011). Photocatalytic disinfection using titanium dioxide: spectrum and

mechanism of antimicrobial activity. *Appl. Microbiol. Biotechnol. 90*, 1847–1868.

56. Fu, F., Dionysiou, D. D., & Liu, H. (2014). The use of zero-valent iron for groundwater remediation and wastewater treatment: a review. *J. Hazard. Mater. 267*, 194–205.
57. Gao, W., Majumder, M., Alemany, L. B., Narayanan, T. N., Ibarra, M. A., Pradhan, B. K., & Ajayan, P. M. (2011). Engineered graphite oxide materials for application in water purification. *ACS Appl. Mater. Interfaces 3*, 1821–1826.
58. Gao, Y., Li, Y., Zhang, L., Huang, H., Hu, J., Shah, S. M., & Su, X.(2012). Adsorption and removal of tetracycline antibiotics from aqueous solution by graphene oxide. *J. Colloid. Interface Sci. 368*, 540–546.
59. Gautam, R. K., Gautam, P. K., Banerjee, S., Soni, S., Singh, S. K., & Chattopadhyaya, M. C. (2015). Removal of Ni (II) by magnetic nanoparticles. *J. Mol. Liq. 204*, 60–69.
60. Ge, F., Li, M. M., Ye, H., & Zhao, B. X. (2012). Effective removal ofheavy metal ions Cd^{2+}, Zn^{2+}, Pb^{2+}, Cu^{2+} from aqueous solutionby polymer-modified magnetic nanoparticles. *J. Hazard. Mater. 211*, 366–372.
61. Ghadim, E. E., Manouchehri, F., Soleimani, G., Hosseini, H., Kimiagar, S., & Nafisi, S. (2013). Adsorption properties of tetracycline onto graphene oxide: equilibrium, kinetic and thermodynamic studies. *PLoS One*, 8, https://doi.org/10.1371/journal.pone.007(9254).
62. Ghaedi, M., Ghayedi, M., Kokhdan, S. N., Sahraei, R., & Daneshfar, A. (2013). Palladium, silver, and zinc oxide nanoparticles loaded on activated carbon as adsorbent for removal of bromophenol red from aqueous solution. *J. Ind. Eng. Chem. 19*, 1209–1217.
63. Ghaedi, M., Ghayedi, M., Kokhdan, S. N., Sahraei, R., & Daneshfar, A. (2013). Palladium, silver, and zinc oxide nanoparticles loaded on activated carbon as adsorbent for removal of bromophenol red from aqueous solution. *J. Ind. Eng. Chem. 19*, 1209–1217.
64. Ghaedi, M., Haghdoust, S., Kokhdan, S. N., Mihandoost, A., Sahraie, R., & Daneshfar, A. (2012). Comparison of activated carbon, multi-walled carbon nanotubes, and cadmium hydroxide nanowire loaded on activated carbon as adsorbents for kinetic and equilibrium study of removal of Safranine *O. Spectrosc. Lett. 45*, 500–510.

65. Ghaedi, M., Shokrollahi, A., Tavallali, H., Shojaiepoor, F., Keshavarz, B., Hossainian, H., Soylak, M., & Purkait, M. (2011). Activated carbon and multiwalled carbon nanotubes as efficient adsorbents for removal of arsenazo (III) and methyl red from wastewater. *Toxicol. Environ. Chem. 93*, 438–449.

66. Giasuddin, A. B., Kanel, S. R., & Choi, H. (2007). Adsorption of humic acid onto nanoscalezerovalent iron and its effect on arsenic removal. *Environ. Sci. Technol. 41*, 2022–2027.

67. Girifalco, L. Hodak, M. & Lee, R. S. (2000). Carbon nanotubes, buckyballs, ropes, and a universal graphitic potential. *Phys. Rev. 62*, 13104–13110.

68. Glassmeyer, S. T., Furlong, E. T., Kolpin, D. W., Cahill, J. D., Zaugg, S. D., Werner, S. L., Meyer, M. T., & Kryak, D. D. (2005). Transport of chemical and microbial compounds from known wastewater discharges: potential for use as indicators of human fecal contamination. *Environ. Sci. Technol. 39*, 5157–5169.

69. Gore, Sane A. (2011). Flame synthesis of carbon nanotubes. Rijeka: *INTECH Open Access Publisher*, 2011.

70. Graydon, J. W., Zhang, X., Kirk, D. W., Jia, C. Q. (2009). Sorption and stability of mercury on activated carbon for emission control. *J. Hazard. Mater. 168*, 978–982.

71. Guesh, K., Mayoral, A., Alvarez, C. M., Chebude, Y., & D´ıaz, I. (2016). Enhanced photo-catalytic activity of TiO_2 supported on zeolites tested in real wastewaters from the textile industry of Ethiopia, *Microporous and Mesoporous Mater. 225*, 88–97.

72. Guo, M., Song, W., Wang, T., Li, Y., Wang, X., & Du, X. (2015). Phenyl functionalization of titanium dioxide nano sheets coating fabricated on a titanium wire for selective solid-phase micro-extraction of polycyclic aromatic hydrocarbons from environment water samples. *Talanta. 144*, 998–1006.

73. Gupta, A. K., &Gupta, M. (2005). Synthesis and surface engineering of iron oxide nanoparticles for biomedical applications. *Biomater. 26*, 3995–4021.

74. Gupta, V. K., Agarwal, S., & Saleh, T. A. (2011). Synthesis and characterization of alumina-coated carbon nanotubes and their application for lead removal. *J. Hazard. Mater. 185*, 17–23.

75. Gupta, V. K., Nayak, A., Agarwal, S., & Tyagi, I. (2014). Potential of

activated carbon from waste rubber tire for the adsorption of phenolics: Effect of pre-treatment conditions. *J. Colloid Interface Sci. 417*, 420–430.

76. Gupta, V. K., Tyagi, I., Sadegh, H., Shahryari-Ghoshekand, R., Makhlouf, A. S. H., & Maazinejad, B. (2015). Nanoparticles as adsorbent; a positive approach for removal of noxious metal ions: a review. *Sci. Technol. Dev. 34*, 195–214.
77. Halling-Sørensen, B., Nielsen, S. N., Lanzky, P., Ingerslev, F., Lu¨tzhøft, H. H., & Jørgensen, S. (1998). Occurrence, fate and effects of pharmaceutical substances in the environment—a review. *Chemosphere. 36*, 357–393.
78. Hamming, L. M., Qiao, R., Messersmith, P. B., & Brinson, L. C. (2009). Effects of dispersion and interfacial modification on the macroscale properties of TiO_2 polymer-matrix nanocomposites. *Composites Sci. and Technol. 69*, 1880–1886.
79. Hemraj-Benny, T., Bandosz, T. J., & Wong, S. S. (2008). Effect of ozonolysis on the pore structure, surface chemistry, and bundling of single-walled carbon nanotubes. *J. Colloid Interface Sci. 317*,375–382.
80. Hendricks, D (2006). Water Treatment Unit Processes: Physical and Chemical, CRC Press. Printed in the USA, ISBN: 9780824706951.
81. Hossain, F., Perales-Perez, O. J., Hwang, S., & Román, F. (2014). Antimicrobial nanomaterials as water disinfectant: Applications, limitations and future perspectives. *Sci. Total Environ.* 466–467, 1047–1059.
82. Huang, Q., Shi, X., Pinto, R. A., Petersen, E. J., & Weber Jr, W. J. (2008). Tunable synthesis and immobilization of zero-valent iron nanoparticles for environmental applications. *Environmental Sc. & Technol. 42*, 8884–8889.
83. Huang, S. H., Liao, M. H., & Chen, D. H. (2003). Direct binding and characterization of lipase onto magnetic nanoparticles. *Biotechnol Progress. 19*, 1095–1100.
84. Iijima, S. (1991). Helical microtubules of graphitic carbon. *Nature. 354*, 56–58.
85. Imamura, K., Yoshikawa, T., Hashimoto, K., & Kominami, H. (2013). Stoichiometric production of aminobenzenes and ketones by photocatalytic reduction of nitrobenzenes in secondary alcoholic suspension of titanium (IV) oxide under metal-free conditions. *Appl. Catal., B:*

Environ. 134–135, 193–197.

86. Iwaszuk, A., Mulheran, P., & Nolan, M. (2013). TiO_2 nanocluster modified-rutile TiO_2 photo-catalyst: A first principles investigation. *J. Materials Chem. A. 1*, 2515–2525.
87. Jain, P., & Pradeep,T. (2005). Potential of silver nanoparticle-coated polyurethane foam as an antibacterial water filter. *Biotech. Bioeng. 90*, 59–63.
88. Janani, S., Rani, K. S. S., Ellappan, P., & Miranda, L. R. (2016). Photodegradation of methylene blue using magnetically reduced graphene oxide bismuth oxybromide composite. *J. Environ. Chem. Eng. 4*, 534–541.
89. Jia, H., & Wang, C. (2012). Adsorption and dechlorination of 2, 4-dichlorophenol (2, 4-DCP) on a multi-functional organo-smectite templated zero-valent iron composite. *Chem. Eng. J. 191*, 202–209.
90. Jiang, Y., Wang, W. N., Biswas, P., & Fortner, J. D. (2014). Facile aerosol synthesis and characterization of ternary crumpled grapheme-TiO_2-magnetite nanocomposites for advanced water treatment. *ACS Appl. Mater. 6*, 11766–11774.
91. Johnson, R. L., Nurmi, J. T., O'Brien Johnson, G. S., Fan, D., O'Brien Johnson, R. L., Shi, Z., Salter-Blanc, A. J., Tratnyek, P. G., & Lowry, G. V. (2013). Field-scale transport and transformation of carboxy methylcellulose-stabilized nano zero-valent iron. *Environ. Sci. Technol. 47*, 1573–1580.
92. Joss, A., Zabczynski, S., Go¨bel, A., Hoffmann, B., Lo¨ffler, D., McArdell, C. S., Ternes, T. A., Thomsen, A., Siegrist, H. (2006). Biological degradation of pharmaceuticals in municipal wastewater treatment: proposing a classification scheme. *Water Res. 40*, 1686–1696.
93. Kang, S., Herzberg, M., Rodrigues, D. F., Elimelech, M. (2008). Antibacterial Effects of Carbon Nanotubes: Size Does Matter. *Langmuir. 24*, 6409–6413.
94. Kang, S., Pinault, M., & Pfefferle, L. D. (2007). Elimelech, M. Single-walled carbon nanotubes exhibit strong antimicrobial activity. *Langmuir. 23*, 8670–8673.
95. Kangwansupamonkon, W., Jitbunpot, W., & Kiatkamjornwong, S. (2010). Photocatalytic efficiency of TiO_2/poly[acrylamide-co-(acrylic acid)] composite for textile dye degradation. *Poly. Degrad. Stab. 95*,

1894–1902.

96. Kheybari, S., Samadi,N., Hosseini, S. V., Fazeli, A., Fazeli, M. R. (2010). Synthesis and antimicrobial effects of silver nanoparticles produced by chemical reduction method. *DARU. 18*, 168–173.
97. Khin, M. M., Nair, A. S., Babu, V. J., Murugan, R., Ramakrishna, S. (2012). A review on nanomaterials for environmental remediation. *Energy Environ. Sci. 5*, 8075–8109.
98. Kim, J., Cote, L. J., & Huang, J. (2012). Two-dimensional soft material: new faces of graphene oxide. *Accounts of chem. Research, 45*, 1356–1364.
99. Kim, J. P., Cho, I. H., Kim, I. T., Kim, C. U., Heo, N. H., & Suh, S. H. (2006). Manufacturing of anti-viral inorganic materials from colloidal silver and titanium oxide. *Revue Roumaine de Chimie. 51*, 1121–1129.
100. Kim, S. H., Lee, S. W., Lee, G. M., Lee, B. T., Yun, S. T., & Kim, S. O. (2016). Monitoring of TiO_2-catalytic UV-LED photo-oxidation of cyanide contained in mine wastewater and leachate. *Chemosphere. 143*, 106–114.
101. Klug, K. L., & Dravid, V. P. (2006). Observation of two-and three-dimensional magnesium oxide nanostructures formed by thermal treatment of magnesium diboride powder. *Appl. Phys. Lett. 81*, 1687–1689.
102. Kolodziejczak-Radzimska, A., & Jesionowski, T. (2014). Zinc oxide – from synthesis to application: a review. *Mater. 7*, 2833–2881.
103. Kotov, N. A. (2006). Materials science: carbon sheet solutions. *Nature. 442*, 254–255.
104. Kumar, K. Y., Muralidhara, H. B., Nayaka, Y. A., Balasubramanyam, J., & Hanumanthappa, H. (2013). Hierarchically assembled mesoporous ZnO nanorods for the removal of lead and cadmium by using differential pulse anodic stripping voltammetric method. *Powder Technol. 239*, 208–216.
105. Kumar, M., & Ando, Y. (2010). Chemical vapor deposition of carbon nanotubes: a review on growth mechanism and mass production. *J. Nanosci. Nanotech. 10*, 3739–758.
106. Kumar, N., Chaurand, P., Rose, J., Diels, L., & Bastiaens, L. (2015). Synergistic effects of sulfate reducing bacteria and zero valent iron on zinc removal and stability in aquifer sediment. *Chem. Eng. J. 260*, 83–89.

107. Kurniawan, T. A., & Lo, W. H. (2009). Removal of refractory compounds from stabilized landfill leachate using an integrated H_2O_2 oxidation and granular activated carbon (GAC) adsorption treatment. *Water Res. 43*, 4079–4091.

108. Kyzas, G. Z., Bikiaris, D. N., Seredych, M., Bandosz, T. J., & Deliyanni, E. A. (2014). Removal of dorzolamide from biomedical wastewaters with adsorption onto graphite oxide/poly (acrylic acid) grafted chitosan nanocomposite. *Bioresour. Technol. 152*, 399–406.

109. Lee, C. L., Lee, H. Y., Tseng, K. H., Hong, P. K. A., & Jou, C. J. G. (2011). Enhanced dechlorination of chlorobenzene by microwave-induced zero-valent iron: particle effects and activation energy, *Environ. Chem. Lett. 9*, 355–359.

110. Lee, J. H., & Yang, Y. S. (2005). Effect of HCl concentration and reaction time on the change in the crystalline state of TiO_2 prepared from aqueous $TiCl_4$ solution by precipitation. *J. European Ceramic Soc. 25*, 3573–3578.

111. Lee, Y. C., Yang, J. W. (2012). Self-assembled flower-like TiO_2 on exfoliated graphite oxide for heavy metal removal. *J. Ind. Eng. Chem. 18*, 1178–1185.

112. Li, B., Cao, H., & Yin, G. (2011). $Mg(OH)_2$@ reduced graphene oxidecomposite for removal of dyes from water. *J. Mater. Chem. 21*, 13765–13768.

113. Li, J., Guo, S., Zhai, Y., & Wang, E. (2009). Nafion-graphene nanocomposite film as enhanced sensing platform for ultra-sensitive determination of cadmium. *Electrochem. Commun. 11*, 1085–1088.

114. Li, N., Zheng, M., Chang, X., Ji, G., Lu, H., Xue, L., Pan, L., & Cao, J. (2011). Preparation of magnetic $CoFe_2O_4$-functionalized graphene sheets via a facile hydrothermal method and their adsorption properties. *J. Solid State Chem. 184*, 953–958.

115. Li, S., Wang, W., Liang, F., & Zhang, W. X. (2017). Heavy metal removal using nanoscale zero-valent iron (nZVI): Theory and application. *J. Hazard. Mater. 322*, 163–171.

116. Li, S., Wang, W., Yan, W., & Zhang, W. X. (2014). Nanoscale zero-valent iron (nZVI) for the treatment of concentrated Cu (II) wastewater: a field demonstration. *Environ. Sci. Processes Impacts. 16*, 524–533.

117. Li, Y., Liu, T., Du, Q., Sun, J., Xia, Y., Wang, Z., Zhang, W., Wang, K., Zhu, H., & Wu, D. (2012). Adsorption of cationic red X-GRL

from aqueous solutions by graphene: equilibrium, kinetics and thermodynamics study. *Chem. Biochem. Eng. 25*,483- 491.

118. Li, Y. H., Ding, J., Luan, Z., Di, Z., Zhu, Y., Xu, C., Wu, D., & Wei, B. (2003). Competitive adsorption of Pb^{2+}, Cu^{2+} and Cd^{2+} ions from aqueous solutions by multiwalled carbon nanotubes. *Carbon. 41*, 2787–2792.
119. Li, Y. H., Wang, S., Luan, Z., Ding, J., Xu, C., & Wu, D. (2003). Adsorption of cadmium (II) from aqueous solution by surface oxidized carbon nanotubes. *Carbon 41*, 1057–1062.
120. Li, Y. H., Wang, S., Wei, J., Zhang, X., Xu, C., Luan, Z., Wu, D., & Wei, B. (2002). Lead adsorption on carbon nanotubes. *Chem. Phys. Lett. 357*, 263–266.
121. Liang, J., Liu, J., Yuan, X., Dong, H., Zeng, G., Wu, H., Wang, H., Liu, J., Hua, S., Zhang, S., & Yu, Z. (2015). Facile synthesis of alumina-decorated multi-walled carbon nanotubes for simultaneous adsorption of cadmium ion and trichloroethylene. *Chem. Eng. J. 273*, 101–110.
122. Liang, P., Shi, T., & Li, J. (2004). Nanometer-size titanium dioxide separation/pre-concentration and FAAS determination of trace Zn and Cd in water sample. *Int. J. Environ. Anal. Chem. 84*, 315–321.
123. Liao, Q., Sun, J., & Gao, L. (2008). Adsorption of chlorophenols by multi-walled carbon nanotubes treated with HNO_3 and NH_3. *Carbon. 46*, 553–555.
124. Lingamdinne, L. P., Koduru, J. R., Roh, H., Choi, Y. L., Chang, Y. Y., & Yang, J. K. (2016). Adsorption removal of Co (II) from wastewater using graphene oxide. *Hydrometallurgy 165*, 90–96.
125. Lissemore, L., Hao, C., Yang, P., Sibley, P. K., Mabury, S., & Solomon, K. R. (2006). An exposure assessment for selected pharmaceuticals within a watershed in Southern Ontario. *Chemosphere. 64*, 717–729.
126. Liu, M., Wang, Y., Chen, L., Zhang, Y., & Lin, Z. (2015). Mg $(OH)_2$ supported nanoscale zero valent iron enhancing the removal of Pb (II) from aqueous solution. *ACS Appl. Mater. Interfaces. 7*, 7961–7969.
127. Lo´pez-Ramo´n, V., Moreno-Castilla, C., Rivera-Utrilla, J., & Radovic, L. R. (2002). Ionic strength effects in aqueous phase adsorption of metal ions on activated carbons. *Carbon. 41*, 2020–2022.
128. Lofrano, G., Carotenuto, M., Libralato, G., Domingos, R. F., Markus, A., Dini, L., Gautam, R. K., Baldantoni, D., Rossi, M., Sharma, S.

K., Chattopadhyaya, M. C., Giugni, M., & Meric, S. (2016). Polymer functionalized nanocomposites for metals removal from water and wastewater: an overview. *Water Res. 92*, 22–37.

129. Lubick, N., & Betts, K. (2008). Silver socks have cloudy lining| Court bans widely used flame retardant. *Environ. Sci. Technol. 42*, 3910–3910.

130. Luo, F., Chen, Z., & Megharaj, M. (2014). Naidu, R. Biomolecules in grape leaf extract involved in one-step synthesis of iron-based nanoparticles. *RSC Adv. 4*, 53467–53474.

131. Luo, T., Cui, J., Hu, S., Huang, Y., & Jing, C. (2010). Arsenic removal and recovery from copper smelting wastewater using TiO_2. *Environ. Sci. Technol. 44*, 9094–9098.

132. Machado, F. M., Bergmann, C. P., Fernandes, T. H. M., Lima, E. C., Royer, B., Calvete, T., & Fagan, S. B. (2011). Adsorption of Reactive Red M-2BE dye from water solutions by multi-walled carbon nanotubes and activated carbon. *J. Hazard. Mater. 192*, 1122–1131.

133. Machida, M., Mochimaru, T., & Tatsumoto, H. (2006). Lead (II) adsorption onto the graphene layer of carbonaceous materials in aqueous solution. *Carbon. 44*, 2681–2688.

134. Madadrang, C. J., Kim, H. Y., Gao, G., Wang, N., Zhu, J., Feng,H., Gorring, M., Kasner, M. L., & Hou, S. (2012). Adsorption behavior of EDTA-graphene oxide for Pb(II) removal. *ACS Appl. Mater. Interfaces. 4*, 1186–1193.

135. Mamadou, S. D., & Savage, N. (2005). Nanoparticles and water quality. *J. Nano. Res. 7*, 325–330.

136. Mateo-Sagasta, J., Raschid-Sally, L. & Thebo, A. (2015). Global wastewater and sludge production, treatment and use. P. Drechsel, M. Qadir and D. Wichelns, Wastewater: Economic Asset in Urbanizing World. *Springer Netherlands*, 15-38.

137. McHenry, M. E., & Laughlin, D. E. (2000). Nano-scale materials development for future magnetic applications. *Acta Materialia. 48*, 223–238.

138. Meng, S., Mansouri, J., Ye, Y., & Chen, V. (2014). Effect of templating agents on the properties and membrane distillation performance of TiO2-coated PVDF membranes, *J. Membr. Sci. 450*, 48–59.

139. Mi, X., Huang, G., Xie, W., Wang, W., Liu, Y., & Gao, J. (2012). Preparation of graphene oxide aerogel and its adsorption for Cu2+

ions. *Carbon. 50*, 4856–4864.

140. Mo, M., Yu, J. C., Zhang, L., & Li, S. K 2005. Self of Assembly of ZnONanorods and Nanosheets into Hollow Microhemispheres and Microspheres. Adv. Mater. *17*(6), 756–760.

141. Mohamed, Y. M., Azzam, A. M., Amin, B. H., & Safwat, N. A. (2015). Mycosynthesis of iron nanoparticles by *Alternaria alternata* and its antibacterial activity. *African J. Biotechnol. 14*, 1234–1241.

142. Moon G., Kim D., Kim H., Bokare A. D., & Choi W. (2014). Platinum-like behavior of reduced graphene oxide as a co-catalyst on TiO2 for the efficient photocatalytic oxidation of arsenite. *Environ. Sci. Technol. Lett. 1*, 185–190.

143. Mostafavi, S. T., Mehrnia, M. R., & Rashidi, A. M. (2009). Preparation of nanofilter from carbon nanotubes for application in virus removal from water. *Desalination. 238*, 271–280.

144. Mou, F., Guan J., Xiao Z., Sun Z., Shi W., & Fan X. A. (2011). Solvent-mediated synthesis of magnetic Fe_2O_3 chest nut like amorphous-core/□--phase-shell hierarchical nano structures with strong As(v) removal capability. *J. Mater. Chem. 21*, 5414–5421.

145. Mu, W., Herrmann, J. M., & Pichat, P. (1989). Room temperature photocatalytic oxidation of liquid cyclohexane into cyclohexanone over neat and modified TiO_2. *Catal. Lett. 3*, 73–84.

146. Mubarak, N. M., Sahu, J. N., Abdullah, E. C., & Jayakumar, N. S. 2013Removal of Heavy Metals from Wastewater Using Carbon Nanotubes. *Sep. Purif. Rev. 43*, 311–338.

147. Mueller, N. C., Braun, J., Bruns, J., Černík, M., Rissing, P., Rickerby, D., & Nowack, B. (2012). Application of nanoscale zero valent iron (NZVI) for groundwater remediation in Europe. *Environ. Sci. Pollut. Res. 19*, 550–558.

148. Navale, G. R., Thripuranthaka, M., Late D. J., & Shinde S. S. (2015). Antimicrobial activity of ZnO nanoparticles against pathogenic bacteria and fungi. *JSM Nanotechnol Nanomed. 3*, 1033.

149. Nepal, D., Balasubramanian, S., Simonian, A. L., & Davis, V. A. (2008). Strong antimicrobial coatings: single-walled carbon nanotubes armored with biopolymers. *Nano Lett. 8*, 1896–1901.

150. Nguyen, A. T., Hsieh, C.-T., & Juang,R. S. (2016). Substituent effects on photo degradation of phenols in binary mixtures by hybrid H_2O_2 and TiO_2 suspensions under UV irradiation. *J. Taiwan Inst. Chem.*

Engineers. *62*, 68–75.

151. Novoselov, K. S., Fal'Ko, V. I., Colombo, L., Gellert, P. R., Schwab, M. G., & Kim, K. (2012). A roadmap for graphene. *Nature*. *490*, 192–200.
152. Novoselov, K. S., Geim, A. K., Morozov, S. V., Jiang, D., Zhang, Y., Dubonos, S. V., Grigorieva, I. V., Firsov, A. A. (2004). Electric field effect in atomically thin carbon films. *Sci*. *306*, 666–669.
153. Ohsaka, T., Shinozaki, K., Tsuruta K., & Hirano K. (2008). Photo-electrochemical degradation of some chlorinated organic compounds on n-TiO_2 electrode. *Chemosphere*. *73*, 1279–1283.
154. Ozmen, M., Can, K., Arslan, G., Tor, A., Cengeloglu, Y., & Ersoz, M. (2010). Adsorption of Cu (II) from aqueous solution by using modified Fe_3O_4 magnetic nanoparticles. *Desalination*. 254, 162–169.
155. Page, K., Palgrave, R. G., Parkin, I. P., Wilson, M., Savin, S. L. P., & Chadwick, A. V. (2007). Titania and silver-titania composite films on glass-potent antimicrobial coatings, *J. Mater. Chem*. *17*, 95–104.
156. Pan, B., Qiu, H., Pan, B., Nie, G., Xiao, L., Lv, L., Zhang, W., Zhang, Q., & Zheng, S. (2010). Highly efficient removal of heavy metals by polymer-supported nanosized hydrated Fe (III) oxides: behavior and XPS study. *Water Res*. *44*, 815–824.
157. Panigrahi, S., Kundu, S., Ghosh, S. K., Nath, S., & Pal, T. (2004). General method of synthesis for metal nanoparticles. *J. Nano. Res*. *6*, 411–414.
158. Patricia E., Nicholas B., Katie J., Dean J., Arthur B., & George C. J. (1999). Preparation and Properties of an Aqueous Ferro fluid. *J. Chem*. 76,943.
159. Pendergast, M. M., & Hoek, E. M. V. (2011). A review of water treatment membrane nanotechnologies. *Energy Environ. Sci*. *4*, 1946–1971.
160. Petersen, E. J., Pinto, R. A., Shi, X., & Huang, Q. (2012). Impact of size and sorption on degradation of trichloroethylene and polychlorinated biphenyls by Nano-scale zerovalent iron. *J. Hazard. Mater*. *243*, 73–79.
161. Pyrzyn´ska, K., & Bystrzejewski, M. (2010). Comparative study of heavy metal ions sorption onto activated carbon, carbon nanotubes, and carbon-encapsulated magnetic nanoparticles. *Colloids Surf. A*. *362*, 102–109.
162. Qu, X., Alvarez, P. J., & Li, Q. (2013). Applications of nanotechnology in water and wastewater treatment. *Wat. Res*. *47*, 3931–3946.

163. Rafique, U., Imtiaz, A., & Khan, A. K. (2012). Synthesis, Characterization and Application of Nanomaterials for the Removal of Emerging Pollutants from Industrial Wastewater, Kinetics and Equilibrium Model. *J. Water Sustain. 2*, 233–244.

164. Rai, P., Gautam, R. K., Banerjee, S., Rawat, V., & Chattopadhyaya, M. C. (2015). Synthesis and characterization of a novel $SnFe_2O_4$@ activated carbon magnetic nanocomposite and its effectiveness in the removal of crystal violet from aqueous solution. *J. Environ. Chem. Eng. 3*, 2281–2291.

165. Rajesh, S., Senthilkumar, S., Jayalakshmi, A., Nirmala, M. T., Ismail A. F., & Mohan, D. (2013). Preparation and performance evaluation of poly (amide–imide) and TiO_2 nanoparticles impregnated polsulfone nano filtration membranes in the removal of humic substances. *Colloids Surf. A. 418*, 92–104.

166. Rao, G. P., Lu, C., & Su, F. (2007). Sorption of divalent metal ions from aqueous solution by carbon nanotubes: A review. *Sep. Purif. Technol. 58*, 224- 231.

167. Rawal S. B., Bera S., Lee D., Jang D. J., & Lee W. I. (2013). Design of visible-light photocatalysts by coupling of narrow band gap semiconductors and TiO_2: effect of their relative energy band positions on the photo-catalytic efficiency. *Catal. Sci. Technol. 3*, 1822–1830.

168. Ren, X., Chen, C., Nagatsu, M., & Wang, X. (2011). Carbon nanotubes as adsorbents in environmental pollution management: A review. *Chem. Eng. J. 170*, 395–410.

169. Robati, D. (2013). Pseudo-second-order kinetic equations for modeling adsorption systems for removal of lead ions using multi-walled carbon nanotube. *J. Nanostruct. Chem. 3*, 55.

170. Roco, M. C. (2003). Nanotechnology: convergence with modern biology and medicine. *Curr. Opin. Biotechnol. 14*, 337–346.

171. Rosal, R., Rodrı´guez, A., Perdigo´n-Melon, J. A., Petre, A., Garcı´a-Calvo, E., Go´mez, M. J., Agu,¨ A. era., & Ferna´ndez-Alba, A. R. (2010). Occurrence of emerging pollutants in urban wastewater and their removal through biological treatment followed by ozonation, *Water Res. 44*,578–588.

172. Sadegh, H., Shahryari-ghoshekandi, R., Agarwal, S., Tyagi, I., Asif, M., & Gupta, V. K. (2015). Microwave-assisted removal of malachite green by carboxylate functionalized multi-walled carbon nanotubes:

kinetics and equilibrium study. *J. Mol. Liq. 206*, 151–158.

173. Sadegh, H., Shahryari-Ghoshekandi, R., Tyagi, I., Agarwal, S., & Gupta, V. K. (2015). Kinetic and thermodynamic studies for alizarin removal from liquid phase using poly-2-hydroxyethyl methacrylate (PHEMA). *J. Mol. Liq. 207*, 21–27.
174. Saif, S., Tahir, A., & Chen, Y. (2016). Green synthesis of iron nanoparticles and their environmental applications and implications. *Nanomater. 6*, 209.
175. Salama, A. (2017). New sustainable hybrid material as adsorbent for dye removal from aqueous solutions. *J. Colloid Interface Sci. 487*, 348–353.
176. San Román, I., Alonso, M. L., Bartolomé, L., Galdames, A., Goiti, E., Ocejo, M., Moragues, M., Alonso, R. M., &Vilas, J. L. (2013). Relevance study of bare and coated zero valent iron nanoparticles for lindane degradation from its by-product monitorization. *Chemosphere. 93*, 1324–1332.
177. Santhosh, C., Velmurugan, V., Jacob, G., Jeong, S. K., Grace, A. N., & Bhatnagar, A. (2016). Role of nanomaterials in water treatment applications: a review. *Chem. Eng. J. 306*, 1116–1137.
178. Sato, T., Qadir, M., Yamamoto, S., Endo, T., & Zahoor, A. (2013). Global, regional, and country level need for data on wastewater generation, treatment, and use. *Agri. Water. Mang. 130*, 1–13.
179. Sean, C. S., & Debora, F. R. (2015). Carbon-based nanomaterials for removal of chemical and biological contaminants from water: A review of mechanisms and applications. *Carbon. 91*, 122–143.
180. Seery, M. K., George, R., Floris, P., & Pillai, S. C. (2007). Silver doped titanium dioxide nanomaterials for enhanced visible light photocatalysis, *J. Photochem. Photobio. A. Chem. 189*, 258–263.
181. Shahryari, Z., Goharrizi, A. S., & Azadi, M. (2010). Experimental study of methylene blue adsorption from aqueous solutions onto carbon nano tubes. *Int. J. Water Res. Environ. Eng. 2*, 16–28.
182. Shahwan, T., Sirriah, S. A., Nairat, M., Boyacı, E., Eroğlu, A. E., Scott, T. B., Hallam, K. R. (2011). Green synthesis of iron nanoparticles and their application as a Fenton-like catalyst for the degradation of aqueous cationic and anionic dyes. *Chem. Eng. J. 172*, 258–266.
183. Shao, D., Hu, J., & Wang, X. (2010). Plasma induced grafting multiwalled carbon nanotube with chitosan and its application for

removal of UO_2^{2+}, Cu^{2+}, and Pb^{2+} from aqueous solutions Plasma Process. *Polym. 7*, 977–985.

184. Shao, D., Jiang, Z., & Wang, X. (2010). SDBS modified XC-72 carbon for the removal of Pb (II) from aqueous solutions. *Plasma Process. Polym. 7*, 552–560.
185. Sharma, S., Ameta, R., Malkani, R. K., & Ameta, S. C. (2013). Photocatalytic degradation of rose Bengal by semiconducting zinc sulphide used as a photocatalyst. *J. Serbian chem. Soc. 78*, 897–905
186. Shen, W., Li, Z., Wang, H., Liu, Y., Guo, Q., & Zhang, Y. (2008). Photocatalytic degradation for methylene blue using zinc oxide prepared by co-deposition and sol-gel methods. *J. Hazard. Mater. 152*, 172–175.
187. Shen, Y., Tang, J., Nie, Z., Wang, Y., Ren, Y., & Zuo, L. (2009). Preparation and application of magnetic Fe_3O_4 nanoparticles for wastewater purification. *Sep. Purif. Technol.* 68, 312–319.
188. Shi, L. N., Zhang, X., & Chen, Z. L. (2011). Removal of chromium (VI) from wastewater using bentonite-supported nanoscale zero-valent iron. *Water Res. 45*, 886–892.
189. Shiklomanov, I. A. (1999). World water resources and their use a joint SHI/UNESCO product. Database. http://webworld.unesco.org/water/ihp/db/shiklomanov/.
190. Šileikaitė A., Prosyčevas I., Puišo J., Juraitis A., & Guobienė A. (2006). Analysis of silver nanoparticles produced by chemical reduction of silver salt solution. *Mater. Sci.* (Medžiagotyra). *12*, ISSN 1392–1320.
191. Singh, G., Rana, D., Matsuura, T., Ramakrishna, S., Narbaitz, R. M., & Tabe, S. (2010). Removal of disinfection byproducts from water by carbonized electro-spunnanofibrous membranes. *Sep. Purif. Technol.* 202–212.
192. Singh, S., Barick, K. C., & Bahadur, D. (2011). Surface engineered magnetic nanoparticles for removal of toxic metal ions and bacterial pathogens. *J. Hazard. Mater. 192*, 1539–1547.
193. Singh, S., Barick, K. C., & Bahadur, D. (2013). Fe_3O_4 embedded ZnOnanocomposites for the removal of toxic metal ions, organic dyes and bacterial pathogens. *J. Mater. Chem. A, 1*, 3325–3333.
194. Singh, S., Barick, K., & Bahadur, D. (2011). Novel and efficient three-dimensional mesoporous ZnO nano-assemblies for environmental remediation. *Int. J. Nanosci. 10*, 1001–1005.

195. Singh, S., Barick, K., & Bahadur, D. (2013). Functional oxide nanomaterials and nanocomposites for the removal of heavy metals and dyes. *Nanomater. Nanotechnol. 3*, 3–20.

196. Song, J., Wang, X., & Chang, C. T. (2014). Preparation and characterization of graphene oxide. *J. Nanomater.*, Volume Article ID 276143, 6 pages http://dx.doi.org/10.1155/2014/276143

197. Sons, W. K., Youk J. H., Lee T. S., & Park, W. H. (2004). Preparation of antimicrobial ultrafine cellulose acetate fibers with silver nanoparticles. *Macromol. Rapid. Commun. 25*, 1632–1637.

198. Stafiej, A., & Pyrzynska, K. (2007). Adsorption of heavy metal ions with carbon nanotubes. *Sep. Purif. Technol. 58*, 49–52.

199. Sun, A. H., Guo, P. J. Li, Z, X., Li, Y. & Cui, P. (2009). Low temperature synthesis of Anatase and Rutile titania nano powders by hydrolysis of $TiCl_4$ using ammonia gas. *J. Alloys Compounds. 481*, 605–609.

200. Taman, R., Ossman, M. E., Mansour M. S., & Farag H. A. (2015). Metal Oxide Nano-particles as an Adsorbent for Removal of Heavy Metals. *J. Adv. Chem. Eng. 5*,3 DOI: 10.4172/2090–4568.1000125.

201. Tang, C., Huang, Y. H., Zeng, H., & Zhang, Z. (2014). Reductive removal of selenate by zero-valent iron: the roles of aqueous Fe2+ and corrosion products, and selenate removal mechanisms. *Water Res. 67*, 166–174.

202. Tang, W. W., Zeng, G. M., Gong, J. L., Liu, Y., Wang, X. Y., Liu, Y. Y., Liu, Z. F., Chen, L., Zhang, X. R., & Tu, D.Z. (2012). Simultaneous adsorption of atrazine and Cu(II) from wastewater by magnetic multi-walled carbon nanotube. *Chem. Eng. J. 211*, 470–478.

203. Tang, Y., Guo, H., Xiao, L., Yu, S., Gao, N., & Wang, Y. (2013). Synthesis of reduced graphene oxide/magnetite composites and investigation of their adsorption performance of fluoroquinolone antibiotics. *Colloids Surf. A. 424*, 74–80.

204. Tawabini, B. S. (2015). Removal of methyl tertiary butyl ether (MTBE) from contaminated water using UV-assisted nano composite materials. *Desalination Water Treat. 55*, 549–554.

205. Tesh, S. J., & Scott, T. B. (2014). Nano-composites for water remediation: a review. *Adv. Mater. 26*, 6056–6068.

206. Theron, S. A., Zussman, E., & Yarin, A. L. (2004). Experimental investigation of the governing parameters in the electrospinning of polymer solutions. *Polymer*, *45*, 2017–2030.

207. Tian, C, Liu R, Liu H, & Qu J. (2013). Disinfection by-products formation and precursors transformation during chlorination and chloramination of highly-polluted source water: significance of ammonia. *Water Res*, *47*, 5901–5910.

208. Tomašević, D. D., Kozma, G., Kerkez, D. V., Dalmacija, B. D., Dalmacija, M. B., Bečelić-Tomin, M. R., Kukovecz, Á., Kónya, Z., & Rončević, S. (2014). Toxic metal immobilization in contaminated sediment using bentonite-and kaolinite-supported nano zero-valent iron. *J. Nanoparticle Res. 16*, 25–48.

209. Tosco, T., Papini, M. P., Viggi, C. C., & Sethi, R. (2014). Nanoscalezerovalent iron particles for groundwater remediation: a review. *J. Clean. Prod. 77*, 10–21.

210. UN, World Water Development Report, Waterfor a Sustainable World, World Water Assessment Programme (WWAP). 2015.

211. UN, World Water Development Report. (2012). World Water Assessment Programme (WWAP).

212. UN, World Water Development Report. (2017). World Water Assessment Programme (WWAP).

213. UNW-DPAC, UN Report: Water and industry in the green economy: Information brief, 2011.

214. Upadhyayula, V. K. K., Deng, S., Mitchell, M. C., & Smith, G. B. (2009). Application of carbon nanotube technology for removal of contaminants in drinking water: A review. *Sci. Total Environ. 408*, 1–13.

215. Upadhyayula, V. K., Ghoshroy, S., Nair, V. S., Smith, G. B., Mitchell, M. C., & Deng, S. (2008). Single-walled carbon nanotubes as fluorescence biosensors for pathogen recognition in water systems. *J. Nanotechno.* Article ID 156358, 5 pages doi:10.1155/2008/156358

216. Varma, R. S. (2012). Greener approach to nanomaterials and their sustainable applications. *Curr. Opin. Chem. Eng. 1*, 123–128.

217. Vaseem, M., Umar, A., & Hahn, Y. B. (2010). ZnO nanoparticles: growth, properties, and applications. In: Umar, A., Hahn, Y. B., editors. Metal oxide nanostructures and their applications. Valencia, CA: *American Scientific Publishers*; 1–36.

218. Vidya, C., Prabha, M. C., & Raj, M. A. (2016). Green mediated synthesis of zinc oxide nanoparticles for the photocatalytic degradation of Rose Bengal dye. *Environ. Nanotechnol. Monit. Manag. 6*, 134–138.

219. Vu, H. C., Dwivedi, A. D., Le, T. T., Seo, S. H., Kim, E. J., & Chang, Y. S. (2016). Magnetite graphene oxide encapsulated in alginate beads for enhanced adsorption of Cr (VI) and as (V) from aqueous solutions: role of crosslinking metal cations in pH control. *Chem. Eng. J, 307*, 220-229.

220. Wang, G., Zhang, S., Xu, X., Li, T., Li, Y., Deng, O., & Gong, G. (2014). Efficiency of nanoscale zero-valent iron on the enhanced low molecular weight organic acid removal Pb from contaminated soil. *Chemosphere. 117*, 617–624.

221. Wang, H., Yuan, X., Wu, Y., Zeng, G., Chen, X., Leng, L., Wu, Z., Jiang, L. & Li, H. (2015). Facile synthesis of amino-functionalized titanium metal-organic frameworks and their superior visible-light photocatalytic activity for Cr (VI) reduction. *J. Hazard. Mater. 286*,187–194.

222. Wang, H., Zhou, A., Peng, F., Yu, H., & Yang, J. (2007). Mechanism study on adsorption of acidified multiwalled carbon nanotubes to Pb(II). *J. Colloid Interface Sci. 316*(2), 277–283

223. Wang, L., Li, J., Jiang, Q., & Zhao, L. (2012). Water-soluble Fe_3O_4 nanoparticles with high solubility for removal of heavy-metal ions from wastewater. *Dalton Trans. 41*, 4544–4551.

224. Wang, Q., Wang, X., Wang, Z., Huang, J., & Wang, Y. (2013). PVDF membranes with simultaneously enhanced permeability and selectivity by breaking the trade-off effect via atomic layer deposition of TiO_2, *J. Membr. Sci. 442*, 57–64.

225. Wang, S., Sun, H., Ang, H.-M., & Tade, M. (2013). Adsorptive remediation of environmental pollutants using novel graphene-based nanomaterials. *Chem. Eng. J. 226*, 336–347.

226. Wang, W., Zhou, M., Jin, Z., & Li, T. (2010). Reactivity characteristics of poly (methyl methacrylate) coated nanoscale iron particles for trichloroethylene remediation. *J. Hazard. Mater. 173*, 724–730.

227. Wang, X., Cai, W., Lin, Y., Wang, G., & Liang, C. (2010a). Mass production of micro/nanostructured porous ZnO plates and their strong structurally enhanced and selective adsorption performance for environmental remediation. *J. Mater. Chem. 20*, 8582–8590.

228. Wang, Z. (2013). Iron complex nanoparticles synthesized by eucalyptus leaves. *ACS Sustain. Chem. Eng. 1*, 1551–1554.

229. Wang, Z., Fang, C., & Megharaj, M. (2014). Characterization of iron–

polyphenol nanoparticles synthesized by three plant extracts and their fenton oxidation of azo dye. *ACS Sustain. Chem. Eng. 2*, 1022–1025.

230. Wang, Z., Wu, D., Wu, G., Yang, N., & Wu, A. (2013). Modifying Fe_3O_4 microspheres with rhodamine hydrazide for selective detection and removal of Hg^{2+} ion in water. *J. Hazard. Mater.* 2013,*244*, 621–627.

231. Welderfael, T., Yadav, O. P., Taddesse, A. M., & Kaushal, J. (2013). Synthesis, characterization and photocatalytic activities of Ag-N-codopedZnO nanoparticles for degradation of methyl red. *Bull. Chem. Soc. Ethi. 27*, 221–232.

232. WHO (World Health Organization), Progress on Drinking Water and Sanitation. World Health Organization. 2014.

233. Xu, D., Tan, X., Chen, C., & Wang, X. (2008). Removal of Pb(II) from aqueous solution by oxidized multiwalled carbon nanotubes. *J. Hazard. Mater. 154*, 407–416.

234. Yan, X. M., Shi, B. Y., Lu, J. J., Feng, C. H., Wang, D. S., & Tang, H. X. (2008). Adsorption and desorption of atrazine on carbon nanotubes. *J. Colloid Interface Sci. 321*, 30–38.

235. Yang, S. T., Chang, Y., Wang, H., Liu, G., Chen, S., Wang, Y., Liu, Y., & Cao, A. (2010). Folding/aggregation of graphene oxide and its application in Cu^{2+} removal. *J. Colloid Interface Sci. 351*, 122–127.

236. Yao, Y., Xu, F., Chen, M., Xu, Z., & Zhu, Z. (2010). Adsorption behavior of methylene blue on carbon nanotubes. *Bioresour. Technol. 101*, 3040–3046

237. Yin, Y., Zhang, G., & Xia, Y. (2002). Synthesis and characterization of MgO nanowires through a vapor-phase precursor method. *Adv. Functional Mater. 12*, 293–298.

238. Ying-Ying, W., & Zhen-Hu, X. (2016). Multi-walled carbon nanotubes and powder-activated carbon adsorbents for the removal of nitrofurazone from aqueous solution. *J. Dispers. Sci. Technol. 37*, 613–624.

239. You, G., Wang, P., Hou, J., Wang, C., Xu, Y., Miao, L., Lv, B., Yang, Y., & Zhang, F. (2017). The use of zero-valent iron (ZVI)-microbe technology for wastewater treatment with special attention to the factors influencing performance: A critical review. *Crit. Rev. Env. Sci. Technol. 47*, 877–907.

240. Yu, J. G., Yu, L. Y., Yang, H., Liu, Q., Chen, X. H., Jiang, X. Y., Chen,

X. Q., & Jiao, F.P. (2015). Graphene nano sheets as novel adsorbents in adsorption, preconcentration and removal of gases, organic compounds and metal ions. *Sci. Total Environ. 502*, 70–79.

241. Yu, X., Tong, S., Ge, M., Zuo, J., Cao, C., & Song, W. (2013). One-step synthesis of magnetic composites of cellulose@ iron oxide nanoparticles for arsenic removal. *J. Mater. Chem. A. 1*, 959–965.
242. Zare-Dorabei, R., Ferdowsi, S. M., Barzin, A., & Tadjarodi, A. (2016). Highly efficient simultaneous ultrasonic-assisted adsorption of Pb (II), Cd (II), Ni (II) and Cu (II) ions from aqueous solutions by graphene oxide modified with 2, 20-dipyridylamine: central composite design optimization. *UltraSono chem. 32*, 265–276.
243. Zare, K., Gupta, V. K., Moradi, O., Makhlouf, A.S. H., Sillanpa¨a¨, M., Nadagouda, M. N., Sadegh, H., Shahryari-ghoshekandi, R., Pal, A., & Wang, Z.J. (2015). A comparative study on the basis of adsorption capacity between CNTs and activated carbon as adsorbents for removal of noxious synthetic dyes: a review. *J. Nanostruct. Chem. 5*, 227–236.
244. Zelmanov, G., & Semiat, R. (2008). Phenol oxidation kinetics in water solution using iron (3)-oxide-based nano-catalysts. *Water Res. 42*, 3848–3856.
245. Zhang, C., Sui, J., Li, J., Tang, Y., & Cai, W. (2012). Efficient removal of heavy metal ions by thiol-functionalized superparamagnetic carbon nanotubes. *Chem. Eng. J. 210*, 45–52.
246. Zhang, Q., Huang, J., Zhao, M., Qian, W., & Wei, F. (2011). Carbon nanotube mass production: principles and processes. *Chem. Sus. Chem. 4*, 864–889
247. Zhang, S. (2014). Preparation of controlled-shape ZnS microcrystals and photocatalytic property. *Ceramics Int. 40*, 4553–4557.
248. Zhao, G., Li, J., Ren, X., Chen, C., & Wang, X. (2011). Few-layered graphene oxide nanosheets as superior sorbents for heavy metal ion pollution management. *Environ. Sci. Technol. 45*, 10454–10462.
249. Zhu, H., Jia, Y., Wu, X., & Wang, H. (2009). Removal of arsenic from water by supported Nano zero-valent iron on activated carbon. *J. Hazard. Mater. 172*, 1591–1596.
250. Zhu, J. J., Liao, X. H., Zhao, X. N., & Hen, H. Y. (2001). Preparation of Silver Nanorods by Electrochemical Methods, *Mater. Lett. 49*, 91–95.
251. Zhu, Y. Q., Hsu, W. K., Zhou, W. Z., Terrones,M., Kroto, H. W., &

Walton, D. R. M. (2001). Selective Co-catalysed growth of novel MgO fishbone fractal nanostructures. *Chem. Phys. Lett. 347*, 337–343.

252. Zhu, Y., Ran, T., Li, Y., Guo, J., & Li, W. (2006). Dependence of the cytotoxicity of multi-walled carbon nanotubes on the culture medium. *Nanotechnol. 17*, 4668–4674.

INDEX